Landscape Ecology: Theory and Practice

Landscape Ecology: Theory and Practice

Editor: Bobby Perez

Callisto Reference
www.callistoreference.com

Callisto Reference,
118-35 Queens Blvd., Suite 400,
Forest Hills, NY 11375, USA

Visit us on the World Wide Web at:
www.callistoreference.com

ISBN: 978-1-64116-267-8 (Hardback)

Printed in China.

Cataloging-in-Publication Data

Landscape ecology : theory and practice / edited by Bobby Perez.
p. cm.
Includes bibliographical references and index.
ISBN 978-1-64116-267-8
1. Landscape ecology. 2. Ecology. I. Perez, Bobby.
QH541.15.L35 L36 2020
577--dc23

Table of Contents

Preface

Landscape ecology is an interdisciplinary field in systems science. The relationships between ecological processes in the environment and ecosystems are studied and improved in landscape ecology. The human impact on landscape structures and functions is investigated in this field. An important principle of landscape ecology is the landscape stability principle, which stresses on the importance of integrity of landscape components in exhibiting resilience to disturbances, promoting total system stability and recovering from disturbances. Some of the vital research areas in this field are land use and land cover change, relating landscape pattern analysis with ecological processes, ecological flows in landscape mosaics, scaling, and landscape conservation and sustainability. This book is compiled in such a manner, that it will provide in-depth knowledge about the theories and practices of landscape ecology. It is a valuable compilation of topics, ranging from the basic to the most complex advancements in this field of study. The readers would gain knowledge that would broaden their perspective about landscape ecology.

Various studies have approached the subject by analyzing it with a single perspective, but the present book provides diverse methodologies and techniques to address this field. This book contains theories and applications needed for understanding the subject from different perspectives. The aim is to keep the readers informed about the progresses in the field; therefore, the contributions were carefully examined to compile novel researches by specialists from across the globe.

Indeed, the job of the editor is the most crucial and challenging in compiling all chapters into a single book. In the end, I would extend my sincere thanks to the chapter authors for their profound work. I am also thankful for the support provided by my family and colleagues during the compilation of this book.

Editor

1

THE INFLUENCE OF ROAD MAINTENANCE SUBSTANCES ON THE GROWTH OF GRAMINACEOUS PLANTS

Agnė KAZLAUSKIENĖ, Vaida BRUKŠTUTĖ

Research Institute of Environmental Protection, Vilnius Gediminas Technical University, Saulėtekio al. 11, 10223 Vilnius, Lithuania

Abstract. The article describes the experimental research on the effect of road maintenance materials on three species of graminaceous plants. The experiment was carried out using technical granular sodium chloride (NaCl) treated with potassium hexacyanoferrate – a road maintenance material most widely applied in wintertime in Lithuania. As an alternative, an organic material obtained on the basis of molasses, patented under the name of *Safecote* in Europe, was selected for the experiment. Three species of graminaceous plants, most frequently used for roadside planting in Lithuania when building new and reconstructing or renovating the existing motor roads, were analysed: perennial rye-grass (*Lolium perenne* L.), meadow fescue (*Festuca pratensis* Huds.), and common meadow-grass (*Poa pratensis* L.). The paper presents experimental findings regarding the effect of road maintenance materials on the parameters of graminaceous plants – the height of the above-ground part and phytomass.

Keywords: graminaceous plant species, perennial rye-grass (*Lolium perenne* L.), meadow fescue (*Festuca pratensis* Huds.), common meadow-grass (*Poa pratensis* L.), molasses, road maintenance salt.

Introduction

During the cold season, approximately 140 000 tons of technical salts are spread on roads to reduce slipperiness in Lithuania. Out of this amount, 70 000 to 80 000 tons are applied on roads of national significance (data of the Lithuanian Road Administration under the Ministry of Transport and Communications) (Baltrėnas, Kazlauskienė 2009a). Technical salts are used to reduce slipperiness in wintertime. They are rather efficient in melting snow and ice but have a negative effect on road components, causing and corrosion of metal parts of motor vehicles as well as r/c structures (bridges, tunnels, etc.); they also have a significant impact on the destruction of cement-concrete structures, and deformation of asphalt-concrete paving, etc. (Kamaitis 2002a, 2002b).

These effects are especially evident in cities as streets are surrounded by numerous buildings and structures, leaving no space for roadside ditches that collect snow and salts cleaned off the roadway. In many such cases, it is impossible to immediately collect and transport snow accumulated on sidewalks. The situation is better with roads of national significance due to many open areas and roadside ditches (Baltrėnas *et al.* 2012).

Significant concentrations of road maintenance salts (up to 3200 mg/*l*) accumulate in snow (Baltrėnas, Kazlauskienė 2009b) and enter the soil together with meltwater carrying heavy metals (Vainalavičiūtė *et al.* 2009; Mikalajūnė, Jakučionytė 2011) and oil products (Žukauskaitė *et al.* 2008; 2012), which negatively affect roadside vegetation (Dobson 1991; Baltrėnas *et al.* 2006; Darginaviciene *et al.* 2008) as well as surface and ground waters (Williams *et al.* 1999) and their fauna (Oškinis, Kaperovičius 2005; Andreikėnaitė *et al.* 2007; Vosylienė *et al.* 2010). As other investigations have shown, soil is alkaline because of cement dust that contains different microelements (Stravinskienė 2011).

By density and quality of its road network, Lithuania equals highly developed states. One of the main aims of road maintenance in winter, aiming to removing snow and ice off road paving and increase the tyre and road paving adhesion coefficient, is to minimise the risk of traffic accidents. Recently, wet salt technology has been applied to protect roads against climatic factors or to de-ice the paving in a very short time. Its effective operation is possible under up to –15 °C (Sakshaug, Vaa 1995; Cornford, Thornes 1996; Žilionienė, Laurinavičius 2007). However,

Corresponding author: Agnė Kazlauskienė
E-mail: agne.kazlauskiene@vgtu.lt

once Lithuania joined the European Union, the requirement has been set to use new generation environmentally friendly materials for road maintenance. With the aim to reduce a negative environmental effect of road maintenance salts, it is expedient to gradually reduce the use of technical salts (in particular in cities) replacing them with alternative new generation substances made on the basis of organic materials (Baltrėnas, Kazlauskienė 2009a).

The best practice, improvement of professional skills, scientific research and studies, and dissemination of their results play an important role in the improvement of road infrastructure (Skrinskas 2012).

This research analyses an organic substance molasses – a by-product of sugar manufacturing processes – to minimise road slipperiness and dustiness. In Europe, this product is patented as *Safocote*. Molasses was presented to the market in Great Britain in 2002. Canada and USA have 14 years of experience using this product, followed by Great Britain, which has been using it for 9 years. In Europe, (Austria, Estonia, Germany, Hungary, Ireland, Norway, Poland, Slovenia, Switzerland and Sweden), it has been in use from 3 to 5 years.

In Vilnius (Lithuania), *Safecote* was primarily used to decrease street slipperiness during the wintertime of 2005/2006. The amount of 18 tons of *Safecote* was used during 2006/2007 (Baltrėnas, Kazlauskienė 2009a).

Safecote added to a NaCl solution helps to stay on a road surface 7 times longer and, therefore, no additional treatment is required. It does not cause corrosion of equipment. If technical road maintenance salts were replaced with molasses substance, it would minimise the impact on concrete, bridges, viaducts and asphalt. Furthermore, it would reduce the number of spreader's working hours as well as the number of trips. No additional capital investment in equipment would be required because *Safecote* can be used with the existing equipment while the required volume of technical salts could be reduced from 30% to 50%.

Safecote should not be used on roads with low traffic intensity as it can increase a slippery layer, causing threat to traffic safety. Molasses substance is viscous and dark colour. When applied in big quantities, it requires additional washing of streets as well as vehicle maintenance in springtime.

Safecote costs more than technical salts but it is less harmful to the environment. The use of technical salts can be decreased adding 10% of *Safecote* to the mixture (Vitaliano 1992; Hanbali 1998; Thornes 2000; Petkuvienė, Paliulis 2009). Such a mixture is an effective solution for ice melting and prevention of slipperiness (Burtwell, Wilson 2002; Burtwell 2004).

Solutions of *Safecote* in water have been used during summer on the streets and gravel roads to reduce dustiness (Bradulienė, Vasarevičius 2010, 2011, 2012, 2014).

Safecote is a relatively new product; therefore, it is expedient to analyse it and its effect on road environment components. In this research, attention is given to graminaceous plants particular to Lithuania (Bradulienė, Vasarevičius 2010, 2011, 2012, 2014).

The impact of *Safecote* on herbaceous vegetation has been episodically studied in Lithuania and foreign countries (Bradulienė, Vasarevičius 2012).

The aim of the research: to analyse how road maintenance materials affect the growth process and main parameters (the height of the above-ground part and phytomass) of graminaceous plants.

1. Methods and materials

Three species of graminaceous plants, most often used for roadside planting in Lithuania when building new or reconstructing and renovating the existing roads, have been selected for this research (Baltrėnas *et al.* 2012). The type of vegetation influences the rate of substrate respiration. Substrates release approximately 20% of the total CO_2 content to the atmosphere (Baltrėnas *et al.* 2011).

The following graminaceous plants have been selected for investigation:

- perennial rye-grass (*darimo* cultivar) – *Lolium perenne* L.;
- meadow fescue (*prana* cultivar) – *Festuca pratensis* Huds.;
- common meadow-grass (*balin* cultivar) – *Poa pratensis* L.

Names of these species have also been mentioned in the article by Jankevičienė (1998).

Also known as molasses, *Safecote* is an organic material, which is a by-product of sugar refining. This organic material contains minerals, including Ca, P, S, K, Na, etc. This material is of dark brown colour, liquid and can be mixed with all technical road maintenance salts and their solutions (Baltrėnas, Kazlauskienė 2009a; Bradulienė, Vasarevičius 2011, 2012, 2014).

The research was carried out with crushed technical sodium chloride, including potassium hexacyanoferrate ($K_4[Fe(CN)_6]$), hereinafter referred to as NaCl. In Lithuania, the highest quantities of this admixture are used on rods in winter season (Baltrėnas, Kazlauskienė 2009b; Baltrėnas *et al.* 2012).

The experiment was carried out as follows: 3 g of seeds of perennial ryegrass, fescue grass and meadow-grass were each separately planted into 3 plastic containers fully filled with 1 kg of substrate.

The substrate of the following composition was used for the research: CaO – 300 mg/*l*, $NH_4 + NO_3$ – 100 mg/*l*, P_2O_5 – 50 mg/*l*, MgO – 80 mg/*l*, Fe_2O_3 – 800 mg/*l*, K_2O – 300 mg/*l*.

The research was carried out in two experiments. In experiment I, substrate had been watered with following solutions:

- 5 g/kg of *Safecote*,
- 5 g/kg NaCl,
- Mixture of 5 g/kg of *Safecote* and NaCl salt.

The mixture of *Safecote* and NaCl salt was composed of: 4.5 g/kg of NaCl and 0.5 g/kg of *Safecote*. The mixture proportion (90% NaCl: 10% *Safecote*) was selected following *Safecote* use instructions to reduce slipperiness. Seeking to reduce the required amounts of NaCl, at least 10% of *Safecote* of NaCl amount should be used.

In experiment II, substrate had been watered with following solutions:

- 3 g/kg of *Safecote*,
- 8 g/kg of *Safecote*,
- 10 g/kg of *Safecote*,
- 8 g/kg of mixture (90% NaCl: 10% *Safecote*) composed of 7.2 g/kg of NaCl and 0.8 g/kg of *Safecote*.

Graminaceous plants of the three mentioned species have also been seeded into uncontaminated substrate for control.

During the entire growth period, plats had been growing at the temperature of +20±2 °C and measured humidity of 52 (±3)% and daylighting of 160 (±10) Lx. Certified equipment METREL Poly MI 6401 had been used to measure air temperature and humidity during the experiment.

The experiment lasted for six weeks because this is the exact time, during which, when accessing substrate, road maintenance materials negatively affect the growth process and main parameters of graminaceous plants under both natural and laboratory conditions. During a longer period, road maintenance materials under natural conditions are washed away by precipitation (rain) and their concentrations are diluted. Under laboratory conditions, a similar process takes place when concentration of materials in substrate is reduced during watering of plants. These are the reasons why it is no value in carrying out the experiment for a longer period.

The bottom of each container had holes to allow a natural water run-off and prevent the accumulation of excessive water. Seeds of each species were simultaneously seeded into clean substrate and substrate contaminated with road maintenance materials (control and experimental plants). Both the control and the experimental plants were grown under the same conditions – at the same temperature and lighting. The experimental and control plants were watered once a week with 50 ml of ionised water (Baltrėnas, Kazlauskienė 2007; Kazlauskienė 2010).

The *height of the above-ground part* of graminaceous plants was measured with a calibrated metal ruler *Preisser* No 11 (with height deviation of 0.026 mm). Five samples of each of the graminaceous species were randomly pulled up from containers to measure the average height of the above-ground part (cm). The height of the above-ground part was measured from the start of roots to the end of the longest leaf (Baltrėnas, Kazlauskienė 2007; Kazlauskienė 2010).

Phytomass of graminaceous plants was measured with the calibrated electronic analytical balance KERN 770–60 (weighing 0.01 g ... 60 g, the error is 0.00001 g). Five samples of each of the graminaceous species were randomly pulled up from containers to determine the average phytomass (mg). The phytomass was determined with the help of an electronic balance, weighing plants without roots, i.e. only the green above-ground part (Baltrėnas, Kazlauskienė 2007; Kazlauskienė 2010).

The height of the above-ground part and the phytomass were measured every week of the experiment, i.e. 6 weeks (Baltrėnas *et al.* 2006; Baltrėnas, Kazlauskienė 2009b; Bradulienė, Vasarevičius 2012).

As illustrated in figures, the error of 10% occurred when measuring the parameters of graminaceous plants.

All data obtained from this experiment is presented as the arithmetic mean of five independent analyses (x±SD, n = 5). Significant differences ($p < 0.05$) were deleted from the data evaluation and the measurements were carried out repeatedly. The average characteristics at typical points were:

$$\bar{c} = \frac{1}{n}\sum_{i=1}^{k} c_i m_i, \qquad (1)$$

where: c_i – value of characteristics at typical points; m_i – probability at the obtaining of value; n – number of experimental days; k – number of different values of the characteristics (Kazlauskienė *et al.* 2013).

The above-mentioned average indicators of plants, were also estimated at the typical point. The standard statistical estimation error of the arithmetic average was equal to approx. 10%. The dependencies between different compound characteristics and all indicators in different samples were determined by statistical evaluation using Mathcad 2001 Professional software.

2. Results

Experiment I. At the end of the 1st week of the experiment, graminaceous plants showed a very weak germination. Some of the plants did not come up at all or hardly germinated; consequently, the analytical measurements of graminaceous plants were taken on the 2nd week of the experiment.

After the 2nd week of the experiment, perennial rye-grass had the highest height of its above-ground part compared to plants of other species. Compared to the control, the height of perennial rye-grass in the substrate containing 5 g/kg of NaCl differed by 1.6 times and that of meadow fescue – by 3.4 times. After the 3rd week of the experiment, the height of control plants of perennial rye-grass reached 11.6 cm. The height of the above-ground part of control and perennial rye-grass in the substrate containing 5 g/kg of NaCl differed by 1.8 times. In one week, the difference of perennial rye-grass increased by 0.2 times. On the 4th week of growth, the height of perennial rye-grass reached 12.7 cm. On the 5th week of the experiment, perennial rye-grass was 14.0 cm. On the 6th week of growth, the height of the above-ground part of perennial rye-grass in the substrate containing *Safecote* was weaker compared to that of control plants (Fig. 1).

During the 3rd week of the experiment, the plants grew significantly; however, since then, common meadow-grass no longer germinated in the substrate with NaCl content. The height of control plants of meadow fescue reached 6.3 cm. Compared to the control, the height of the above-ground part of meadow fescue in the substrate containing 5 g/kg NaCl differed by 2.1 times. In one week, the difference of meadow fescue decreased by 1.3 times. On the 3rd week of growth, the height of plants in the substrate contaminated with *Safecote* considerably increased. In the substrate contaminated with NaCl and *Safecote*, plants were also higher compared to those containing NaCl alone: perennial rye-grass – 1.4 times, meadow fescue – 2.3 times. On the 4th week of growth, the height of meadow fescue reached 7.5 cm. The above-ground part of plants in the substrate contaminated with *Safecote* was the highest on the 4th and 5th weeks of the experiment. On the 5th week of the experiment, meadow fescue was 8.6 cm high. Compared to control plants, meadow fescue continued to grow 1.2 times taller.

Analysis of the weekly plant increment shows that the biggest increment of the above-ground part of plants was recorded on the 3rd week of growth, when perennial rye-grass increased by 1.9 cm in the substrate contaminated with 5 g/kg of NaCl, and 4.1 cm in the substrate containing the mixture of NaCl and *Safecote*. Meadow fescue grew by 1.2 and 2.7 cm, respectively. The increase of graminaceous plants in the substrate contaminated with NaCl was more even compared to the substrate containing *Safecote* (Fig. 2).

The height of the above-ground part of plants in the substrate contaminated with NaCl was the smallest, while common meadow-grass did not germinate at all. On the 3rd week of the experiment, the height of control plants of common meadow-grass reached 2.8 cm. On the 4th week of growth, the height of common meadow-grass reached 3.5 cm. On the 5th week of the experiment, common meadow-grass was 4.8 cm high. In the substrate containing NaCl, common meadow-grass did not germinate while in the substrate contaminated with the mixture of 5 g/kg of *Safecote* and NaCl salt grew evenly and increased by 0.4 cm per week on average (Fig. 3).

During the experiment, phytomass of perennial rye-grass was changing depending on contamination of the substrate. After two weeks of growth, phytomass of perennial rye-grass control plants was 8.9 mg; that of plants in the substrate containing 5 g/kg of NaCl – 7.5 mg; in the substrate containing 5 g/kg of *Safecote* – 8.6 mg; and in the substrate contaminated with the mixture of 5 g/kg NaCl/*Safecote* – 7.1 mg. Other weeks of the experiment showed a clear dependence of phytomass on the specificity on substrate contamination. The weakest phytomass of perennial rye-grass was in the substrate containing 5 g/kg NaCl. The phytomass of plants in the substrate

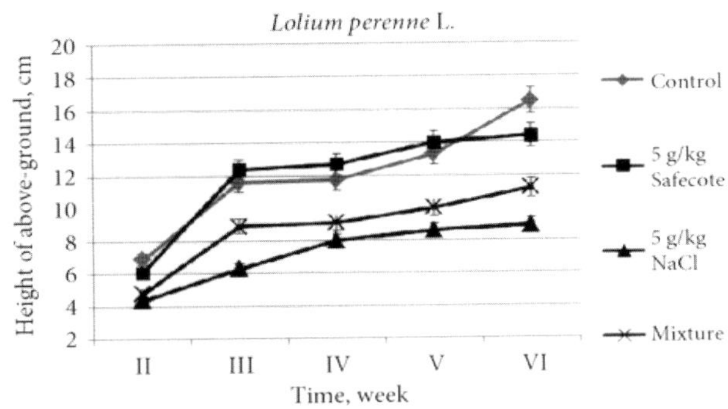

Fig. 1. Change in the height of the above-ground part of perennial rye-grass (*Lolium perenne* L.) (Experiment I)

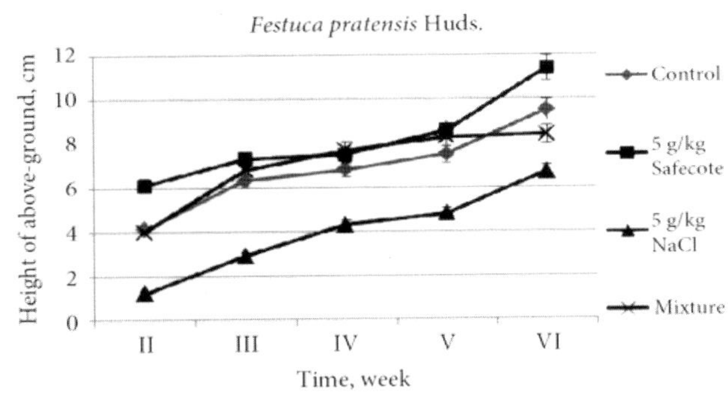

Fig. 2. Change in the height of the above-ground part of meadow fescue (*Festuca pratensis* Huds.) (Experiment I)

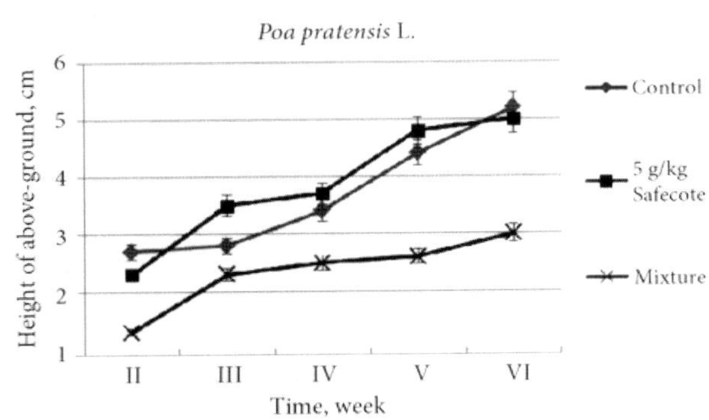

Fig. 3. Change in the height of the above-ground part of common meadow-grass (*Poa pratensis* L.) (Experiment I)

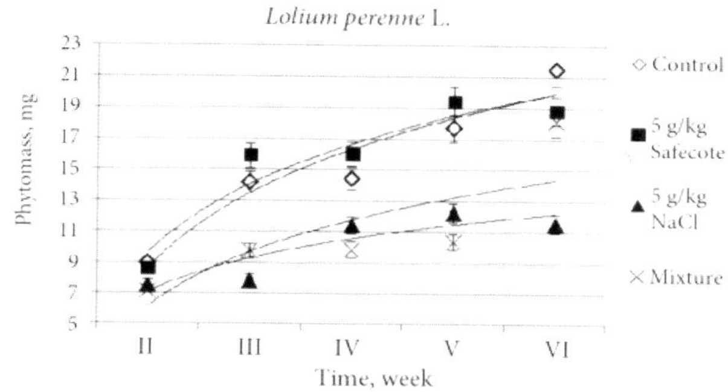

Fig. 4. Change in phytomass of perennial rye-grass (*Lolium perenne* L.) (Experiment I)

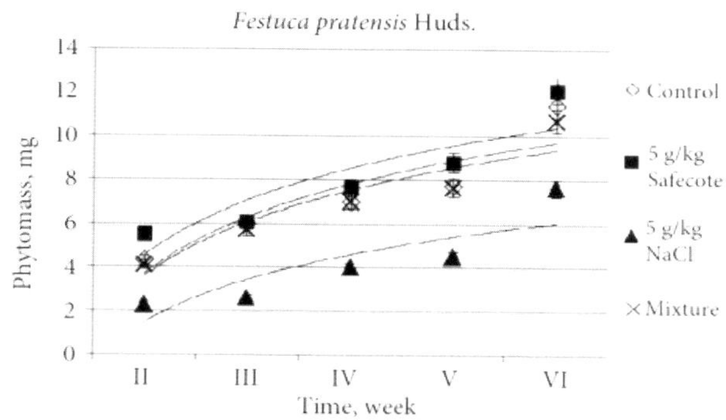

Fig. 5. Change of in phytomass of meadow fescue (*Festuca pratensis* Huds.) (Experiment I)

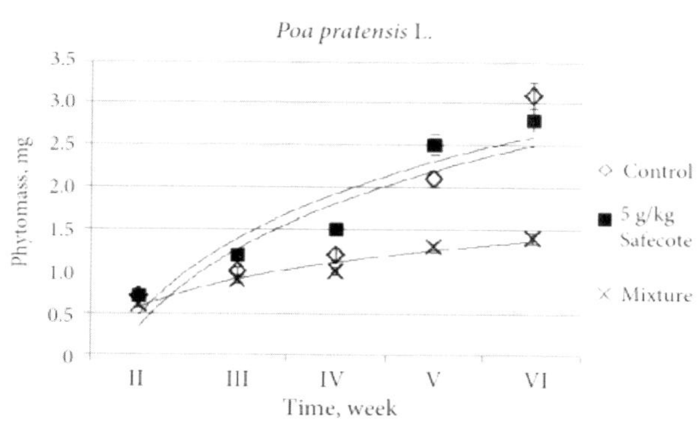

Fig. 6. Change in phytomass of common meadow-grass (*Poa pratensis* L.) (Experiment I)

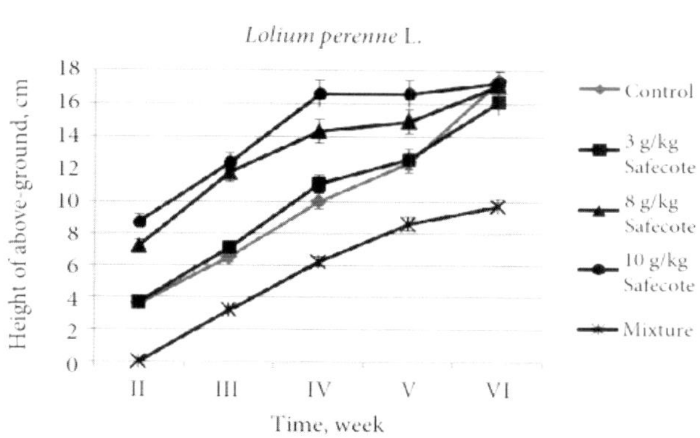

Fig. 7. Change in phytomass of perennial rye-grass (*Lolium perenne* L.) (Experiment II)

contaminated with the mixture of 5 g/kg NaCl and *Safecote*, was bigger compared to the substrate containing only NaCl but by 4 mg smaller than that of control plants. Tor five weeks, in the substrate contaminated only with 5 g/kg *Safecote*, phytomass of perennial rye-grass was on average 1.7 mg bigger than that of control plants but after week 6 it decreased by 2.7 mg (Fig. 4).

Phytomass values of meadow fescue in substrates contaminated only with *Safecote* and the mixture of *Safecote* and NaCl were similar. As Figure 5 shows, phytomass of plants in the substrate contaminated with the mixture of 5 g/kg NaCl and *Safecote* was somewhat smaller than that of control plants, while in the substrate contaminated only with *Safecote*, it was inconsiderably bigger compared to control plants. A significant difference in phytomass of perennial rye-grass in the substrate containing 5 g/kg NaCl and the control substrate, i.e. 3.8 mg, was recorded after growth weeks 3 and 6.

During the first four weeks of the experiment, phytomass of common meadow-grass in differently contaminated substrates was changing evenly. From week 4 to week 6, a significant difference in phytomass change was recorded: in the substrate contaminated with the mixture of 5 g/kg of NaCl and *Safecote*, plants grew by only 0.4 mg, while in the control substrate – 2.1 mg, and in the substrate containing 5 g/kg of *Safecote* – 1.3 mg. Just as in the case of perennial rye-grass, phytomass of common meadow-grass in the substrate contaminated only with *Safecote* was bigger than that of control plants (Fig. 6). During the experiment, common meadow-grass has not germinated in the substrate containing 5 g/kg of NaCl. Within the growth period, phytomass of common meadow-grass grass was the smallest compared to perennial rye-grass and meadow fescue.

Experiment I showed that when added into the substrate, *Safecote* had a positive effect on the growth process of graminaceous vegetation.

Experiment II. As mentioned in the methodology, only two *Safecote* concentrations – 5 g/kg and 0.5 g in the mixture with NaCl – were added to the substrate in the first part of the experiment. To confirm experimental findings, it was decided to continue the experiment by analysing at least several more concentrations of *Safecote* (3 g/kg, 8 g/kg, 10 g/kg) and one more high concentration mixture with NaCl (7.2 g/kg NaCl and 0.8 g/kg *Safecote*).

After the 2nd week, perennial rye-grass had the longest above-ground part compared to plants of the other 2 species, i.e. common meadow-grass and meadow fescue.

After the 2nd week of growth, it has become clear that the higher is *Safecote* concentration, the bigger is the height of the above-ground part of plants. After week 2, all plants growing in the substrate contaminated only with *Safecote* were taller than the control plants. After week 2, none of the plant species germinated in the substrate contaminated with the mixture of 8 g/kg NaCl and *Safecote*. Compared to the control substrate, the height of the above-ground part of perennial rye-grass in the substrate containing 10 g/kg of *Safecote*, differed by 1.9 times (Fig. 7).

The plants significantly grew on the experiment's 3rd week. The height of control plants of perennial rye-grass reached 6.5 cm, that of fescue-grass – 3.4 cm, common meadow-grass – 2.3 cm. The above-ground part height of meadow fescue differed by 2.4 times in substrate containing 10 g/kg of *Safecote*, when compared to control substrate. Perennial rye-grass reached the height of 3.2 cm In substrate contaminated with the mixture of 8 g/kg NaCl and *Safecote*, fescue-grass – 1.5 cm while common meadow-grass did not germinate even after the 3rd week (Figs 7, 8).

Compared to the control substrate on the 3rd week of growth, the height of the above-ground part of common meadow-grass in the substrate containing 10 g/kg of *Safecote* differed by 2.2 times. On weeks 4 and 5 of the experiment, the height of plants was the greatest in the substrate containing 10 g/kg of *Safecote*: on the 4th week, the height of perennial rye-grass reached 16.6 cm, that of meadow fescue – 10.8 cm, common meadow-grass – 7.2 cm; on the 5th week perennial rye-grass was 16.6, fescue-grass 12.7 cm and common meadow-grass 12.0 cm tall (Figs 7, 8, 9).

On the 6th week of growth, perennial rye-grass in differently contaminated substrates showed nearly the same height of its above-ground part, former differences disappeared. The highest perennial rye-grass plants were in control pots and reached 17.4 cm. However, the difference in heights of fescue-grass and common meadow-grass plants in differently contaminated substrates continued to remain significant. The height of fescue-grass and common meadow-grass in all substrates contaminated with *Safecote* was greater than that of control plants that were watered only with ionised water (Figs 7, 8, 9).

In the substrate contaminated with the mixture of 8 g/kg NaCl and *Safecote*, perennial rye-grass and meadow fescue germinated only on the 3 week of the experiment, while common meadow-grass has not germinated during the entire experiment. Every week germinated plants showed an average increase of: perennial rye-grass – 2.4 cm, meadow fescue – 1.4 cm. After 6 weeks of growth, the height of the above-ground part of perennial rye-grass reached 9.7 cm and that of fescue-grass – 5.5 cm. During the experiment, the height of the above-ground part of plants in the substrate contaminated with the mixture of NaCl and *Safecote* was smaller than that of control plants (Figs 7, 8, 9).

Phytomass of perennial rye-grass was changing diversely depending on contamination of the substrate. Phytomass of control plants was evenly increasing by the average of 4 mg per week. Phytomass of plants in the substrate with 3 g/kg *Safecote* content and the substrate containing the mixture of 8 g/kg NaCl and *Safecote* was changing similarly to that of control plants but plants in the substrate contaminated with the mixture of NaCl and *Safecote* had the weakest phytomass. Compared to control plants, it was 0.1 mg less on the 2nd growth week, 3 mg – 3rd week, 0.7 mg – 4th week, 1.9 mg – 5th week, and 1.5 mg – 6th week. Plants in substrates containing 8 g/kg *Safecote* and 10 g/kg *Safecote* had bigger phytomass compared to control plants (Fig. 10).

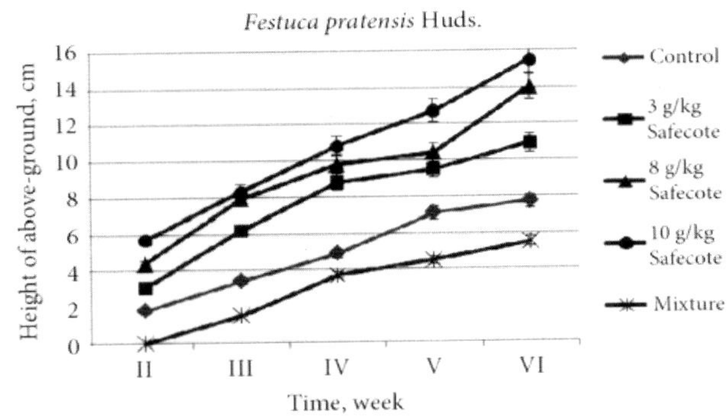

Fig. 8. Change of meadow fescue (*Festuca pratensis* Huds.) height of above-ground (Experiment II)

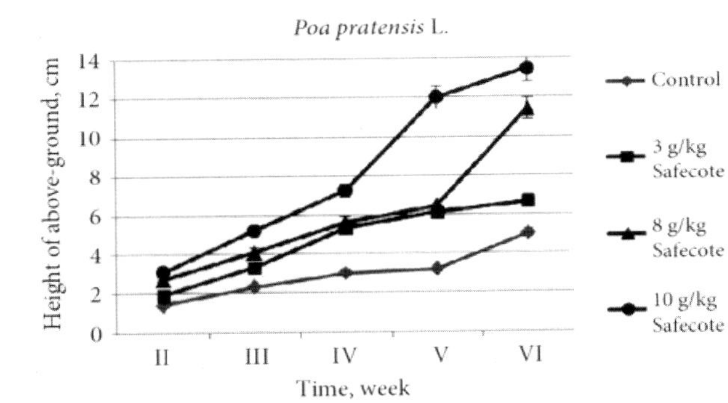

Fig. 9. Change the above-ground height of common meadow-grass (*Poa pratensis* L.) (Experiment II)

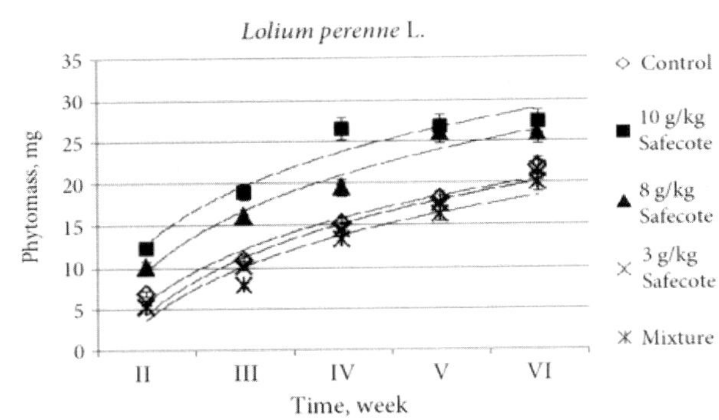

Fig. 10. Change in phytomass of perennial rye-grass (*Lolium perenne* L.) (Experiment II)

As illustrated in Figure 11, the biggest variation in the height of above-ground part of perennial ryegrass was in week when temperature was changed approximately 2 °C. Correlation coefficient of variation of above-ground part of perennial ryegrass and temperature was equal to 0.62.

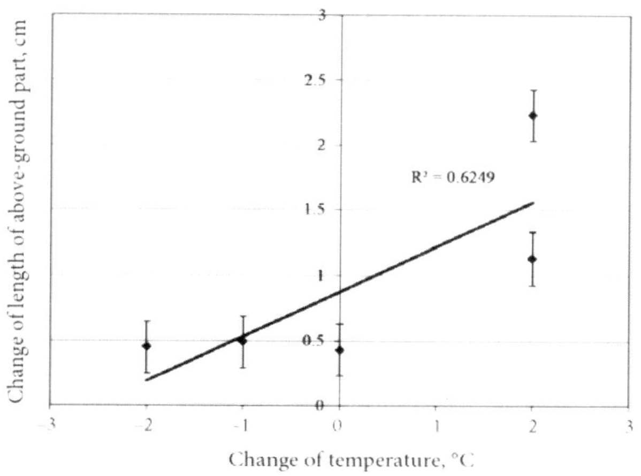

Fig. 11. Variation of the height of the above-ground part of perennial rye-grass (*Lolium perenne* L.) in relation to temperature (Experiment II)

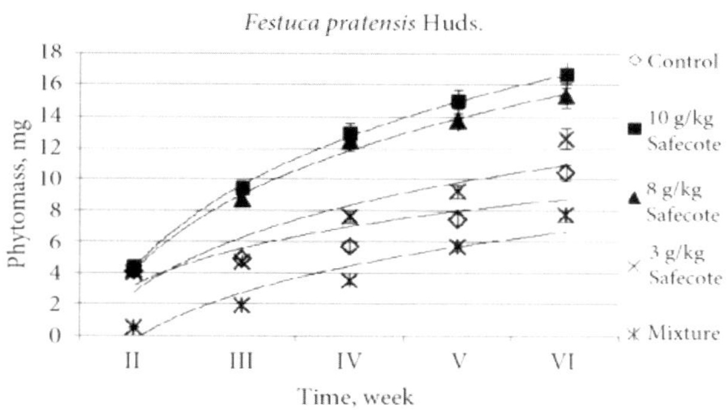

Fig. 12. Change in phytomass of meadow fescue (*Festuca pratensis* Huds.) (Experiment II)

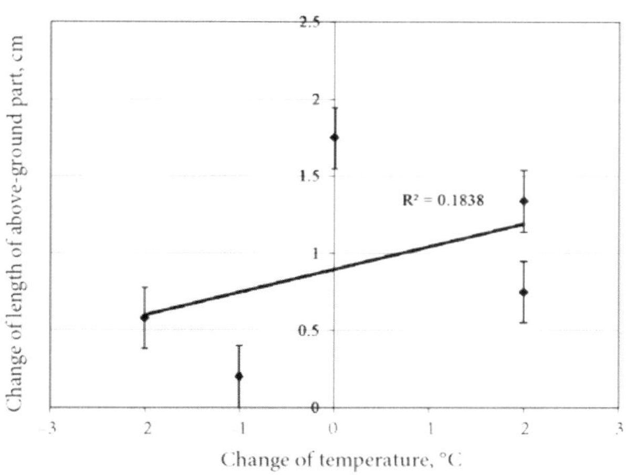

Fig. 13. Variation of the height of the above-ground part of meadow fescue (*Festuca pratensis* Huds.) in relation to temperature (Experiment II)

Obvious differences in phytomass of meadow fescue have been recorded in the substrate with *Safecote* content. The higher was *Safecote* concentration in the substrate, the bigger was phytomass of plants. Phytomass of plants in the substrate contaminated with the mixture of 8 g/kg NaCl and *Safecote*, was weaker than that of control plants: after 2 weeks – 3.5 mg, after 3 weeks – 2.9 mg, after 4 weeks – 2.2 mg, after 5 weeks – 1.7 mg, after 6 weeks – 2.7 mg (Fig. 12).

As illustrated in Figure 13, the biggest variation in the height of the above-ground part of fescue grass was recorded on the week when the temperature was changed to 0 °C. The correlation coefficient of variation of the above-ground part of fescue grass and temperature was equal to 0.18.

During the first weeks of the experiment, phytomass of common meadow-grass was changing gradually and after the 4th week, it amounted to the average of 2.0 mg. Common meadow-grass in the substrate with 10 g/kg *Safecote* content was distinguished by somewhat bigger phytomass. From week 4 to week 6, a big difference in the change of phytomass was recorded: in substrates with 8 g/kg and 10 g/kg of *Safecote* content, plants increased by 7.1 mg and 8.2 mg, respectively. Phytomass of common meadow-grass, just as that of perennial rye-grass, was bigger in pots with higher concentration of *Safecote* (Fig. 14).

As illustrated in Figure 15, the biggest change in the height of the above-ground part of meadow-grass was recorded on the week when the temperature was changed to 0 °C. The correlation coefficient of variation of the above-ground part of meadow-grass and temperature was 0.01.

3. Discussion

J. Bradulienė and S. Vasarevičius (2012) investigated the impact of different concentrations of *Safecote* and *Safecote* mixture with $CaCl_2$ on timothy-grass

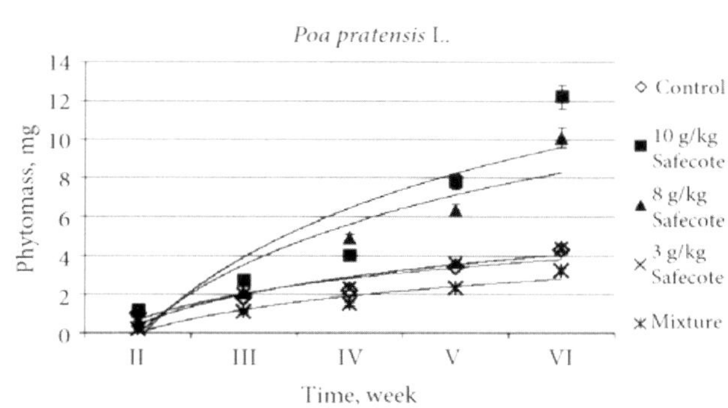

Fig. 14. Change in phytomass of common meadow-grass (*Poa pratensis* L.) (Experiment II)

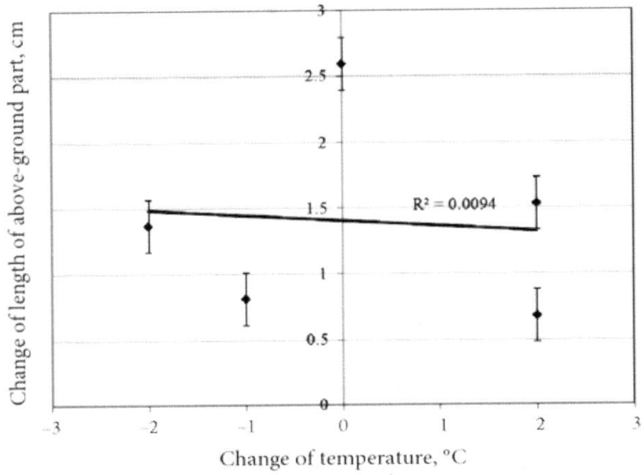

Fig. 15. Variation of the height of the above-ground part of common meadow-grass (*Poa pratensis* L.) in relation to temperature (Experiment II)

(*Phleum pratense* L.), white clover (*Trifolium repens* L.) and crimson clover (*Trifolium incarnatum* L.). These researchers found that the height of herbaceous plants above the ground has a linear dependency on *Safecote* concentrations in the soil, i.e. the greater is the concentration of *Safecote*, the higher part of herbaceous plants was above the ground.

The comparison of results with those of earlier investigations regarding the toxic impact of NaCl on herbaceous vegetation (Baltrėnas *et al.* 2006; Baltrėnas, Kazlauskienė 2007, 2009b; Baltrėnas *et al.* 2012) established that a negative influence of NaCl on vegetation parameters can be reduced by changing some part of NaCl to *Safecote*, e.g. making mixtures. The research results show that technical salts have an adverse effect on the growth of grass plants by slowing down the growth of the above-ground part of plants and limiting their phytomass.

Conclusions

1. Findings of the experiment show that perennial rye-grass had the biggest and meadow grass – the smallest phytomass among the analysed graminaceous plants grown in contaminated substrate.

2. During the experiment, it was determined that perennial rye-grass reached the highest, while meadow grass – the lowest height of the above-ground part of plants.

3. Findings of the experiment also showed that perennial rye-grass had the highest resistance to the toxic effect of salts while common meadow-grass – the lowest as it did not germinate in substrates with big concentrations (5–8 g/kg) of NaCl.

4. *Safecote* infiltrated into the soil had a positive impact on the growth of herbaceous vegetation, e.g. it accelerated the growth of the part of plants above the ground and increased their biomass. The higher was *Safecote* concentration in the soil (in this research, the highest was 10 g/kg), the better it improved the height of herbaceous plants above the ground and added to their phytomass.

5. It has been experimentally proved that the content of NaCl in the substrate has a negative effect on the growth process of graminaceous vegetation, i.e. it retards the increment of the above-ground part and reduces phytomass.

6. Herbaceous vegetation grown in the soil with *Safecote*, sometimes showed higher parameters compared to control plants (even till 50%); consequently, mineral compounds of *Safecote* ingredients can be used as soil fertilisers to improve condition of plants. When inserted into the substrate, *Safecote* had a positive effect on the growth process of graminaceous vegetation, i.e. accelerated the increment of the above-ground part of plants and increases their phytomass. The higher was *Safecote* concentration in the substrate, the greater was the height of the above-ground part and phytomass of graminaceous vegetation. In some cases, graminaceous plants in the substrate with *Safecote* content showed better parameters than those in the control substrate.

7. When NaCl application was reduced by 10%, replacing it with the same amount of *Safecote*, the substrate condition was improved with mineral substances and favourable conditions for plant development were created.

References

Andreikėnaitė, L.; Bakšienė, J.; Vosylienė, M. Z. 2007. Studies of micronuclei and other nuclear abnormalities in blood of rainbow trout (Oncorhynchus mykiss) treated with heavy metal mixture and road maintenance salts, *Acta Zoologica Lituanica* 17(3): 213–219. http://dx.doi.org/10.1080/13921657.2007.10512834

Baltrėnas, P.; Kazlauskienė, A. 2007. Grass vegetation dynamics in soil contaminated with salt, *Ekologija* 53(3): 58–63.

Baltrėnas, P.; Kazlauskienė, A. 2009a. Sustainable ecological development reducing negative effects of road maintenance salts, *Technological and Economic Development of Economy* 15(1): 178–188. http://dx.doi.org/10.3846/1392-8619.2009.15.178-188

Baltrėnas, P.; Kazlauskienė, A. 2009b. Selection of grass vegetation species resistant to road maintenance salts, in J. P. Navarro-Aviñó (Ed.). *Phytoremediation: the green salvation of the world.* Research Signpost, 1–24.

Baltrėnas, P.; Kazlauskienė, A.; Mikalajūnė, A. 2012. *Aplinkos apsauga keliuose:* monografija [Roads Environment Protection. Monograph]. Vilnius: Technika. 384 p.

Baltrėnas, P.; Kazlauskienė, A.; Zaveckytė, J. 2006. Experimental investigation into toxic impact of road maintenance salt on grass vegetation, *Journal of Environmental Engineering and Landscape Management* 14(2): 83–88.

Baltrėnas, P.; Pranskevičius, M.; Lietuvninkas, A. 2011. Investigation and evaluation of carbon dioxide emissions from soil in Neris Regional Park, *Journal of Environmental Engineering and Landscape Management* 19(2): 115–122. http://dx.doi.org/10.3846/16486897.2011.580917

Bradulienė, J.; Vasarevičius, S. 2010. Žvyrkelių dulkėtumo mažinimo tirpalu „*Safecote*" ir jo mišiniu su $CaCl_2$ eksperimentiniai tyrimai [Experimental investigation into reduction of gravel road dustiness using "*Safecote*" solution and its mixture with $CaCl_2$], *Journal of Environmental Engineering and Landscape Management* 18(2): 145–153. http://dx.doi.org/10.3846/jeelm.2010.17

Bradulienė, J.; Vasarevičius, S. 2011. Žvyrkelio dulkėjimo ir aplinkos sąlygų sąryšio statistinis įvertinimas, dulkėjimui mažinti naudojant tirpalą melasos pagrindu [Statistical evaluation of the relationship between gravel road dustiness and environmental conditions using molasses solution for reducing dustiness], *Mokslas – Lietuvos ateitis – Science –Future of Lithuania* 3(5): 5–11. http://dx.doi.org/10.3846/mla.2011.079

Bradulienė, J.; Vasarevičius, V. 2012. The impact on grass vegetation of materials reducing gravel road dustiness, *The Baltic Journal of Road and Bridge Engineering* 7(4): 288–296. http://dx.doi.org/10.3846/bjrbe.2012.38

Bradulienė, J.; Vasarevičius, V. 2014. Investigations of gravel road dust minimization by applying different salt solutions, *Polish Journal of Environmental Studies* 23(3): 943–948.

Burtwell, M. 2004. Assessment of Safecote: new deicer product, in *Proceeding of 6th International Symposium on Snow Removal and Ice Control Technology,* 7–9 June 2004, Spokane, Washington, 529–540.

Burtwell, M.; Wilson, M. 2002. Safecote de-icer in UK winter maintenance, *Highways* 71(10): 5–13.

Cornford, D.; Thornes, J. E. 1996. A comparison between, spatial winter indices and expenditure on winter road maintenance in Scotland, *International Journal of Climatology* 16: 339–357. http://dx.doi.org/10.1002/(SICI)1097-0088(199603)16:3<339::AID-JOC40>3.0.CO;2-U

Darginaviciene, J.; Pasakinskiene, I.; Maksimov, G.; Rognli, O. A.; Jurkoniene, S.; Sveikauskas, V.; Bareikiene, N. 2008. Changes in plasmalemma K^+Mg^{2+}–ATPase dephosphorylating activity and H^+ transport in relation to freezing tolerance and seasonal growth of *Festuca pratensis* Huds, *Journal of Plant Physiology* 165: 825–832. http://dx.doi.org/10.1016/j.jplph.2007.07.009

Dobson, M. C. 1991. *De-icing salt damage to trees and shrubs.* Forestry commission, Bulletin 101, HMSO, London. 64 p.

Hanbali, R. M. 1998. The economic impact of winter road maintenance on road users, *Transportation Research Record* 1442: 429–439.

Jankevičienė, R. 1998. *Botanikos vardų žodynas.* Vilnius: Botanikos instituto leidykla. 524 p.

Kamaitis, Z. 2002a. Damage to concrete bridges due to reinforcement corrosion. Part I. Site investigations, *Transport* 17(4): 137–142.

Kamaitis, Z. 2002b. Damage to concrete bridges due to reinforcement corrosion: part II-design considerations, *Transport* 17(5): 163–170.

Kazlauskienė, A. 2010. Žolinių augalų augimo dinamika natrio formiatu užterštame dirvožemyje [The dynamics of growing herbaceous plants in the soil contaminated with sodium formate], *Mokslas – Lietuvos ateitis – Science – Future of Lithuania* 2(5): 36–40. http://dx.doi.org/10.3846/mla.2010.088

Kazlauskienė, A.; Valentukevičienė, M.; Ignatavičius, G. 2013. Sustainability assessment of heavy metals and road maintenance salts in sweep sand from roadside environment, *Technological and Economic Development of Economy* 19(2): 224–236. http://dx.doi.org/10.3846/20294913.2013.796500

Mikalajūnė, A.; Jakučionytė, L. 2011. Investigation into heavy metal concentration by the gravel roadsides, *Journal of Environmental Engineering and Landscape Management* 19(1): 89–100. http://dx.doi.org/10.3846/16486897.2011.557474

Oškinis, V.; Kasperovičius, T. 2005. Impact of road maintenance salts on water ecosystems according to diatom flora investigation, *Journal of Environmental Engineering and Landscape Management* 13(1): 51–55.

Petkuvienė, J.; Paliulis, D. 2009. Experimental research of road maintenance salts and molasses ("Safecote") corrosive impact on metals, *Journal of Environmental Engineering and Landscape Management* 17(4): 236–243. http://dx.doi.org/10.3846/1648-6897.2009.17.236-243

Sakshaug, K.; Vaa, T. 1995. *Road salting and traffic safety – the effect of road salting on accidents and driving speed.* Report No. STF63 A95005, SINTEF transport engineering, Norway. 59 p.

Skrinskas, S. 2012. International and scientific activities of the Lithuanian road administration are the source of experience, *The Baltic Journal of Road and Bridge Engineering* 7(3): 169–172. http://dx.doi.org/10.3846/bjrbe.2012.23

Stravinskienė, V. 2011. Pollution of "Akmenės cementas" vicinity: alkalizing microelements in soil, composition of vegetation species and projection coverage, *Journal of Environmental Engineering and Landscape Management* 19(2): 130–139. http://dx.doi.org/10.3846/16486897.2011.579449

Thornes, J. E. 2000. Road salting – an international benefit/cost review, in *Proceeding of the 8th World Salt Symposium*, 7–11 May 2000, Hague, the Netherlands 2: 787–791. Elsevier.

Vainalavičiūtė, R.; Kazlauskienė, A.; Baltrėnas, P.; Jankaitė, A. 2009. Investigation of particulate matter and chloride concentrations in the environment of gravelled roads, *Geologija* 51(1–2): 1–11. http://dx.doi.org/10.2478/v10056-009-0001-5

Vitaliano, D. F. 1992. An economic assessment of the social costs of highway salting and the efficiency of substituting a new de-icing material, *Journal of Policy Analysis and Management* 2(3): 397–418. http://dx.doi.org/10.2307/3325069

Vosylienė, M. Z.; Kazlauskienė, A.; Mikalajūnė, A. 2010. Alterations in selected haematological parameters of rainbow trout (*Oncorhynchus mykiss*) exposed to complex mixtures of contaminants, *Environment Protection Engineering* 36(4): 95–104.

Williams, D. D.; Williams, N. E.; Cao, Y. 1999. Road salt contamination of groundwater in a major metropolitan area and development of a biological index to monitor its impact, *Water Research* 34(1): 127–138. http://dx.doi.org/10.1016/S0043-1354(99)00129-3

Žilionienė, D.; Laurinavičius, A. 2007. De-icing experience in Lithuania, *The Baltic Journal of Road and Bridge Engineering* 2(2): 73–79.

Žukauskaitė, A.; Jakubauskaitė, V.; Belous, O.; Ambrazaitienė, D.; Stasiškienė, Ž. 2008. Impact of heavy metals on the oil products biodegradation process, *Waste Management & Research* 26: 500–507. http://dx.doi.org/10.1177/0734242X08089838

Žukauskaitė, A.; Jakubauskaitė, V.; Ambrazaitienė, D.; Zabukas, V.; Paulauskienė, T. 2012. The impact of chemical additives on the process of biodegradation of oil products, *Journal of Environmental Engineering and Landscape Management* 20(1): 17–26. http://dx.doi.org/10.3846/16486897.2011.633336

Agnė KAZLAUSKIENĖ is a researcher of the Research Institute of Environmental Protection, Doctor of Technological Sciences (Environmental Engineering), Dep. of Environmental Protection, Vilnius Gediminas Technical University (VGTU). Author of 14 scientific publications and 1 invention. Research interests: environmental pollution, road maintenance substances, environmental monitoring.

Vaida BRUKŠTUTĖ. Master of Science (Environmental Engineering), Dep. of Environmental Protection, Vilnius Gediminas Technical University (VGTU). Bachelor of Science (Environmental Engineering), VGTU, 2007. Research interests: environmental protection, pollution prevention.

2

SUBJECTIVE EVALUATION OF THE ENVIRONMENTAL QUALITY IN CHINA'S INDUSTRIAL CORRIDORS

Jian XU[a,b], Mei ZHANG[c], Long SHAO[a], Jian KANG[b,a]

[a] *School of Architecture, Harbin Institute of Technology, Harbin 150001, China*

[b] *School of Architecture, University of Sheffield, Sheffield S10 2TN, UK*

[c] *School of East Asia Studies, University of Sheffield, Sheffield S10 2TD, UK*

Abstract. Based on 270 questionnaire surveys in 8 cities of 5 industrial corridors in China, this study aims to examine the effects of industry construction on the evaluation of environmental pollution, natural environment, built environment, personal perception and development and policy. The results show that the evaluations on environmental pollution and landscape design are both below the medium level, but the evaluations of the living comfort and safety are both above the medium level. Further analysis, females usually give lower evaluation scores than males, and age and health situations are negatively related to the evaluation results; People indicate a great desire to reduce the environmental pollution and protect the natural environment. Moreover, the landscape was analysed using colour extraction techniques based on video recording, there are significant correlations between industrial pixel ratio and evaluation results of air quality, vegetation pixel ratio and evaluation results of river water quality, and public facilities pixel ratio and evaluation results of comfort levels.

Keywords: China, industrial corridor, attitude, subjective evaluation, environmental quality.

Introduction

Globally, at present, the industrial distribution experiences an evolution process from decentralized to centralized pattern, particularly in developing countries (Ellison, Glaeser 1999). Industrial cities are rapidly developing into a pattern of "industrial corridors", which are defined as a number of cities, with a maximum spacing of 150 km, say, that are connected by roads or railways that constitute an integrated industrial area and share economic development goals (Xu *et al.* 2015). Based on the statistics of 51 industrial corridors in China, they are normally 100–300 km in length, 30–100 km in width, and the number of cities in an industrial corridor ranges from 3 to 20. This is particularly the case currently in East Asian countries, India, the southwest America, Africa and some other manufacturing-dominant countries.

Over the past decade, as "world's factory floor", China has had a 10.42% average growth rate in Gross Domestic Product (GDP), to which manufacturing contributed 41% (National Bureau... 2014). Such large-scale industrial production activities for economic development significantly led to the serious environmental pollution/contamination: The air pollution index (AQI, including ozone, fine particulate matter, nitrogen dioxide, carbon monoxide, sulphur dioxide and total reduced sulphur compounds) of over 75% of the total number of cities is below health-based standards (HEI International... 2004), rapid increase in cancer incidence in industrial corridor areas (Chen *et al.* 1998) and PM10 and smaller particles (Quan *et al.* 2011) lead to an estimated 2.4 million premature deaths every year (Zhang *et al.* 2010). Half of China's water resources are considered to be unfit to be treated to make it safely drinkable (Zhou *et al.* 2014). Industrial solid waste (ISW) constitutes 75.9% of total waste and was rapidly increasing (Wang *et al.* 2008). Moreover, industrial noise pollution seriously jeopardises the health of workers and nearby residents (Nelson *et al.* 2005). The environmental pollution of industrial corridors influences both natural environment and built environment and the overall environmental quality (European Environment Agency 2012) is affected. Environmental quality is a set of properties and characteristics of the environment, either generalised or local, as they impinge on human beings and other organisms. It is a measure of the condition of an environment relative to

Corresponding author: Jian Kang
E-mail: j.kang@sheffield.ac.uk

the requirements of one or more species and or to any human need or purpose (Johnson *et al.* 1997).

With the attention on worldwide industrial growth, it is equally important to study and understand citizens' attitudes towards the industrial corridor environment. A number of subjective studies on the urban environment (Zhang, Kang 2009), campus environment (Shi *et al.* 2005), residential areas (Wu *et al.* 1995), leisure satisfaction and quality of life of urban central districts (Ngai 2005) have been performed. However, the industrial environmental quality has not been subjectively evaluated.

Therefore, this study aims to subjectively evaluate the environmental quality of industrial corridors of China through a recent survey of 8 cities in 5 industrial corridors. According to a framework of subjective and objective evaluation method of environmental quality (Marans 2003), seventeen key research questions have been analysed, including environmental pollution, natural environment, built environment, personal perception and development and policy. First, this paper reveals the general public's attitude by analysing the preliminary results obtained regarding interviewees' subjective evaluation, and comparisons have been made with other cities like Beijing and some other industrial countries in Europe and America. Then, the correlations of all environmental and social factors are explained. Furthermore, we explain the correlations between contextual and demographical data and the relationship between the objective proportion of different pixels and the subjective results, understand the interaction of subjective evaluation and objective landscape elements. The conclusions can be used to understand people's subjective feelings under the fast industry development situation. Moreover, this paper puts forward some ideas which could be useful for industrial park planning and design.

1. Methodology

1.1. Study areas

To date, there are 51 industrial corridors supported by the Chinese government. Figure 1 shows the distribution of the provincial industrial corridors in China based on a systematic study of the provincial government documents (National Bureau... 2013). 87% of the provinces have one or more industrial corridors, then the cities which have industrial corridors are divided into five types according to the starting year of the development, which cover three consecutive "five-year plan" periods (Casey, Koleski 2011). Given the significant differences in industry type between north and south China, this paper concentrates on north China, and five industrial corridors were selected by stratified random sampling method based on the classification of starting year (Taleb 2005), and then most representative cities within the industrial corridors were selected. The eight selected cities are shown in Figure 1. Although the city population density varied, the population density of industrial areas was approximately the same. Thirty questionnaires were distributed in each city. For the sampled cases, the start years span over a relatively wide time range. The plan areas of the five industrial corridors range from 192 to 837 square kilometres. In addition, the GDP ranges from 4.8 to 10 billion dollars up to 2010.

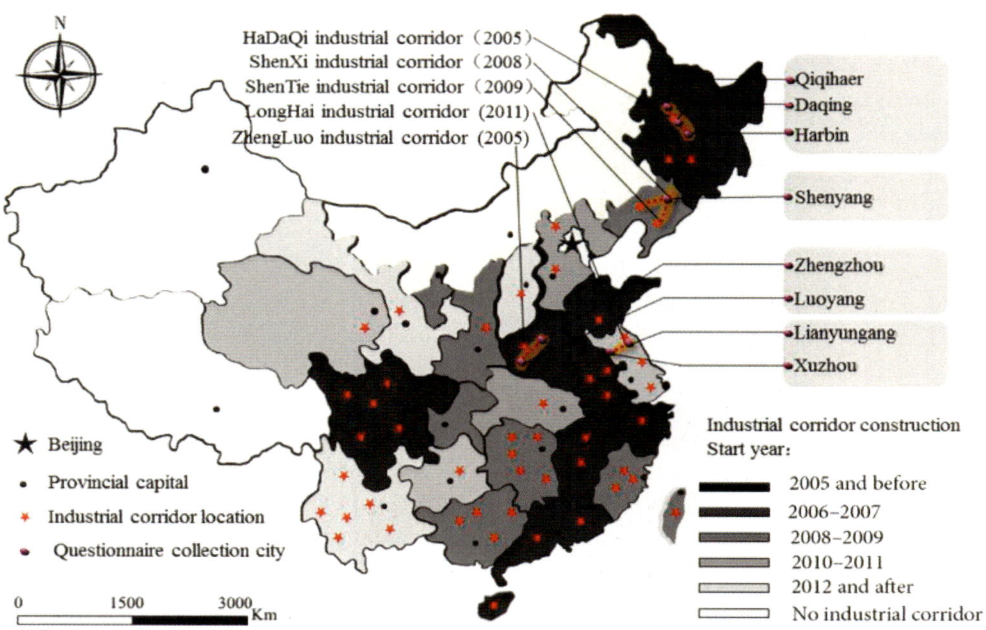

Fig. 1. Provincial industrial corridor distribution of China, including the construction start year and the case study cities, where the numbers in brackets after the industrial corridor refer to the start year of the industrial corridor development

1.2. Landscape characteristics

A video method was used to record and analyse the landscape characteristics in the main roads of industrial zones where questionnaire survey was made. Videos were recorded at noon in clear weather in May 2013, so the shadow colour effect of the environment was minimised. The videos were recorded at a constant normal run speed (9 km/hr) on the main road by motor vehicles, where the questionnaires were distributed in 5 industrial corridors. The average recording distance was 4 km (minimum sample distance $\geq$ 2 km). With this method, 45 videos were recorded in 8 typical cities of 5 industrial corridors, and the total of video recording distance was more than 180 km and the typical characteristics of the sampled main roads of the industrial zones in five industrial corridors are shown in Figure 2.

To identify different landscape/architectural elements (industrial, vegetation, public facilities), the video recording was analysed by developing a MATLAB program for colour identification. First, the distance from the shooting points, the architecture of all of the industrial corridor videos and the proportion of the video images occupied by sky were adjusted to the same values. Then, the video was split into images that comprehensively consider the shooting speed, depth of field and video frame rate. Based on these criteria, to ensure the image continuity as much as possible, an image was selected and extracted every ten frames. To effectively restrain the noise, we used median filter processing. First, we selected multiple objects and 24 * 24 samples for the training set from the centralised images; then, we extracted the colour histogram in the RGB colour space. Examples of actual video capture images and the colour characteristics of the "landscape elements of industrial corridors are shown in Figure 3, where the industrial pixels include industrial buildings, office buildings, fences and other industrial production facilities; the vegetation pixels include lawns and trees of the environmental greening and green sculpture; and the public facilities include roadways and sidewalks. The overall results are shown in Table 1. Further analysis shows that the industry has a correction coefficient of −0.817 ($p \leq 0.01$) with the vegetation, and the vegetation is also correlated with public facilities, with a coefficient of 0.667 ($p \leq 0.05$), using Spearman's rank correlation coefficient.

Table 1. Pixel proportion (percentage) of each landscape element of industrial corridors.

City	Industrial pixels (factory building)	Vegetation pixels (trees and grass)	Public facilities pixels (roadways and side-walks)	The total percentage (except for the sky)
Zhengzhou	31.27%	14.22%	16.28%	61.77%
Luoyang	30.14%	16.28%	14.66%	61.08%
Xuzhou	21.54%	15.14%	11.92%	48.60%
Lianyungang	18.03%	27.24%	17.33%	62.50%
Shenxi	23.02%	16.78%	15.17%	54.97%
Shentie	26.10%	17.85%	12.02%	55.97%
Harbin	22.98%	18.79%	21.08%	62.85%
Daqing	19.08%	23.07%	18.35%	60.60%
Qiqihaer	19.42%	19.07%	17.32%	55.81%
Mean sore	23.51%	18.71%	16.02%	58.24%

A: LongHai industrial corridor
B: ZhengLuo industrial corridor
C: ShenXi industrial corridor
D: ShenTie industrial corridor
E: HaDaQi industrial corridor
1: Xuzhou, Chengdong industrial zone (in city), Changan Avenue to Xinxie Avenue
2: Xuzhou, Jinshanqiao industrial park (in city), Jinqiao road to Pantaoshan road
3: Gongyi, Xihecun industrial zone (out of city), 237 pro-vincial road
4: Zhengzhou, Xicunzhen industrial zone (out of city), National road G310
5: Gongyi, LongWei industrial economic zone (in city), Heluo road
6: Shenyang, TieXi industrial development zone (in city), Kaifa road
7: Shenyang, Shenbei industrial zone(in city),Shenbeixi road
8: Daqing, Daqing industrial economic zone (out of city), Ha-Da high way
9: Harbin, Pingfang industrial economic zone (in city), Songhua road to Dalianbei road

Fig. 2. Typical characteristics of the main sampling road of the industrial zones in the five industrial corridors

Fig. 3. Contrast images of the actual video-captured images and colour characteristics of landscape elements of industrial corridors

1.3. Questionnaire

The questions included environmental pollution, natural environment, built environment, personal perception and development and policy. These five aspects are not only complete interpretation of environmental quality evaluation index (Johnson *et al.* 1997), but also correspond to the five subsystems of "system dynamics model of industrial growth and landscape ecology" (Xu *et al.* 2015).

Environmental pollution involves the conditions of industrial corridor in terms of air, water, solid waste and noise pollution, based on the environmental elements morphological classification, which corresponds with the items included in two recent surveys of Beijing (Wong 2003; Zhang, Kang 2009). Natural environment includes a number of indicators. In general, "flora and fauna", relates to biosphere diversity and also landscape ecological situation (Goudie 2013; Wu 2013; Hobbs 1997). From the objective perspective, the number and survival situation of animals reflects the continuity of the ecological corridors, also reflecting the protection of the natural environment (Mörtberg, Wallentinus 2000). Consequently, the questionnaire was designed based on two topics: "the living situation of animals/birds/insects (bees, butterflies, dragonflies, etc.)" and "landscape greening". Built environment questions were designed to address human activities, public service facilities (Forsberg, von Malmborg 2004). Landscape design includes landscape sculpture design (includes pavilion, gallery, landscape architecture, art sculpture and other garden elements with the function of art and use), waterscape design and application and building height/density. Public service facilities are associated with the degree of public evaluation with recreational physical-activity facilities (fixed facilities and devices that can be used to develop sports fitness, teaching, training and competition and are open to the public) and leisure squares. Moreover, personal perception is an important part of environmental quality, which includes comfort levels (contain both psychic comfort and physical comfort, e.g. housing, catering, transport, infrastructure, health and inner feelings etc.) and security situation (industrial, construction, transportation, etc.) (Andersson *et al.* 2003). For these topics, five-point linear scales were used to indicate the five degrees of evaluation results listed in Table 1: –2, Not satisfied at all; –1, Not satisfied; 0, Medium; 1, Satisfied; and 2, Very satisfied.

Moreover, four questions regarding development and policy were designed at the end of the questionnaire. This set of questions appropriately reflects the public's willingness to slow the construction of an industrial zone to reduce environmental pollution and the destruction of natural environment, and indicates the degree of the public's willingness to complain to the government regarding industrial environmental problems that they face if there is a simple and convenient method to do so. An additional question concerned the public's willingness to donate 0.5% of their monthly income to improve the industrial environment. Five linear scales of agreement were generally used.

To avoid the randomness of questionnaire data, the sample size, statistical analysis of reliability and validity of questionnaires have been carried out, and it was shown that the sample capacity can meet the requirements (Taleb 2005).

1.4. Survey procedure and interviewees

The surveys were performed as face-to-face interview in public spaces as in previous studies (Yang, Kang 2005). In the 8 cities (9 sites), 270 interviews were performed, and 30 questionnaires were distributed in each location with the help of 6 university students; 59.6% of the 270 interviews were administered inside whilst 40.4% were administered outside in urban industrial parks. The interviewees were selected in the case-study sites. In terms of gender, by using simple random sampling method (Taleb 2005), there were 69.3% males and 30.7% females, which generally conforms to the general proportions of males and females in such industrial areas of north China. For each age category, as shown in Figure 4a, there were groups of people aged 10–20, 20–30, 30–40, 40–50, 50–60 and over 60 years; the population age distribution curve is generally close to the Chinese age distribution in 2010 (National Bureau... 2010). The interviewees' residence situations were divided into three conditions, as shown in Figure 4b. Five types of areas in each industrial zone were chosen in which to issue the questionnaires, and the proportions of interviewees in different locations are shown in Figure 4c. Regarding the interviewees' occupations, 8 categories were considered: factory workers, factory managers, famers, business people, teachers, retirees and others.

1.5. Conceptual framework

Corresponding to the questions above, and also with reference to subjective and objective evaluation of environmental quality in previous studies (Marans 2003), the framework of this study is shown in Figure 5. This research is divided into two parts, the subjective questionnaire survey as its main part and objective environmental conditions as a minor part. The research method was designed to show the actual environmental situation of the industrial corridors clearly, and to explore the relationships between the objective environmental factors and the subjective evaluation. At the result analysis level, the conceptual framework includes subjective responses (questionnaire survey), objective conditions (analysis of landscape characteristics pixels), statistical analysis of the correlations among subjective factors (relationships among the 5 categories of factors), and, the effect of demographical/contextual factors and the effect of landscape features. The results will contribute to the understanding of actual conditions in the subjective and objective aspects of China's industrial corridors. Moreover, the results are useful for understanding the evaluation of people with different demographical and contextual background, which in turn can contribute to the improvement of the environment to satisfy different groups of people. Furthermore, by understanding the relationships between subjective

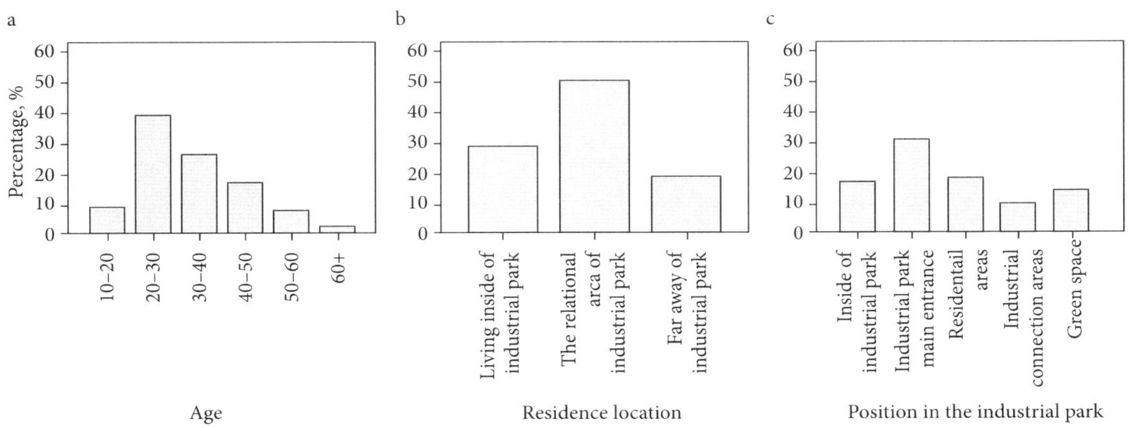

Fig. 4. The distribution of interviewees' age, residence location, and interview position in the industrial parks

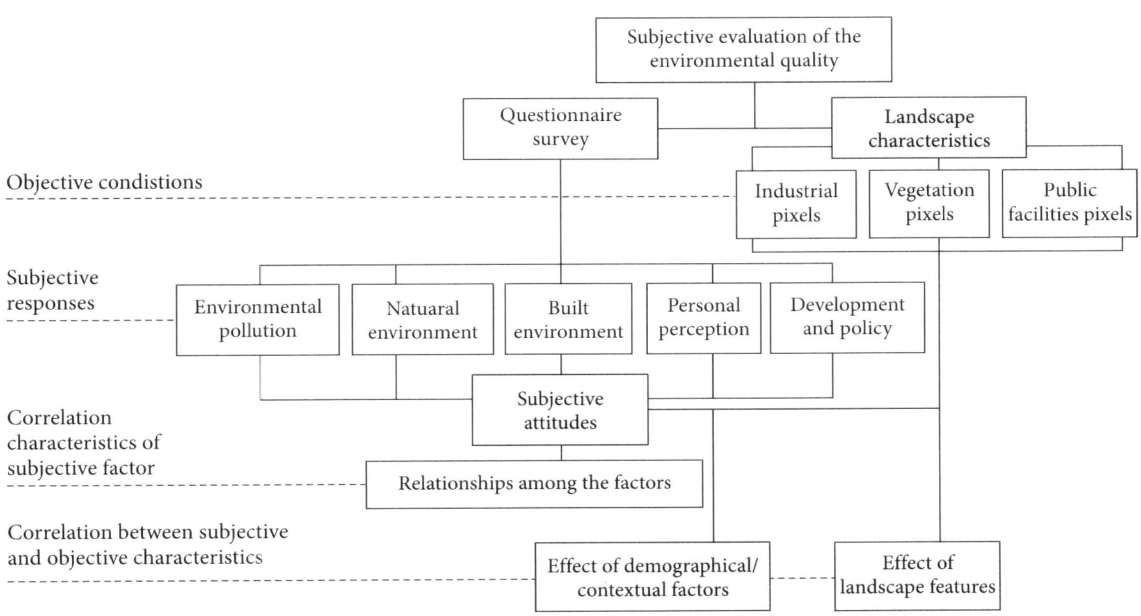

Fig. 5. The framework of the subjective evaluation of the environmental quality of industrial corridors

evaluation results and objective environmental conditions, better planning systems can be established, and this is also useful for developing prediction models on industrial environment system (Xu *et al.* 2015).

2. Subjective attitudes

The results of the statistical analysis of the 13 subjective evaluation questions in China's industrial corridors are shown in Table 2.

2.1. Environmental pollution

In terms of air quality, the mean score is –0.16. Compared with the survey results obtained for a typical seriously polluted city, Beijing (Zhang, Kang 2009) (–0.8, after conversion), indicates that people in this polluted city generally feel worse. In other words, although industrial pollution is the second most common cause of air pollution (Goyal *et al.* 2006), the public's satisfaction with the environment of the industrial corridor is not the lowest that has been reported.

For river water quality, the mean score of evaluation results is –0.16, which is considerably higher than that of a typical polluted city, Beijing (–0.9) (Zhang, Kang 2009). Untreated and insufficiently treated industrial wastewater discharge are the principal drivers of drinking water pollution in cities (Wang *et al.* 2008). Indeed, it has been reported that more than three-quarters of the water flowing through Chinese cities is unsuitable for drinking because of industrial pollution (Zhang *et al.* 2010).

For the ISW treatment, the mean score of evaluation results regarding industrial corridor public is 0.09; in Beijing, the score for litter contamination is –0.7 (Zhang, Kang 2009). It should be noted that the ISW evaluation of industrial corridor tends to be above the medium level, possibly because transportation, landfill, incineration and other treatment of ISW often stay out of the public interests, although ISW has had a huge and long term impact on the environment. Whereas in the city, the garbage recycling process is more relevant to every citizen. It is noted that, currently, 2.7% of the world's ISW is directly discharged into the environment from China; this number is rapidly increasing: compared with 1989, the quantity of ISW generated in 2002 increased by 66% (Huang *et al.* 2006). Therefore, this important environmental problem must be addressed in the industrial corridor development.

Noise has a mean score –0.31, and the SD is 1.001. Although the effects of noise on workers' hearing and work efficiency have been intensively studied (Barron 2002), environmental noise pollution is also an important part of the overall environment (Yang, Kang 2005). Compared with Beijing, –0.9 (Zhang, Kang 2009), the city square environment is subjectively considered to be noisier than the industrial corridors. A possible reason for this

Table 2. Interviewees' evaluation of the industrial corridor environmental quality

Digital codes and questions			Not satisfied at all –2	Not satisfied –1	Medium 0	Satisfied 1	Very satisfied 2	Mean score	SD (standard deviation)
Environmental pollution	1. Air quality		26.7%	33.7%	20.4%	15.6%	3.7%	–0.16	1.141
	2. River water quality		14.1%	26.7%	24.1%	31.5%	3.7%	–0.16	1.128
	3. Industrial solid waste (ISW) treatment		7.4%	30.0%	28.9%	31.5%	2.2%	–0.09	0.998
	4. Noise		12.2%	31.5%	33%	21.5%	1.9%	–0.31	1.001
Natural environment	5. Environmental greening		11.1%	24.1	38.9%	24.8%	1.1%	–	0.972
	6. Animals\birds\insects living situation		31.9%	17.8%	27.8%	19.6%	3.0%	–0.56	1.208
Built environment	Landscape design	7. Landscape sculpture design	30.7%	20.7%	25.6%	20.0%	3.0%	–0.56	1.202
		8. Waterscape design and application	38.1%	25.9%	19.6%	14.1%	2.2%	–0.84	1.149
		9. Building height/ density	2.6%	32.6%	38.1%	24.8%	1.9%	–0.09	0.864
	Public service facilities	10. Physical-activity facilities	7.0%	23.7%	24.8%	34.4%	10.0%	–0.17	1.113
		11. Leisure square	5.9%	24.4%	21.9%	36.3%	11.5%	0.23	1.121
Personal perception	12. Comfort levels		2.6%	7.0%	21.5%	55.6%	13.3%	0.70	0.881
	13. Security situation		4.8%	4.1%	23.7%	42.2%	25.2%	0.79	1.022

result is that industrial parks are generally located in relatively quiet areas, and the noise sources are mainly from industries and traffic, whereas in urban open public spaces there are much more noise types.

In summary, as shown in Table 2, people are generally unsatisfied with the pollution/contamination situation. It is interesting to note that this result corresponds to the results of two surveys that were carried out in Beijing (Wong 2003; Zhang, Kang 2009), where the top three environmental problems are air, water and noise. It should be noted that Beijing showed even lower environment evaluation scores than the industrial corridors (Zhang, Kang 2009). The subjective evaluation of air, river water, solid waste in industrial corridors are below the medium level in general, so that the eco-industrial park concept needs to be more implemented.

2.2. Natural environment

In terms of the living situation of animals/birds/insects, the mean score of evaluation results is –0.56. These results of living situation of animals/birds/insects reflect the total number of animals and migration situation, and demonstrate the health condition of ecosystem from the perspective of biology (Nasi *et al.* 2009). Ecological corridors are believed to facilitate movement between connected patches of a habitat, which increases gene flow, promotes the re-establishment of locally extinct populations, and increases species diversity in otherwise isolated areas (Tilman, Kareiva 1997). The mean score of landscape greening demonstrates that landscape greening of industrial corridors remains to be improved, to ensure the sustainable development of regional ecosystem.

It can be seen that, in terms of both living situation of animals/birds/insects and environmental greening of the natural environment, the evaluation results are under the medium level, suggesting that the construction of the industrial corridors has affected the natural environment of the region. Moreover, there was a possibility that the animal migration corridors and green corridors were cut off, seriously affecting the ecosystem (Mörtberg, Wallentinus 2000).

2.3. Built environment

As shown in Table 2, the scores of landscape sculpture are below the medium level. Moreover, the mean score of evaluation results associated with an artificial waterscape is the lowest among all elements of landscape design. In terms of building height/density, the mean score of evaluation results is notably close to neutral.

With respect to public service facilities, the average evaluation score of physical-activity facilities (public fitness equipment) is under the medium level (–0.17), whereas the average evaluation score of leisure square is above the medium level (0.23).

One possible reason for the lowest score of waterscape design and application is the limitations in waterscape design with cold weather (Ruying 2007) for the Shen Xi, Shen Tie and Ha Da Qi industrial corridors. Moreover, these results show that people have higher artistic requirements for environmental design beyond function; The neutral evaluation of building height/density suggests that the building planning conforms to people's expectation of the industrial corridor environment; In addition, the widespread demand in sports facilities need to be filled in industrial corridors.

2.4. Personal perception

Personal perception is the way a person forms judgements and makes conclusions concerning the characteristics and motives of others or surrounding environment (Gifford 2007). For the 2 relevant aspects of this survey, which are comfort levels and security situation, the evaluation mean scores are 0.7 and 0.79, respectively, which could be perceived as that most industrial corridor areas are relatively comfortable and safe with few accidents.

2.5. Development and policy

For the two questions concerning "Do you think it is right to slow down the construction of industrial parks in order to reduce environmental pollution?" and "Do you think it is right to slow down industrial park construction in order to protect the natural environment?", the agreement mean scores are 1.50 and 1.49, respectively. More details are shown in Table 3. This result is significantly higher than that obtained by the survey conducted in Beijing for "slowing down the construction of industrial parks to reduce environmental pollution" in 2003, in which 40% of the respondents indicated that the top priority should be economic development and the next highest priority should be to protect the environment in both rural and urban areas (Wong 2003). In another survey including the same question and conducted in Beijing in 2009, the average score was 0.6, and the number of "agree" and "strongly agree" responses accounted for 43.9% and 20.6% of the total responses, respectively (Zhang, Kang 2009). One possible reason for these findings is that people tend to gradually gain awareness of environmental pollution and natural environment protection over time. These results indicate that people in industrial polluted area have a greater awareness of the need to protect environment.

For the question "For complaining to the government for industrial environment, do you think there is a simple way?", the agreement mean score is 1.21, which is much higher than that obtained by the survey conducted in Beijing (Zhang, Kang 2009), in which the mean score was only 0.6. Again, there are two possible reasons for this result: change over the past 5 years and the greater willingness of people in industrial corridors to complain. It is noted that before 2000, access to information about complaining to the government and local environmental protection bureaus was notably difficult (Chan, Wong 1994); however, since 2005, public environment hotlines have been established in every major city in China.

The agreement mean score of the item "would you like to give monthly donation for improving industrial environment (0.5% of personal income)?" is 1.35, which is higher than that gathered in the Beijing survey, 0.7 (Zhang, Kang 2009). A similar study on "willing to pay more for environmentally friendly products" show that, in 1989, 67% of Americans stated that they were willing to pay 5–10% more for ecologically compatible products. By 1991, environmentally conscious individuals were willing to pay between 15–20% more for green products. By 2001, 79% of their samples were willing to pay 40% more for green products (Laroche *et al.* 2001). Therefore, similarly to the above trajectory in industrial history, from pollution to governance and the recovery process, the increase in pollution leads to an increase in awareness of environmental protection, and people are willing to do more to protect the environment.

3. Relationships among the factors

Because the subjective evaluation of the industrial corridor environment is a complex affair that depends on

Table 3. Interviewees' evaluation of development and policy

Digital codes and questions	Strongly disagree	Disagree	Neutral	Agree	Strongly agree	Mean score	SD
	−2	−1	0	1	2		
14. Slow down the construction of industrial park to reduce the pollution of the environment	0 (0%)	1 (0.4%)	25 (9.3%)	83 (30.7%)	161 (59.6%)	1.50	0.678
15. Slow down the industrial park construction to protect the natural environment.	0 (0%)	1 (0.4%)	24 (8.9%)	87 (32.2%)	158 (58.5%)	1.49	0.672
16. Complain to the government for industrial environment, if there is a sample way.	0 (0%)	3 (1.1%)	41 (15.2%)	122 (45.2%)	104 (38.5%)	1.21	0.734
17. Monthly donation for improving industrial environment (0.5% of personal income)	1 (0.4%)	2 (0.7%)	31 (11.5%)	103 (38.1%)	133 (49.3%)	1.35	0.741

Table 4. Correlation coefficients and significance levels of the environmental quality factors, where the digital codes are shown in Table 2 and Table 3.

	Environmental pollution				Natural environment		Built environment					Personal perception		Development and policy			
							Landscape design			Public service facilities							
	1	2	3	4	5	6	7	8	9	10	11	12	13	14	15	16	17
1	1	0.363**	0.281**	0.266**	0.103	0.100	0.105	0.118	0.041	0.225**	0.190**	0.117*	−0.090	0.069	0.059	−0.056	0.070
2		1	0.236**	0.163**	0.040	0.001	−0.028	0.044	−0.014	0.098	0.017	0.014*	0.042	0.096	0.089	0.060	0.133
3			1	0.250**	−0.031	0.051	0.018	0.031	−0.022	0.048	0.093	0.067	−0.025	0.045	0.007	0.037	−0.032
4				1	0.020	0.073	0.006	−0.054	0.022	0.157**	0.147*	0.078	0.016	0.058	0.018	−0.075	0.047
5					1	0.309**	0.283**	0.215**	0.173**	0.052	0.083	0.084	−0.004	−0.083	0.008	−0.092	−0.043
6						1	0.858**	0.664**	0.424**	0.203**	0.212**	0.109	−0.038	−0.097	−0.074	−0.101	−0.090
7							1	0.720**	0.403**	0.166**	0.196**	0.142*	−0.032	−0.087	−0.108	−0.164**	−0.102
8								1	0.500**	0.142*	0.164**	0.062	−0.010	−0.098	−0.146*	−0.096	−0.113
9									1	0.129*	0.110	−0.013	0.002	−0.070	−0.059	−0.073	−0.041
10										1	0.426**	0.087	0.102	−0.112	−0.059	−0.083	−0.056
11											1	0.097	0.050	−0.125*	−0.089	−0.021	−0.119*
12												1	−0.083	−0.082	−0.088	−0.258**	−0.037
13													1	.064	−0.004	0.138*	0.038
14														1	0.618**	0.235**	0.849**
15															1	0.232**	0.675**
16																1	0.203**
17																	1

Note: the marks * and ** indicate significant differences or correlations; * represents $p \leq 0.05$, and ** represents $p \leq 0.01$.

various factors, it is essential to study the correlations among those factors in terms of the Spearman correlation coefficient and significance level, Table 4 shows that the factors in each category are mostly closely related, but there are also interesting correlations among different categories.

The environmental pollution category has 6 correlations with built environment, among which air quality (1) is correlated to physical-activity facilities (10) and leisure squares (11), which is expected. Moreover, air quality and river water quality are both correlated to comfort levels (12).

There is a significant correlation between 3 factors of the natural environment category and 2 factors of the landscape design category. This result is identical to Hobbs' view of future landscape and the future of landscape ecology, the connectivity of natural environment is the basis of landscape ecological sustainable healthy development (Hobbs 1997). Based on the analysis of automatic linear modelling, the importance of landscape sculpture design, waterscape design and application, and animals\ birds and insects living situation to environmental greening are 0.043, 0.052 and 0.106, respectively. To further illustrate the correlation, the regression scatter diagrams between environmental greening (5) and the other 3 factors (6–8) are shown in Figure 6.

For the built environment category, interesting correlations with other factors are observed. Regarding public service facilities, both physical-activity facilities (10) and leisure squares (11) are related to air quality (1), noise (2), living conditions of animals/birds/insects (6), landscape sculpture design (7) and waterscape design and application (8), which can be concluded that improving the quality of the air, reducing noise, increasing the connectivity of greening and increasing the quantity and quality of artificial landscape will improve the movement desire of the public. In particular, physical-activity facilities and air quality have the highest correlation coefficient of 0.225**. Furthermore, building height/density (9) is positively correlated to physical-activity facilities, but there is no significant correlation with leisure squares, which shows that the people in industrial parks need more physical-activity facilities rather than square spaces. Regarding the relationship with development and policy, the evaluation result associated with leisure squares (11) is negatively correlated to environmental protection awareness (14) and donation willingness (17).

The comfort levels (12) of industrial corridors is what the public always focus on. The results in Table 4 show that air and water qualities are two strong influencing factors, again indicting they are the bottom line for ensuring basic health, and landscape sculpture design also plays an important role in creating comfort levels. In other words, strengthening the construction of a pavilion, corridor, or seat design and increasing the number of rest areas and the number of sculpture in a given region are useful in improving comfort levels. To further analyse the effect of other factors on the comfort level, automatic linear modelling is used, as shown in Figure 7, in which the comfort levels (target variable) and 15 other factors (predictive variable) are considered. It is observed that the ranking of the factors influencing the comfort levels is gender, location, protection awareness of natural environment, occupation, health condition and age.

For the development and policy category, it should be noted that the awareness to protect the environment is only significantly negatively correlated to the evaluation result of leisure squares, possibly because of the popular square dance, which represents a type movement therapy in China that has developed over the past 20 years (Sigel 1994). Moreover, the awareness to protect the natural environment is negatively correlated to waterscape design and application, although the importance of waterscape design and application to the awareness to protect the natural environment is only 0.176 based on the automatic linear modelling. Indeed, waterscape design is the highest form of landscape design. In general, people with higher design requirements for waterscapes have a greater desire to reduce environmental pollution and protect the natural environment (Litton *et al.* 1971).

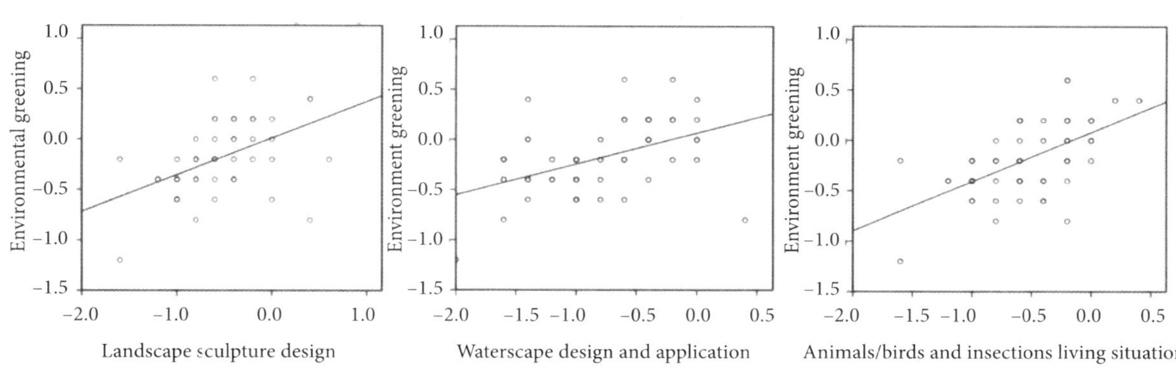

Fig. 6. Regression scatter diagrams between environmental greening and other factors

Complaining is an important aspect of environmental protection. The results in Table 4 show that complaining is correlated to landscape sculpture design, comfort levels and security situation, which suggests that the people who demand landscape sculpture are more demanding of a better living environment and are more willing to complain. It should be noted that complaint behaviour is negatively correlated to comfort levels but positively related to security situation, which suggests that beyond a certain level of comfort and safety, there is a desire to complain. Based on further analysis using automatic linear modelling, if the correlation coefficient increases to 0.20 (identical research method to Fig. 7), except above factors, environmental greening and sports facilities also could influence the willingness of complaining, and the importance are 0.021 and 0.019, respectively.

The willingness to donate monthly is negatively correlated to the evaluation result associated with a leisure square. Further analysis using automatic linear modelling shows that donation is also positively correlated to comfort levels and security situation with an importance level of 0.028 and 0.096, respectively, which suggests that those who like outdoor activities and feel more comfortable and safe are more willing to donate.

4. Effect of demographical/contextual factors

It is also essential to examine the effect of demographical/contextual factors on subjective evaluation, the results of which are shown in Table 5 based on Spearman correlations.

4.1. Demographical factors

In terms of gender, the difference between males and females is significant with respect to air quality (1), landscape sculpture design (7), waterscape design and application (8), personal perception (12, 13) and development and policy (14–17). As shown in Figure 8a, for air quality, landscape sculpture design and waterscape design and application, females feel more unsatisfied than males. However, males feel more comfortable than females, females feel safer than males. This result is similar to that in a previous study, where it was shown that the comfort demand of females is higher, while males have a higher sense of crisis in open air (Sasaki *et al.* 2000). For the four questions on development and policy, the results are shown in Figure 8b, which are similar to the finding of a previous study that "females would pay more for green products" (Laroche *et al.* 2001) and females have higher willingness to protect environment. Moreover, the result shows that males' capacity and tolerance of environment are higher than those of females.

Age correlates to environmental pollution (1–4), environmental greening (5), physical- activity facilities (10) and willingness to donate (17), and the importance levels are 0.708 (average value), 0.426, 0.913 and 0.107, respectively based on the automatic linear modelling. As shown in Figure 8c, people who are older than 50 feel less satisfied than younger people regarding air quality (1) and water quality (2). However, for the ISW treatment (3) and noise (4), there is no linear relationship between age and subjective evaluation. Figure 8d shows that the

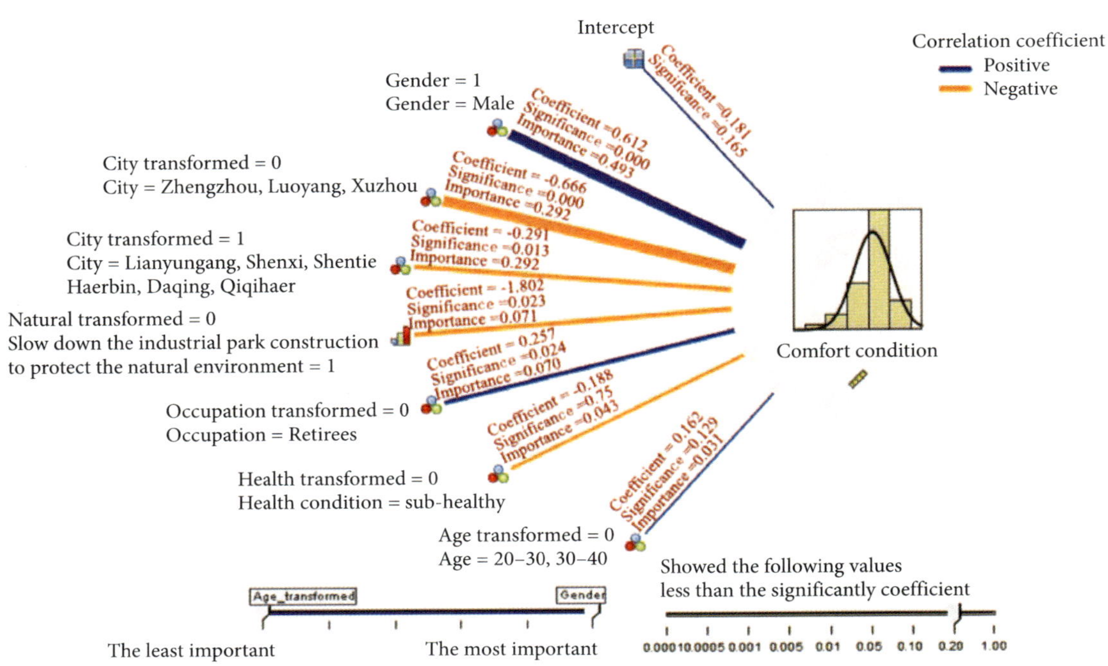

Fig. 7. The correlation between comfort levels and all other factors, based on automatic linear modelling

distribution curves of environmental greening (5) and physical-activity facilities (10) are similar; In addition, the figure shows that people over 50 feel more unsatisfied than the other age categories. This indicates that the elderly pay more attention to physical health, and are more concerned about environmental pollution and natural environmental quality, whereas those in the 10–20-year-old group feel more satisfied than the other age groups. Regarding willingness to donate, people who are aged 50–60 years show less willingness, possibly because people in this age group live under China's biggest economic pressures, they have financial burdens for their parents, children and grandchildren.

Occupation presents an obvious correlation with air quality (1), noise (4), natural environment (5, 6) and landscape design (7, 8), public serves facilities (10, 11), willingness to donate (17), with an importance level of 0.812, 0.673, 0.259, 0.542, 0.361 and 0. 947 (average value) respectively based on the automatic linear modelling. As shown in Figure 8e, retirees have the highest demand for fresh air and quietness; For noise, it is interesting to note that teachers feel more satisfied than others; As shown in Figure 8f, students and retirees, who like outdoor activities the most, feel the most unsatisfied with the landscape; the people who feel the most dissatisfied about the living conditions of animals/birds/insects (6) also feel dissatisfied with physical-activity facilities (10); regarding willingness to donate (17), factory managers, teachers and students, all of whom have better education than others, are the most willing, where the importance level of the factors above is with the range of 0.250–0.470 based on the automatic linear modelling.

The healthy condition correlates to air and river water quality (1, 2), environmental greening (5), landscape sculpture design (7), building height/density (9), public serves facilities (10, 11) and awareness of protecting the natural environment (15). As shown in Figs 8g and 8h, ill and sub-healthy people feel more unsatisfied with the abovementioned aspects, possibly because they have a higher demand for these factors. Correspondingly, they are more willing to protect the natural environment.

4.2. Contextual: location

As shown in Table 5, industrial corridor location correlates to the ISW treatment (3), natural environment (5, 6), landscape design (7, 8), physical-activity facilities (10), security situation (13) and willingness to donate (17). Figure 9a shows the mean scores obtained for environment greening (5), animals\birds\insects living situation (6), landscape sculpture design (7) and waterscape

Table 5. Spearman correlation coefficients among demographical/contextual factors considering location and people's subjective evaluation, with significance levels (2-tailed); mean differences in terms of the subjective evaluation between males and females and between residents inside/outside the city, with significance levels (t-test, 2-tailed). Digital codes are shown in Table 2 and Table 3

Question			Demographical				Contextual: location			
			Gender	Age	Occu-pation	Healthy condition	Industrial corridor	Inside/ outside of the city	Position in industrial park	Residence location
Environmental pollution		1	0.117*	−0.133*	−0.135*	−0.095*	0.008	0.001	0.002	−0.134*
		2	0.031	−0.126*	−0.003	−0.103*	0.089	0.006	0.054	−0.048
		3	0.011	0.125*	−0.010	−0.004	−0.182**	0.134*	0.081	0.033
		4	0.036	0.021*	−0.141*	−0.059	0.053	0.165**	−0.102*	−0.046
Natural environment		5	0.042	−0.118*	−0.184**	−0.361**	−0.111*	0.069	−0.058	−0.115*
		6	0.060	−0.061	−0.268**	−0.078	−0.116*	0.082	−0.050	−0.092
Built environ-ment	Landscape design	7	0.220**	−0.044	−0.236**	−0.122*	−0.158**	0.094	−0.039	−0.090
		8	0.139*	−0.001	−0.145**	−0.089	−0.231**	0.016	0.016	0.039
		9	0.042	−0.071	−0.093	−0.127*	−0.070	0.062	0.020	−0.069
	Public service facilities	10	0.005	−0.136*	−0.228**	−0.055*	0.123*	0.059	−0.049	−0.031
		11	0.004	0.039	−0.127*	−0.135*	0.037	0.008	−0.065	0.049
Personal perception		12	0.351**	−0.079	−0.108*	−0.084	0.015	0.059	−0.024	0.025
		13	0.365**	−0.034	−0.125*	−0.097	−0.142**	0.085	−0.039	0.017
Development and policy		14	0.155**	−0.067	−0.024	−0.006	−0.022	0.171**	−0.100	−0.044
		15	0.199**	−0.080	−0.047	0.118*	0.076	0.133*	−0.070	−0.125*
		16	0.610**	0.060	0.143	0.056	0.013	0.061*	−0.003	−0.016
		17	0.123*	−0.118*	0.002**	−0.025	0.153**	0.187**	−0.086	−0.047

design and application (8) among the five industrial corridors. Overall, only the environmental greening of the Shen Xi industrial corridor can satisfy the public, with a mean evaluation score of 0.067. Moreover, all people feel dissatisfied with the living conditions of animals/birds/insects, landscape sculpture design and waterscape design and application. In this respect, the worst situation occurs in the Ha Da Qi industrial corridor. In general, the above-mentioned results suggest that the entire landscape of the industrial corridors should strengthen design and management. As shown in Figure 9b, in terms of the ISW treatment (3), physical-activity facilities (10), and donation issues, there are considerable differences among the industrial corridors, possibly due to different construction times, completion degree of construction, scale, and regional differences of industrial corridors. For security situation (13), all of the mean evaluations of the industrial corridors indicate that the respondents are "satisfied".

The survey locations of the cities (inside/outside of the city) significantly affect the evaluation result with ISW treatment, noise, development and policy aspects (17). As shown in Figure 9c, for the ISW treatment (3), the mean evaluation score inside the city is –0.25, but the mean evaluation score outside of the city is 0.083. For noise (4), it is interesting to note that the results show that people in the city consider their surroundings to be less noisy (–0.174) than do people outside the city (–0.505). One reason for this finding may be that most interview locations outside

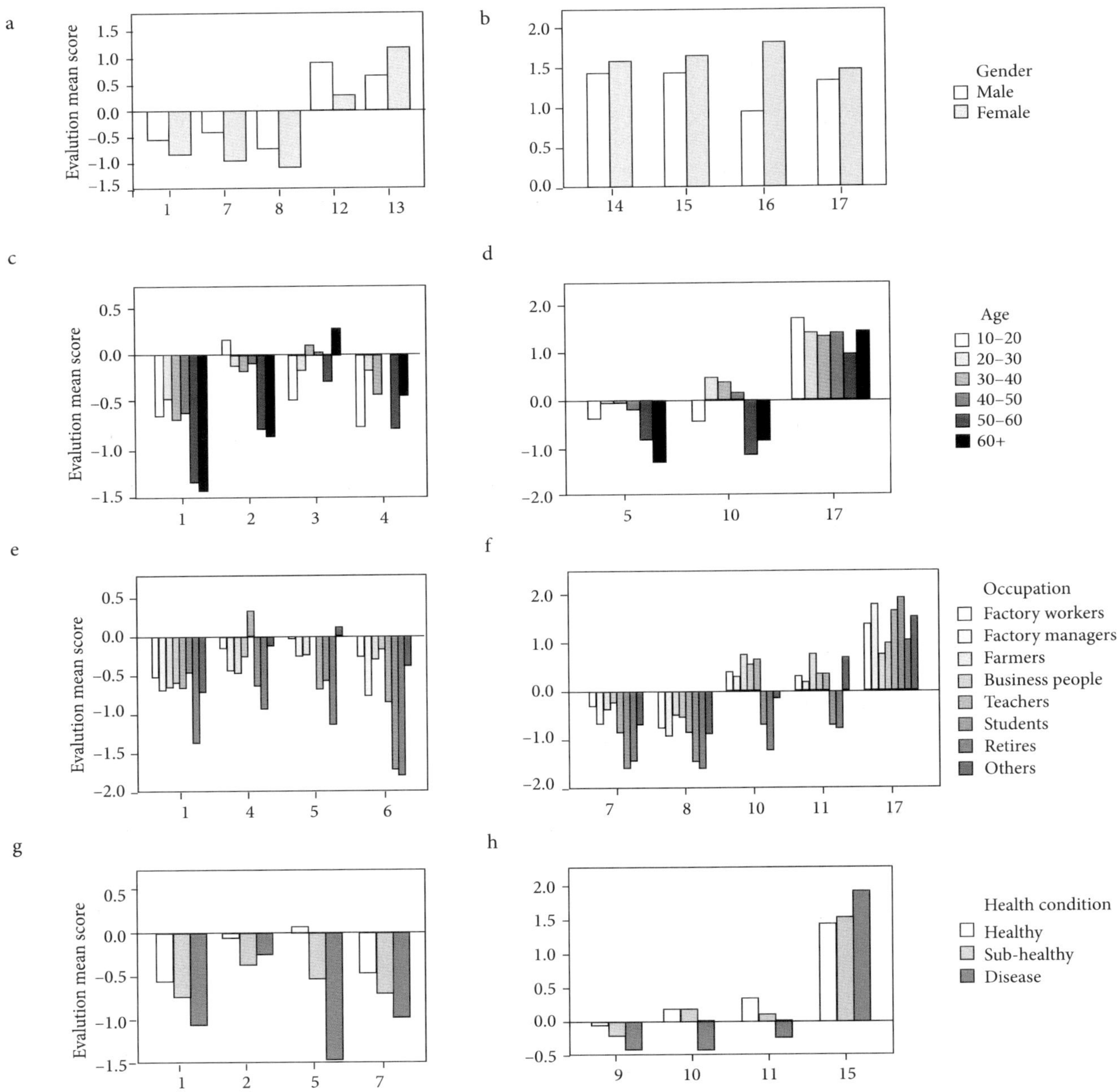

Fig. 8. Mean evaluation scores with different demographical factors, with the digital codes shown in Table 2 and Table 3.

the city are located on a highway, or the environment outside the city is relativity quiet, which raises awareness of industrial and traffic noise. For the 4 questions regarding development and policy (14–17), as shown in Figure 9d, the environmental protection consciousness of people interviewed inside a city is clearly greater than that of people interviewed outside of a city.

Figure 9e shows that the evaluation result on noise conditions (4) is associated with different positions in an industrial park, and the evaluation score of industrial connection areas and green space are lower than others. According to the measurement, traffic noise is present in 90% of the time, with a peak value reaching 72.7 dB, due to heavy trucks (Yong *et al.* 2012), which is higher than the environmental noise standards of urban regional (China) (State environmental... 1999), and also considerably higher than the comfortable level for people to have conversation (Yang, Kang 2005).

Residence location correlates to air quality (1), environmental greening (5) and environmental awareness to protect the natural environment (15). As shown in Figure 9f, for air quality and environmental greening, people who live inside an industrial park are more satisfied than others. Two possible reasons for this finding are the relatively low building density in such areas and adaptability (Sundstrom, E., Sundstrom, M. G. 1986). In addition, people who live far away are less willing to protect the natural environment, possibly because this industrial environment does not change their living environment immediately, leading a lack of sense of responsibility.

4.3. Effect of landscape features

Based on the analysis described in 2.2, the respective proportions of industrial elements, vegetation elements, and public-facility elements were calculated by the aforementioned MATLAB program using videos of the industrial corridors; the relationships between the landscape features and subjective evaluation are shown in Table 6. It is interesting to note that the industrial pixels is correlated to the subjective evaluation score of air quality and the

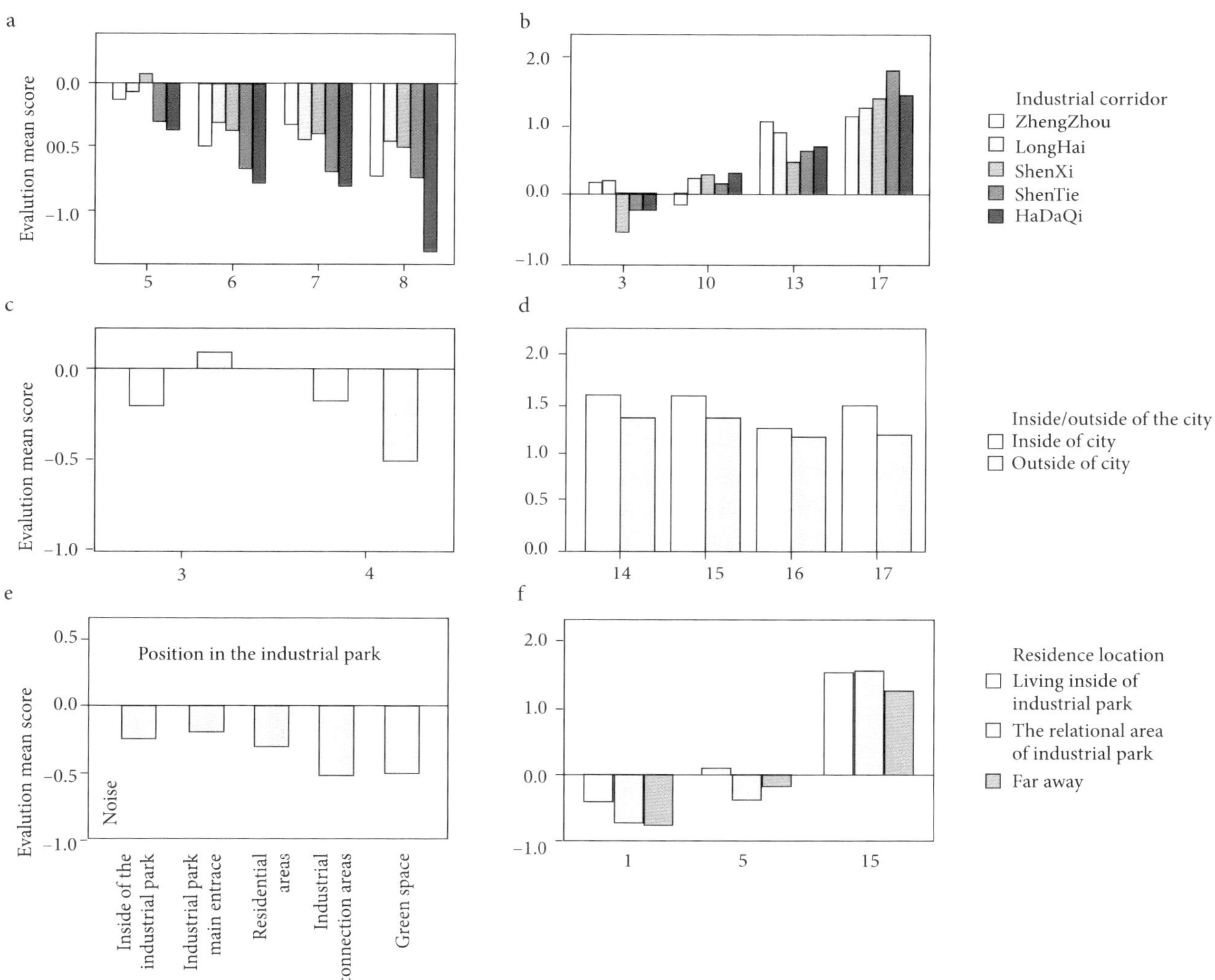

Fig 9. Mean evaluation scores with different contextual factors

Note: The digital codes shown in Table 2 and Table 3

building height/density with correlation coefficients of −0.733* and 0.210*, respectively. This clearly shows that industrial activity has affected the perceptions of air quality. Moreover, the vegetation pixels is positively correlated to the perceptions of river water quality, environmental greening and fauna, with correlation coefficients of 0.672*, 0.233* and 0.427*, respectively. Furthermore, the public facilities pixels is correlated with the perceptions of sports and comfort levels, with correlation coefficients of 0.636* and 0.050*, respectively, which demonstrates that building more public facilities and landscape sculpture can increase people's desire for sports and comfort levels. These correlations suggest that there might be a strong effect of visual characteristics on the subjective evaluation of environmental quality. Similar interactions have also been observed in other studies, such as visual-audio interactions (Yang, Kang 2005).

In terms of the development and policy aspect, Table 6 indicates that the proportions of industrial pixels and vegetation pixels are correlated with people's willingness to complain and the proportion of public facilities pixels is correlated with people's willingness to donate. These results suggest that people's willingness to complain derives from the imbalance between industrial development and natural environment. Moreover, if people find themselves in a better public infrastructure situation, their willingness to donate will be encouraged.

Table 6. Relationships between the landscape features and subjective evaluation. The digital codes are shown in Table 2–3, (* represents $p \leq 0.05$).

Questions			Industrial pixels	Vegetation pixels	Public facilities pixels
Environmental pollutions		1	−0.733*	0.533	−0.083
		2	−0.387	0.672*	0.471
		3	0.100	−0.300	−0.200
		4	−0.418	0.234	0.075
Natural environment		5	0.067	0.233*	−0.433
		6	0.042	0.427*	−0.544
Built environment	Landscape design	7	0.360	−0.460	-0.310
		8	−0.042	−0.210	-0.479
		9	0.210*	0.210	-0.269
	Public service facilities	10	−0.469	0.577	0.636*
		11	−0.170	0.094	0.145
Personal perception		12	0.083	−0.417	0.050*
		13	0.217	−0.500	−0.317
Development and policy		14	0.515	−0.194	−0.169
		15	−0.393	0.544	−0.033
		16	0.429*	−0.403*	0.311
		17	−0.234	0.577	0.318*

Conclusions

This study explores the subjective evaluation of China's industrial corridor environment based on field surveys and public attitudes towards environmental pollution.

In summary, although the evaluation results of environmental quality are generally at the medium level in China's industrial corridors, the evaluation result indicated by the environmental pollution evaluation of the industrial corridors is even higher than that observed for a typical polluted city, Beijing. In addition, the public generally feels satisfied with the comfort levels and security situation in the industrial corridors. Moreover, compared with Beijing, with respect to development and policy, people who evaluated lower score of pollution and natural environment, are more willing to complain to the government, and have a greater desire to contribute to improving the environment, but compared with developed countries, the consciousness of personal dedication to protect the environment is still relatively low.

Based on the relationship between all of the evaluation factors in the industrial corridors, it is observed that people have a greater demand for physical-activity facilities than leisure squares; Air and river water quality, landscape sculpture design and environmental greening have the greatest effect on comfort levels; the people who have higher requirements for waterscape design have a stronger desire to reduce environmental pollution and protect the natural environment; People, based on the premise of ensuring comfort and safety, desire to complain to the government regarding the environment; Moreover, people who like outdoor activities feel much safer and are more willing to donate.

In addition, this study expounds the effects of personal characteristics, location attribute and landscape factors on subjective evaluation. Regarding gender, females are more sensitive for the environment and landscape and are more positive about protecting the environment than males are; In terms of age and health condition, people who are over 50 years of age and ill and sub-healthy people are more sensitive to environmental problems; With respect to occupation, students and retirees have a greater demand for physical-activity facilities; Factory managers, teachers and students, who generally have a better education than others, have greater desire to donate for improving industrial environment; In terms of the willingness to donate, with respect to contextual factors, it is interesting to note that the interviewees who were located within an industrial park inside a city perceived less noise and indicated a stronger awareness of protecting the environment than did the interviewees located outside of a city; Moreover, it is interesting to note that evaluation of noise is related to survey location in industrial parks. Furthermore, people who live far away are less willing to protect the natural environment of industrial corridors.

In terms of the relationship between the landscape characteristics pixels and subjective evaluation, it has been revealed that the industrial pixels directly relate to the subjective evaluation of air quality, and vegetation pixels directly relate to the subjective evaluation of water quality. Moreover, the proportions of industrial pixels and vegetation pixels are related to people's willingness to complain. Furthermore, if people find themselves in a better public infrastructure situation, their willingness to donate will be encouraged.

China has a considerable contribution to worldwide pollution, and it would be interesting to understand Chinese people's subjective evaluation of industrial environmental quality, which would be relevant to the pollution control and environmental improvement. Moreover, it is also important to understand the attitude of people who work or live in the rapid developing industrial corridors in China, as reference to other developing countries and developed countries. Furthermore, the results of this research would be useful for making comparisons with the developed countries when they were in the similar historical process.

This research also puts forward some ideas which could be relevant to environment planning and design of industrial corridors and industrial parks.

It is noted that this research scope is industrial corridors of north China, so that the conclusions may not be generalised in general situation in China.

Acknowledgments

The authors acknowledge the financial support of the National Natural Science Foundation of China "Heritage corridor conservation of Industrial cultural landscape of the Chinese eastern railway" (No. 51278141).

References

Andersson, C.; Hellervik, A.; Lindgren, K.; Hagson, A.; Tornberg, J. 2003. Urban economy as a scale-free network, *Physical Review* E 68: 036124.
http://dx.doi.org/10.1103/PhysRevE.68.036124

Barron, R. F. 2002. *Industrial noise control and acoustics*. CRC Press. http://dx.doi.org/10.1201/9780203910085

Casey, J.; Koleski, K. 2011. *Economic U-C, Commission SR. Backgrounder: China's 12th five-year plan*. US-China Economic and Security Review Commission.

Chan, H. S.; Wong, K. K. 1994. Environmental attitudes and concerns of the environmental protection bureaucrats in Guangzhou, People's Republic of China: implications for environmental policy implementation, *International Journal of Public Administration* 17: 1523–1554.
http://dx.doi.org/10.1080/01900699408524953

Chen, V. W.; Andrews, P. A.; Wu X. C.; Correa, C. N.; Fontham, E. 1998. Cancer incidence in the industrial corridor: an update, *The Journal of the Louisiana State Medical Society* 150(4): 158–167.

HEI International Scientific Oversight Committee. 2004. Health effects of outdoor air pollution in developing countries of Asia: a literature review. Boston: Health Effects Institute (Special Report No. 15).

European Environment Agency. 2012. Glossary: Environmental Quality (definition) [online], [cited 01 Jan 2015]. Available from Internet: http://www.eea.europa.eu/. Copenhagen: Environmental Terminology and Discovery Service.

Ellison, G.; Glaeser, E. L. 1999. *The geographic concentration of industry: does natural advantage explain agglomeration?*. Harvard Institute of Economic Research.

Forsberg, A.; von Malmborg, F. 2004. Tools for environmental assessment of the built environment, *Building and Environment* 39(2): 223–228.
http://dx.doi.org/10.1016/j.buildenv.2003.09.004

Gifford, R. 2007. *Environmental psychology: Principles and practice*. Colville: Optimal books.

Goyal, S.; Ghatge, S.; Nema, P.; Tamhane, S. 2006. Understanding urban vehicular pollution problem vis-a-vis ambient air quality–case study of a megacity (Delhi, India), *Environmental Monitoring and Assessment* 119: 557–569.
http://dx.doi.org/10.1007/s10661-005-9043-2

Goudie, A. S. 2013. *The human impact on the natural environment: past, present, and future*. John Wiley and Sons.

Hobbs, R. 1997. Future landscapes and the future of landscape ecology, *Landscape and Urban Planning* 37: 1–9.
http://dx.doi.org/10.1016/S0169-2046(96)00364-7

Huang, Q.; Wang, Q.; Dong, L.; Xi, B.; Zhou, B. 2006. The current situation of solid waste management in China, *Journal of Material Cycles and Waste Management* 8: 63–69.
http://dx.doi.org/10.1007/s10163-005-0137-2

Johnson, D.; Ambrose, S.; Bassett, T.; Bowen, M.; Crummey, D.; Isaacson, J.; Johnson, D.; Lamb, P.; Saul, M.; Winter-Nelson, A. 1997. Meanings of environmental terms, *Journal of Environmental Quality* 26(3): 581–589.
http://dx.doi.org/10.2134/jeq1997.00472425002600030002x

Laroche, M.; Bergeron, J.; Barbaro-Forleo, G. 2001. Targeting consumers who are willing to pay more for environmentally friendly products, *Journal of Consumer Marketing* 18(6): 503–520.
http://dx.doi.org/10.1108/EUM0000000006155

Litton, R. B.; Tetlow, R. J.; Sorensen, J.; Beatty, R. A. 1971. *An aesthetic overview of the role of water in the landscape*. New York: Water Information Centre.

Mörtberg, U.; Wallentinus, H. G. 2000. Red-listed forest bird species in an urban environment–assessment of green space corridors, *Landscape and Urban Planning* 50(4): 215–226. http://dx.doi.org/10.1016/S0169-2046(00)00090-6

Marans, R. W. 2003. Understanding environmental quality through quality of life studies: the 2001 DAS and its use of subjective and objective indicators, *Landscape and Urban Planning* 65(1): 73–83.
http://dx.doi.org/10.1016/S0169-2046(02)00239-6

Nasi, R.; Koponen, P.; Poulsen, J. G.; Buitenzorgy, M.; Rusmantoro, W. 2009. *Impact of landscape and corridor design on primates in a large-scale industrial tropical plantation landscape*, in: Plantation forests and biodiversity: Oxymoron or opportunity?. Springer, 181–202.

National Bureau of Statistics of China. 2010. *China Statistical Yearbook 2010*. Beijing: China Statistics Press.

National Bureau of Statistics of China. 2013. *China Statistical Yearbook 2013*. Beijing: China Statistics Press.

National Bureau of Statistics of China. 2014. *China Statistical Yearbook 2014*. Beijing: China Statistics Press.

Nelson, D. I.; Nelson, R. Y.; Concha-Barrientos, M.; Fingerhut, M. 2005. The global burden of occupational noise-induced hearing loss, *American Journal of Industrial Medicine* 48: 446–458. http://dx.doi.org/10.1002/ajim.20223

Ngai, V. T. 2005. Leisure satisfaction and quality of life in Macao, China, *Leisure Studies* 24: 195–207. http://dx.doi.org/10.1080/02614360412331313502

Quan, J.; Zhang, Q.; He, H.; Liu, J.; Huang, M.; Jin, H. 2011. Analysis of the formation of fog and haze in North China Plain (NCP), *Atmospheric Chemistry and Physics* 11: 8205–8214. http://dx.doi.org/10.5194/acp-11-8205-2011

Ruying, G. 2007. Discussion on waterscape design of northern cities based on the water-saving gardens, *Journal of Anhui Agricultural Sciences* 35: 11094.

Shi, J.; Ge, J.; Hokao, K. 2005. Campus lifestyle and its relationship with residential environment evaluation: a case study of hangzhou city, China (Architectural/Urban Planning and Design), *Journal of Asian Architecture and Building Engineering* 4: 323–330. http://dx.doi.org/10.3130/jaabe.4.323

Sigel, J. G. 1994. Dance movement therapy in China, *American Journal of Dance Therapy* 16: 53–56. http://dx.doi.org/10.1007/BF02359241

Sundstrom, E.; Sundstrom, M. G. 1986. *Work places: The psychology of the physical environment in offices and factories*. CUP Archive.

Sasaki, R.; Yamada, M.; Uematsu, Y.; Saeki, H. 2000. Comfort environment assessment based on bodily sensation in open air: relationship between comfort sensation and meteorological factors, *Journal of Wind Engineering and Industrial Aerodynamic* 87(1): 93–110. http://dx.doi.org/10.1016/S0167-6105(00)00018-0

State Environmental Protection Administration of China. 1999. Environmental noise standards of urban regional (China) GB3096-1993.

Taleb, N. 2005. *Fooled by randomness: The hidden role of chance in life and in the markets*. Random House Incorporated.

Tilman, D.; Kareiva, P. M. (Eds). 1997. *Spatial ecology: The role of space in population dynamics and interspecific interactions*. Vol. 30. Princeton University Press. ISBN 9780691016528

Wang, M.; Webber, M.; Finlayson, B.; Barnett, J. 2008. Rural industries and water pollution in China, *Journal of Environmental Management* 86: 648–659. http://dx.doi.org/10.1016/j.jenvman.2006.12.019

Wong, K. K. 2003. The environmental awareness of university students in Beijing, China, *Journal of Contemporary China* 12: 519–536. http://dx.doi.org/10.1080/10670560305472

Wu, J. J. 2013. *Landscape ecology. Ecological system*. Springer, 179–200.

Wu, S. X.; Li, J. P.; Huo, Y.; Zhang, S. M. 1995. Evaluation of factors affecting the living and environmental quality of resident areas, *Acta Scientiae Circumstantiae* 3.

Xu, J.; Kang, J.; Shao, L.; Zhao, T. Y. 2015. System dynamic modelling of industrial growth and landscape ecology in China, *Journal of Environmental Management* 161: 92–105. http://dx.doi.org/10.1016/j.jenvman.2015.06.026

Yang, W.; Kang, J. 2005. Acoustic comfort evaluation in urban open public spaces, *Applied Acoustics* 66: 211–229. http://dx.doi.org/10.1016/j.apacoust.2004.07.011

Yong, T.; Li, Y. F.; Li, Z. L. 2012. Study on scheme comparison of acoustic highway environment protection measures, *Shanxi Architecture* 5: 87.

Zhang, J.; Mauzerall, D. L.; Zhu, T.; Liang, S.; Ezzati, M.; Remais, J. V. 2010. Environmental health in China: progress towards clean air and safe water, *The Lancet* 375: 1110–1119. http://dx.doi.org/10.1016/S0140-6736(10)60062-1

Zhang, M., Kang, J. 2009. Subjective evaluation of urban environment: a case study in Beijing, *International Journal of Environment and Pollution* 39: 187–199. http://dx.doi.org/10.1504/IJEP.2009.027151

Zhou, Y.; Hu, S. T.; Xi, B, Su J.; Hao, F.; Wu, J. 2014. Status and challenges of water pollution problems in China: learning from the European experience, *Environmental Earth Sciences* 1–12.

Jian XU, BSc, MSc (North China University of Forest), is a final year PhD candidate, jointly at the School of Architecture, University of Sheffield, UK, and the School of Architecture, Harbin Institute of Technology. Her research field is landscape, planning and architectural technology. She has published 10 journal and conference papers.

Mei ZHANG, BA (Beijing University), PhD (University of Cambridge), is a Lecturer at the University of Sheffield, UK. Her field is environmental sustainability in East Asian, evidenced by over 40 publications.

Long SHAO, BSc, MSc, PhD (Harbin Institute of Technology) is a Professor at the School of Architecture, Harbin Institute of Technology, China. His research field is landscape. He has published 40 journal and conference papers.

Jian KANG, BEngArch MSc (Tsinghua University, Beijing), PhD (University of Cambridge), has been Professor at the University of Sheffield, UK, since 2003. His field is built environment sustainability, evidenced by 70+ prestigious engineering-consultancy projects, 80+ funded research projects, and 800+ publications.

3

EFFECTS OF URBAN GREEN BELTS ON THE AIR TEMPERATURE, HUMIDITY AND AIR QUALITY

Chunyang ZHU[a], Peng JI[b], Shuhua LI[c]

[a]*Huazhong Agricultural University, No. 1, Shizishan Street, Hongshan District, 430070 Wuhan Hubei Province, China*

[b]*Heilongjiang Bayi Agricultural University, No. 5, Xinfeng Road, Longfeng District, 163319 Daqing Heilongjiang Province, China*

[c]*School of Architecture, Tsinghua University, Haidian District, 100084 Beijing, China*

Abstract. As urbanization increases, designing green space that offers ecological benefits is an increasingly important goal of urban planning. As a linear green space in an urban environment, green belts lower air temperature, increase relative humidity, and improve air quality. To quantify the ecological effects of urban green belts and to identify a critical width for effective urban green belts, we analysed the width of urban green belts in terms of their effects on air temperature (T), relative humidity (RH), concentration of negative air ions (NAI) and bacteria rate (BR). The air T, RH and NAI from 8:00 to 18:00 and BR at 9:00 over seven days were investigated on six widths of green belts (0–10 m, 10–20 m, 20–30 m, 30–40 m, 40–50 m and over 50 m) along the west Fourth Ring Road of Beijing in April, July, October and December 2009. We found that (1) the T-RH benefits increased with the width of the green belts, and the 6 m belt had the smallest effect on T-RH, followed by the 16 m and 27 m belts, whereas the effect was obvious with the 34 m belt and conspicuous and stable with the 42 m belt (approximately 80% green coverage) ($P < 0.05$); (2) the critical width reference value of urban green belts for an obvious effect on the increase in NAI concentration was approximately 42 m (approximately 80% green coverage) ($P < 0.05$) and the NAI concentration increased with the width of green belts even in July; and (3) the positive effect on the decrease in the BR was greater than the negative effect, the BR decreased with the green belt width and the changes in the BRs were stable with the 34 m belt. The results of this study may help urban planners and designers achieve urban green space designs that optimize ecological effects and cultural benefits.

Keywords: green belt, width, temperature-humidity (T-RH), negative air ions (NAI) concentration, bacteria rate (BR), ecological environment benefit.

Introduction

Rapid urbanization is accompanied by high concentrations of urban populations and built-up areas in many countries, creating a need to plan green space for greater ecological effectiveness (Fitzsimons *et al.* 2012; Smith 1993). As a linear green space in an urban environment, green belts are appropriate for various types of urban areas, and they contribute significantly to urban green space in terms of size, function, and location (Rohling 1998; Schiller 1997; Wu *et al.* 2012). Green belts can function as an urban ecological corridor, introducing natural airflow into inner urban areas from the suburbs to help counter the urban heat island effect (Gunn 2007). If green belts along rivers are sufficiently wide, they can improve the microclimate and impede the formation of large areas of heat blocks (Hathway, Sharples 2012). In addition to the vegetation, the thermal characteristics of urban green spaces are affected by factors such as the area and shape of the zone (Cao *et al.* 2010). With their highly marginalized characteristics, green belts decrease the urban heat island effect to a greater degree than green space that are circular, rectangular or other forms (Wang, Lin 2006). Generally, research efforts have focused on the temperature-humidity benefits of green belts in areas near rivers. The width of riverine green belts has been shown to affect the decrease in temperature and the increase in humidity in spring, summer and autumn; the effect is strongest in summer (Ji *et al.* 2013). In addition, significant research has focused on establishing the distribution of heatislands and the status of urban greening using Remote Sensing and Geographic Information Systems (Cao *et al.* 2010; Feyisa *et al.*

Corresponding author: Shuhua Li
E-mail: 155175563@qq.com

2014; Mackey *et al.* 2012; Maimaitiyiming *et al.* 2014). The results have shown that an urban corridor system can have a clear influence on the thermal environment over a distance of 300 m (Du 2007). In addition, green belts play an important role in improving air quality (Islam *et al.* 2012). The concentration of negative air ions and bacteria rates are important indices for evaluating air quality (Tikhonov *et al.* 2004) and play an important role in improving environmental indicators, but related research has mainly focused on forest vegetation communities (Liang *et al.* 2014).

Quantifying the ecological effects of urban green belts is important for urban green space planning and construction. For example, understanding the exact extent to which a specific green belt design would impact urban ecological attributes and whether these attributes increase with the increasing width of the green belt could help create a tool that city planners and designers could use to manage urban ecological attributes more efficiently. In this study, we investigated the effects of green belts on temperature (T), relative humidity (RH), and the concentration of negative air ions (NAI) and bacteria (BR) using individual measurements.

The aims of the study are as follows: (1) to quantify the effects of urban green belts on T, RH, NAI and BR throughout the season and (2) to identify a critical width for effective urban green belts based on their ability to decrease T, increase RH, increase NAI and decrease BR.

1. Data and methods

1.1. Study site

The experimental area is situated at a latitude of 39° 57′ north and a longitude of 116° 14′ east, and the experimental area has a warm, semi-humid climate; it is cold and dry in winter and hot and rainy in summer. The annual average temperature is approximately 11.8 °C. The lowest temperature is in January, with an average temperature of –4.7 °C, and the highest temperature is in July, with an average temperature of 26.1 °C. The annual precipitation is 626 mm; the peak is in summer with 75% of the annual precipitation in the period from June to August. The annual evaporation is approximately 1800–2000 mm, which is greater than the precipitation.

The structure of Beijing's green space system is "ring-shaped" and "wedge-shaped", and the "ring-shaped" structure has a heterogeneous distribution along the six ring roads in Beijing (Fig. 1). The study area is located in the west Fourth Ring Road, and the belt plots were located in a linear green space along the west Fourth Ring Road with a green cover of 70–90% approximately 12 km from city centre (Tiananmen Square). The length of the study area was approximately 2.5 km (Fig. 1). All plots had similar plant communities and structure, which ensures consistent environmental conditions, and the widths varied according to the gradient. Six plots of different widths (0–10 m, 10–20 m, 20–30 m, 30–40 m, 40–50 m, and over 50 m) were selected for evaluation. Each plot had 70–80% green cover; one side was adjacent to a walkway while the other side was adjacent to the roadside. The control was also located along the west Fourth Ring Road; it was required to be bare, hard ground not affected by vegetation, and its surrounding environment was similar to that of the plots (Fig. 2). The meteorological conditions and traffic flow counts during the testing periods at the study site are given in Table 1. The meteorological data, including environmental T, RH, wind speed and pressure, were measured at five time intervals in April, July, October and December. The highest average air T for the five time intervals appeared in July, followed by April, October and December, while the peak value was observed at noon; the average RH was highest in July, followed by October, April and December, and the maximum and minimum values were detected in the morning and at noon, respectively; the highest average pressure was observed in December, followed by April, October and July, and the peak value was detected in the morning. In addition, the average wind speed was less than 0.4 m/s, and the ranges varied from 0.2 m/s to 0.9 m/s across four seasons at the testing points. There was no obvious difference of the traffic flow at the study site between seasons, and there were peaks in both the morning and evening of approximately

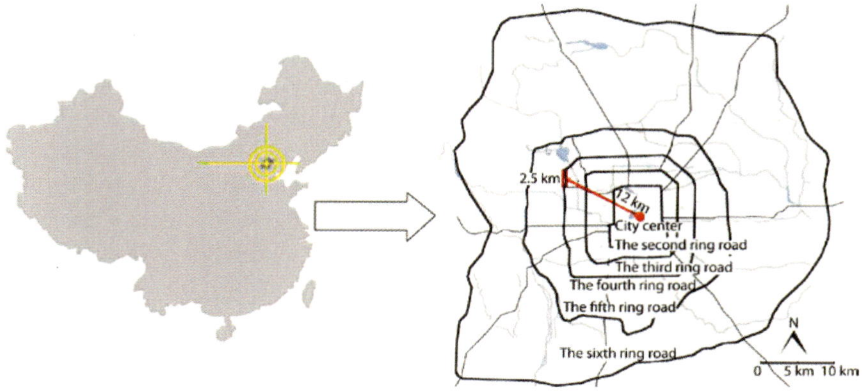

Fig. 1. Location of six green belts along the west Fourth Ring Road of Beijing

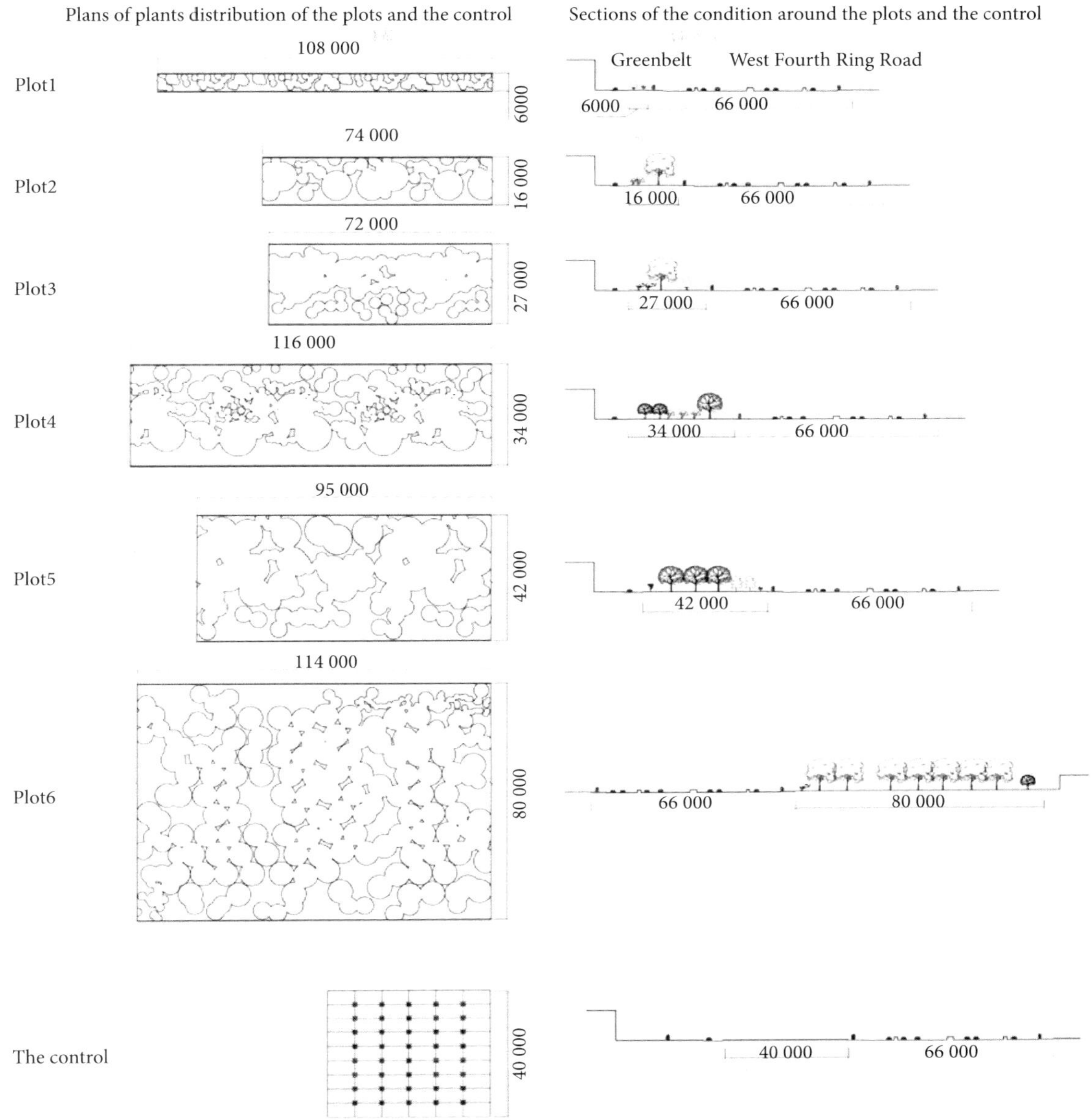

Fig. 2. Diagrams of plants distribution within the green belts and sections of the areas around the green belts

Table 1. Meteorological conditions and traffic flow counts during the testing periods at the study site

Experimental period	Time intervals	T(°C)	RH (%)	Wind speed ($m \cdot s^{-1}$)	Pressure (hPa)	Traffic flow counts ($Veh \cdot h^{-1}$)	Weather
April	8:00–10:00	16.2	33.4	0.2	1011.1	14,000–19,000	Clear
	10:00–12:00	23.3	22.1	0.2	1011.3		
	12:00–14:00	27.7	20.6	0.5	1009.7		
	14:00–16:00	25.5	19.7	0.9	1008.2		
	16:00–18:00	23.6	22.7	0.2	1007.9		
	Average	23.3	23.7	0.4	1009.6		
July	8:00–10:00	32.0	58.3	0.3	999.5	13,000–19,000	Clear
	10:00–12:00	36.0	37.2	0.2	999.9		
	12:00–14:00	38.3	33.2	0.5	998.2		
	14:00–16:00	39.1	32.7	0.7	995.8		
	16:00–18:00	36.6	43.4	0.2	995.6		
	Average	36.4	41.0	0.38	997.8		

End of Table 1

Experimental period	Time intervals	T(°C)	RH (%)	Wind speed ($m \cdot s^{-1}$)	Pressure (hPa)	Traffic flow counts ($Veh \cdot h^{-1}$)	Weather
October	8:00–10:00	14.6	46.8	0.4	1009.8	14,000–19,000	Clear
	10:00–12:00	20.1	29.7	0.5	1010.0		
	12:00–14:00	23.6	24.5	0.5	1008.4		
	14:00–16:00	24.2	23.7	0.4	1007.1		
	16:00–18:00	21.4	29.6	0.2	1006.7		
	Average	20.8	30.9	0.4	1008.4		
December	8:00–10:00	–3.7	21.6	0.4	1024.8	14,000–20,000	Clear
	10:00–12:00	2.9	11.2	0.3	1024.8		
	12:00–14:00	4.2	9.0	0.3	1018.3		
	14:00–16:00	6.1	7.9	0.6	1020.0		
	16:00–18:00	0.4	14.0	0.2	1020.0		
	Average	2.0	12.7	0.36	1021.6		

20,000 $Veh \cdot h^{-1}$. There were 13,000–14,000 $Veh \cdot h^{-1}$ at noon and 15,000–18,000 $Veh \cdot h^{-1}$ during other time intervals. Vehicle speed was less than 40 km/h during peak traffic flow.

The basic data from the plant communities are given in Table 2.

In Table 2, the ratio of canopy density had been measured by the canopy projection method, an accurate method for the determination of small-scale vegetation. First, the trees were positioned before the horizontal distance from the edge of the crown to the trunk was measured in different directions. Second, the vertical projection diagram was drawn on grid-shaped paper using the measured data, the repeated calculation of overlapping trees was eliminated, and the total areas of the crown projection were calculated. Finally, the canopy projection was calculated with the ratio of area of crown vertical projection/area of green.

Table 2. Basic characteristics of the six green belts

Green belt width (m)	Plant Constitution	Planting density (plants/m^2)	Canopy Density (%)
6	*Sabina chinensis, Prunus triloba*	0.095	70–80
16	*Sophora japonica, Sabina chinensis, Lonicera maackii*	0.036	70–80
27	*Sophora japonica, Sabina chinensis, Prunus triloba, Forsythia suspensa*	0.035	70–80
34	*Salix matsudana cv.* Pendula, *Sabina chinensis, Lonicera maackii, Forsythia suspensa, Ligustrum×vicaryi*	0.039	70–80
42	*Salix matsudana cv.* Umbraculifera, *Fraxinus chinensis, Sophora japonica, Pinus tabulaeformis, Sabina chinensis, Prunus ceraifera cv.* Pissardii	0.033	70–80
80	*Salix matsudana cv.* Pendula, *Robinia pseudoacacia, Forsythia suspensa*	0.030	75–85

1.2. Methods

To test T and RH, mesh points were set in the plots. Each plot was divided into 4 sections with replications; each section was evenly divided into 8 test points longitudinally and 3 replications transversely. 40 points were selected as the control. T and RH were measured 1.5 m above the ground.

To accurately measure T and RH, the T and RH probes were located in the shadow of the crown or observer to avoid the inaccurate data caused by direct solar radiation.

To test the NAI concentration, the stationing method was used with the T and RH distribution. NAI was measured at 1.5 m above the ground around the east, west, north and south, respectively.

In the paper, the q and CI were the evaluation index of the NAI concentration, as shown in the following formula:

$$q = \frac{n^+}{n^-}; \quad (1)$$

where $n+$ is the positive ion rates; $n-$ is the negative ion rates; and

$$CI = \frac{n^-}{1000q}. \quad (2)$$

To test the BR, the plate Medium Composed Method was used. The petri dish number was calculated after 5 min in the air, followed by culturing at 37 °C for 48 hours.

The stationing method was used with the T and RH distribution, and BRs were measured 1.5 m above the ground.

The BR per cubic meter was calculated by the Ormeilians formula. The formula is as follows:

$$\mathrm{BR/m^3} = \frac{5000N}{A \times t}, \tag{3}$$

where N, A and t represent the BRs in plates, area of the Petri dish (cm²), and time exposed in the air (min), respectively.

The experimental period was in April, July, October and December of 2009. The data on air T, RH and NAI concentrations as well as the meteorological conditions were calculated in sunny and windless weather from 8:00 to 18:00 every day for 3 days in six plots; measured every 2 hours and recorded for 2 minutes of each test point at 8:00, 10:00, 12:00, 14:00 and 16:00 at each time interval; and the BR was calculated at 9:00.

1.3. Testing instrument

T and RH were detected by a temperature and humidity recorder (FUSO 8829, Japan).The device had a measuring range of –40 to 85 °C and 0–100% RH with a resolution of 0.1 °C and 0.1% RH. The test requirement for measurements was to stay within the plot and control for 5 minutes before testing to ensure a stable T and RH.

The NAI concentration was analysed by an anion concentration tester (ZL1-DLY-5G, USA) that had a measuring range of 10–1.999×10⁹ ions/cm³, a resolution of 10 ions/cm³, and an error ≤±10%. The ion mobility was 0.15 cm²/ V·sec, the linear flow rate of air sampling was 180 cm/sec, the response time constant was approximately 15 sec, and the rated working environment had a temperature range from –0 to 40 °C and a relative humidity ≤90%.

The BR was tested using an agar plate medium in a petri dish. The agar plate medium consisted of peptone, glucose, and beef extract.

1.4. Data processing

The data were analysed using SPSS 17.0 and Eviews 7.0 software. Descriptive statistical analyses were performed to evaluate the average, variability, standard deviation and regression for each parameter measured and the autocorrelation processing for the T and RH measured. ANOVA was performed according to the 3 days replications design. The means were compared using the Duncan's multiple range test at the 0.05 level of probability.

2. Results

2.1. Effects of different widths of green belts on air temperature and relative humidity

2.1.1. Air temperature and relative humidity changes over five time intervals

The air T and RH gap for each plot was analysed compared with the control for all five time intervals respectively.

The effect on T changed gradually with time according to the comparison of the six plots over the five time intervals (Fig. 3). In July and October, the peaks were at 12:00–14:00 and 14:00–16:00. In April, the peaks were at 12:00–14:00 and 16:00–18:00. The effects differed by season and across the five time intervals. In December, all of the green belts had increased T during cooler parts of the day, that is, at 8:00–10:00 and 16:00–18:00, whereas

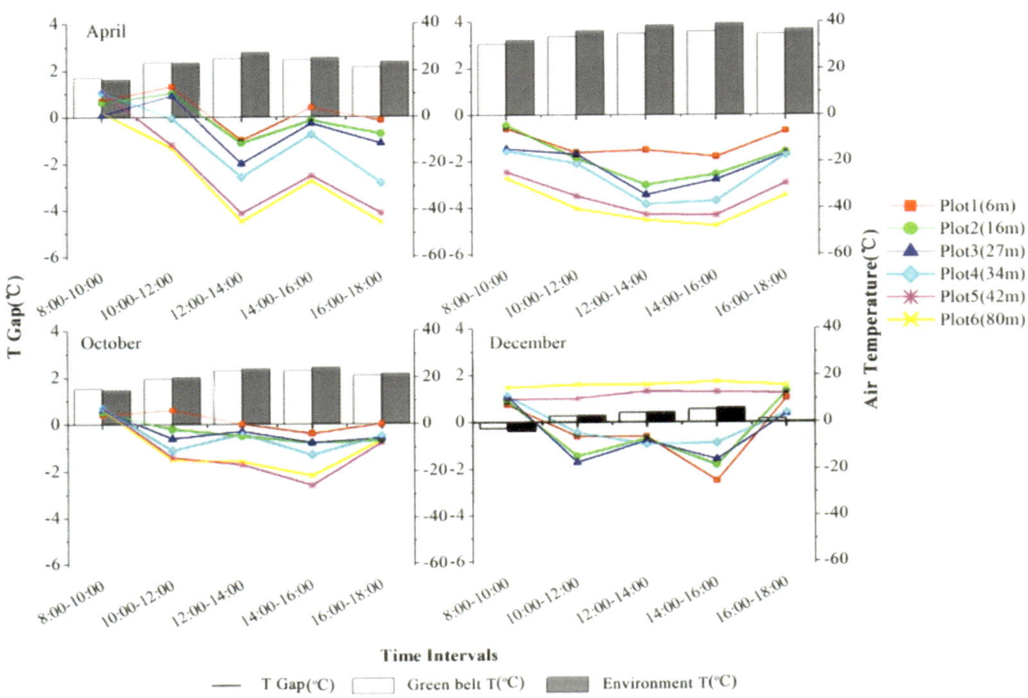

Fig. 3. T changes of green belts over five time intervals

plots 5 and 6 showed increased T for all time intervals. For example, the data collected at 14:00–16:00 from July indicate the most significant environmental benefits of lower T. The T in plot 1 was 1.8 °C lower than that of the control; the T in plots 2 and 3 was 2.6–2.8 °C lower; the T of plot 4 was 3.7 °C lower; and the T of plots 5 and 6 was 4.4–4.8 °C lower. This result indicates that the T effect was stronger in wider plots. The amount of T reduction was smaller at 16:00–18:00, indicating that the T-lowering effect of the green belt was not very strong for the ambient temperature decrease compared with that of 14:00–16:00.

The increase in RH was most conspicuous in all six plots at 8:00–10:00 and 16:00–18:00 in April and at 8:00–10:00 and 10:00–12:00 in July. In October, the maximum difference was at 14:00–16:00, whereas in December, the effect was not obvious for any time interval because the vegetation had lost its leaves (Fig. 4). In July, the increase in RH in plots 4–6 was greater than in plots 1–3. The greatest increase was in plot 6, which was 15% more humid than the control group. In plot 5, the RH was 4% higher than in plot 4. The effect on the increase in RH was most evident in plots 5 and 6 (Fig. 4).

The results indicated that width exerted a definite effect on the decrease in T and increase in RH for all plots in April, July and October. A clear effect of width was observed in plot 4, where green belt effects were stronger than in plots 1–3 at 12:00–14:00 and 14:00–16:00. A conspicuous, stable effect was evident in plots 5 and 6; the effect in plot 5 was as great as that in plot 6 from 10:00 to 16:00.

2.1.2. Changing average air temperature and relative humidity over 3 days

The 3-day average air T and RH gap for each plot was analysed compared with the control for the five time intervals.

The benefit as measured by the 3-day average T increased with the width of the six plots during April, July and October, and the order was plot 6 > plot 5 > plot 4 > plot 3 > plot 2 > plot 1 (Fig. 5). The variance analysis(ANOVA) showed a significant difference in the 3-day average T in the six plots compared with the control group using Duncan's post hoc tests ($P < 0.05$). The T benefit of plot 1 was notable despite being the narrowest. By contrast, the green belts did not produce an obvious effect until a certain width was attained. In April and July, the differences in air T between plots 5 and 6 were not significant despite the greater width of plot 6. There were no significant differences between plots 5 and 4, but there were significant differences between plot 5 and plots 1, 2, and 3. There were no significant differences among plots 1, 2, 3 and 4. In December, there was no significant difference between plots 5 and 6. The T benefit was most obvious in July, when plot 4 was 2.6 °C cooler than the control and plots 5 and 6 were 3.5 °C and 3.9 °C cooler, respectively. From this analysis, the critical width reference value for an urban green belt to effect an obvious decrease in T was approximately 34 m (plot 4), with an optimal width reference value of approximately 42 m (plot 5).

As observed from the 3-day average RH for all time intervals (Fig. 6), there were some influences on the increase in RH in all the plots (plot 6 > plot 5 > plot 4 > plot 3 > plot 2 >

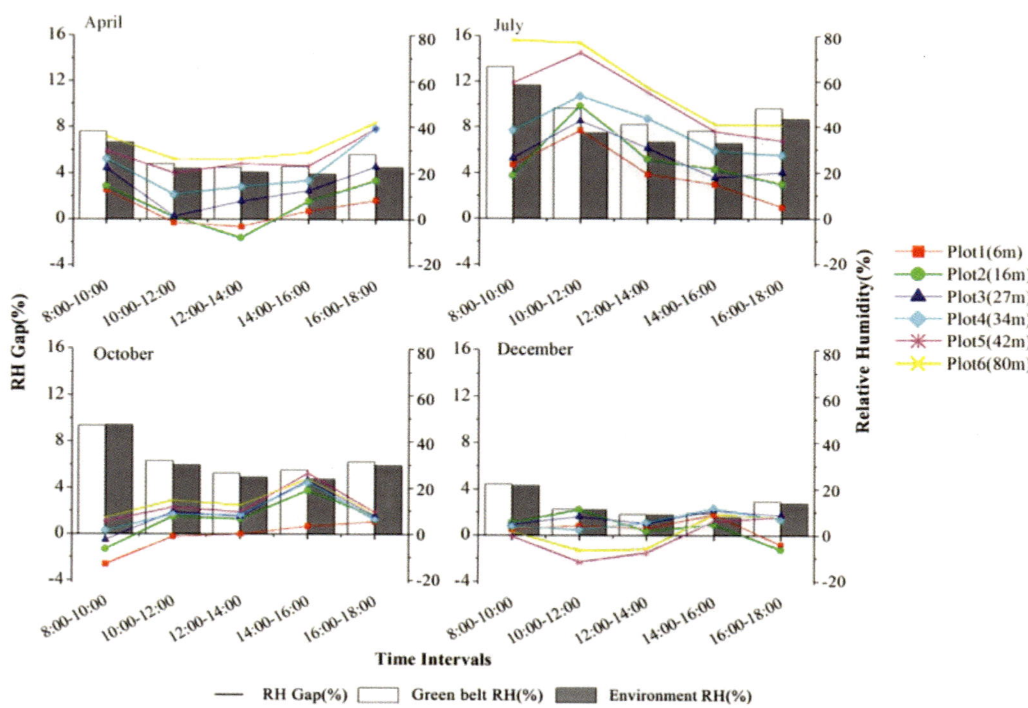

Fig. 4. RH changes of green belts over five time intervals

plot 1) during April, July and October. The variance analysis indicated that the differences in the 3-day average RH among the six green belts and the control group were significant based on Duncan's post hoc tests ($P < 0.05$). There was an obvious benefit to RH, and significant differences were detected at certain times: in April and July, there were no significant differences between plots 5 and 6 or between plots 4 and 5; however, there were significant differences between plot 5 and plots 1, 2, and 3. In October and December, there were no obvious differences among the six plots. The RH benefit was most obvious in July, when plot 4 was 7.6% more humid than the control, and plots 5 and 6 were 10.2–11.6% more humid than the control. According to this analysis, the critical width reference value for an urban green belt to clearly raise RH was approximately 34 m (plot 4), and the optimal width reference value was approximately 42 m (plot 5). The 3-day average T and RH showed that the benefits increased with the width of the green belt. In April, July and October,

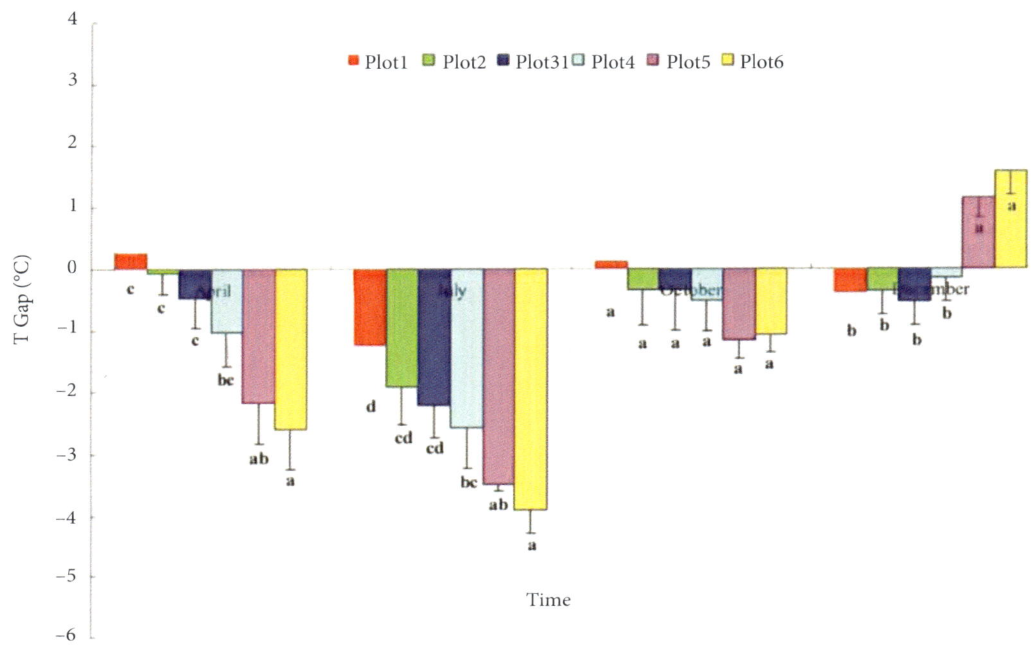

Different letters indicate significant differences among green belts with different widths at 0.05 level.

Fig. 5. The 3-day average of air T in the green belts

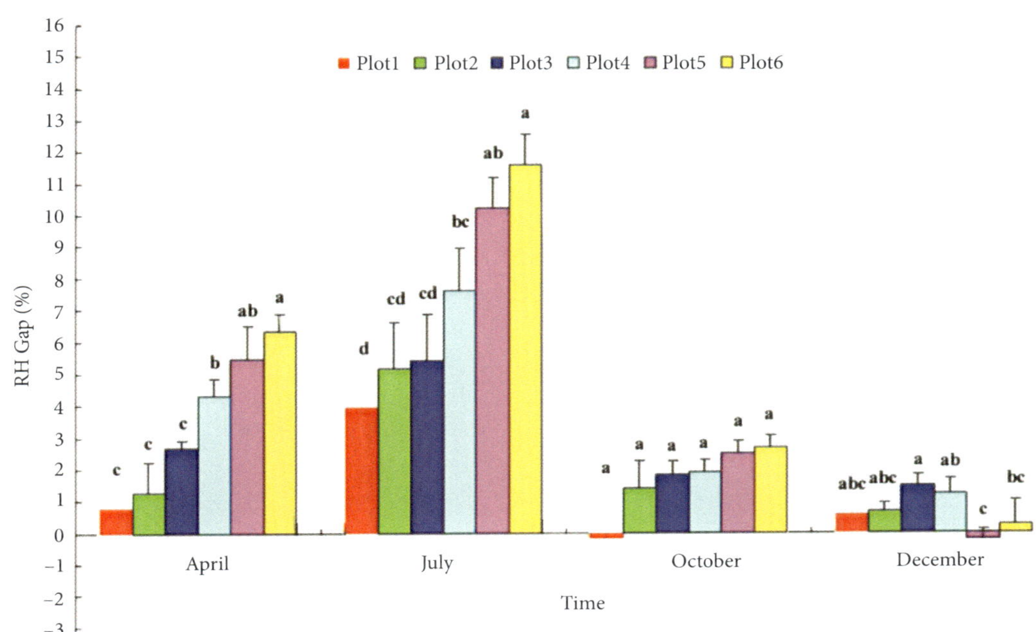

Different letters indicate significant differences among green belts with different widths at 0.05 level.

Fig. 6. The 3-day average of air RH in the green belts

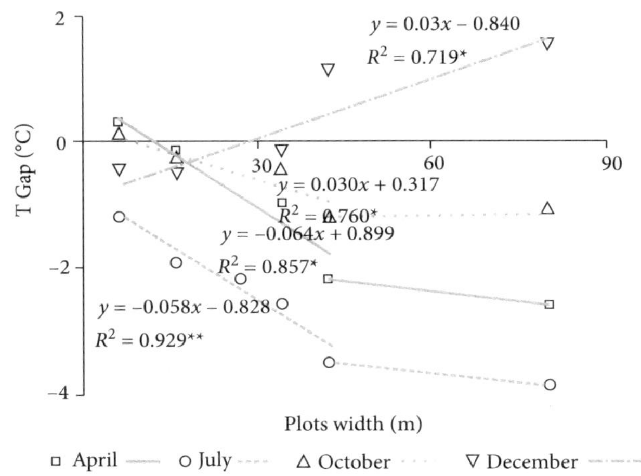

Fig. 7. Linear correlation analysis of average T in green belts

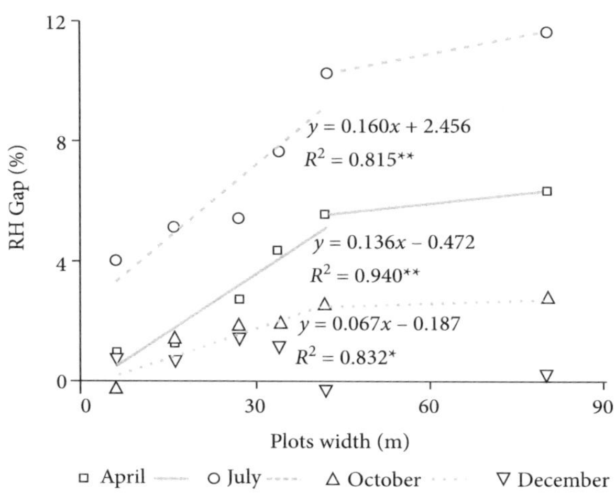

Fig. 8. Linear correlation analysis of average RH in green belts

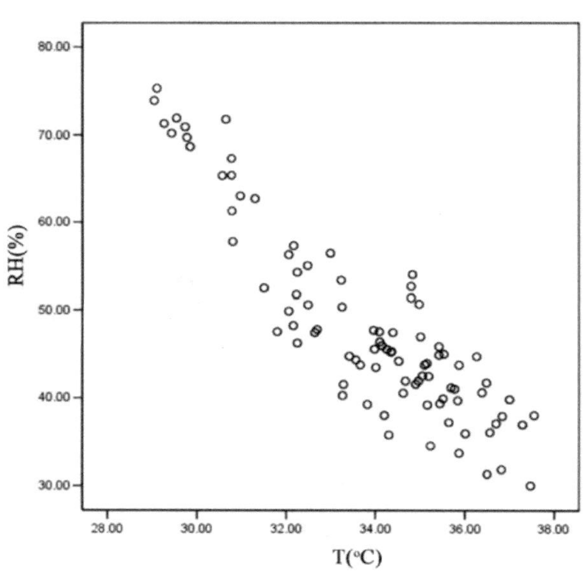

Fig. 9. Scatter diagram of the correlation trend of air T and RH

the width of plot 1 had little effect on T and no obvious effect on RH. A moderate effect on T-RH benefits was observed in plots 2–4, whereas a conspicuous, stable effect was observed in plots 5 and 6. In December, there was an obvious effect of the plots on the increase in T in the green belt, with plots 5 and 6 showing the greatest effect.

From the linear correlation analyses of the average T and RH (Fig. 7; Fig. 8), the T and RH benefits in plot 5 were similar to those in plot 6 and higher than those in plot 4 even though the width of plot 5 was not markedly different from that of plot 4. Both, however, are distinctly different from that of plot 6. It can be concluded that the optimal width reference value of an urban green belt for the effect on the decrease in T and the increase in RH was that of plot 5.

2.1.3. Regression analysis of air temperature and relative humidity

Air temperature is related to the relative humidity in the green space. However, relationships between T and RH are different in different types of green space, especially in the green belt with a marginal characteristic (Wang, Lin 2006; Cohen *et al.* 2012).

A regression analysis was applied to the average T and RH data (90 readings) over 3 days for the six plots (bivariate) in July, when the environmental benefits were at their most significant. From the scatter diagram of the correlation trend (Fig. 9), there was a significant linear correlation between T and RH. The correlation coefficient between T and RH was –0.898** (** represents significance, $P < 0.01$) from the analysis of the correlation coefficient matrix. The Spearman rank correlation coefficient was also determined by analysing a matrix of the nonparametric correlation coefficients to be –0.848** ($P < 0.01$).

In the regression model, F was determined to be 366.086 ($P < 0.01$), so the model was statistically significant. From an analysis of residual statistics, the results showed a normal distribution. Using the generalised difference method in Eviews, DW = 1.83, du < DW < 4-du, there was no autocorrelation of the model when the first observed value underwent Prais-Winsten transformation. In the regression equation, Y (RH) = 184.721– 4.038 × X(T) was based on the regression analysis between T and RH in the six green belts (Table 3).

2.2. Effect of green belt width on negative air ion concentration

2.2.1. Negative air ion concentration changes for the five time intervals

The difference in NAI concentration between each plot and the control was analysed over five time intervals.

Figure 10 shows that the maximum and minimum NAI concentrations differed among the five time intervals by season. The NAI concentrations in April and July were similar for the plots and control when the maximums were at 8:00–10:00 (355 and 466 ions/cm^3, respectively) and the minimums were at 12:00–14:00 (240 and 189 ions/cm^3, respectively). In October, the maximum occurred at 12:00–14:00 (246 ions/cm^3), and the minimum occurred at 16:00–18:00 (161 ions/cm^3). In December, when the leaves fell, the NAI concentration showed little variation.

Table 3. Autocorrelation processing of the regression model of air T and RH in the six green belts

Dependent Variable: Y – 0.2859 * Y (–1)				
Method: Least Squares				
Variable	Coefficient	Standard Error	t-Statistic	Probability
C	131.9096	6.554161	20.12609	0.0000
X – 0.2859 * X (–1)	–4.037758	0.270770	–14.91213	0.0000
R-squared	0.718785	Mean dependent variable		34.44522
Adjusted R-squared	0.715552	S.D. dependent variable		8.644510
Standard Error of regression	4.610431	Akaike info criterion		5.916735
Sum squared residue	1849.278	Schwarz criterion		5.972660
Log likelihood	–261.2947	F-statistic		222.3716
Durbin-Watson statistic	1.830497	Probability (F-statistic)		0.000000

The width of the green belt was an important factor in the NAI concentration. In July, the NAI concentration was the highest, and all plots had a similar trend. The NAI concentrations in plots 4–6 were higher than in plots 1–3, and plot 6 had the highest NAI concentration, with the maximum NAI concentration at 8:00–10:00 (1032 ions/cm^3) and the minimum at 12:00–14:00 (220 ions/cm^3).

2.2.2. Changing average negative air ions concentration over 3 days

The 3-day average NAI concentration for the five time intervals was analysed for the plots and the control.

The 3-day average NAI concentration (Fig. 11) showed that the NAI concentration benefit increased with the width of the plots during April, July and October, and the order was plot 6 > plot 5 > plot 4 > plot 3 > plot 2 > plot 1. The NAI concentrations in wider plots were 1.5–2.5 times higher than those of the narrower plots. Based on Duncan's post hoc tests ($P < 0.05$), in April, there was no significant difference in the 3-day averages of NAI between plots 5 and 6, but there were significant differences between plots 5 and 6 and other plots. In July, there was a significant difference between plot 6 and the other plots and no obvious difference between plots 4 and 5, but there was a significant difference between plots 4 and 5 and plots 1–3. In October, there was no significant difference between plot 6 and plots 4 and 5 or between plots 4 and 5 and plot 3, but there was a significant difference between plots 6 and 3. However, there was a significant difference between plots 4 and 5 and plots 1 and 2. In December,

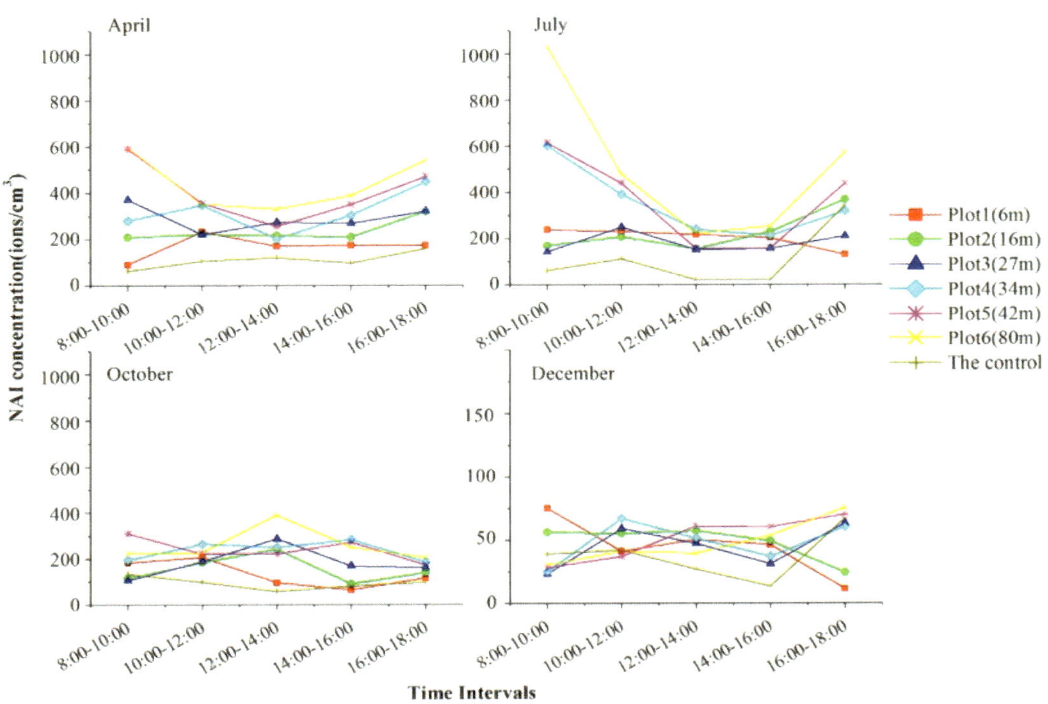

Fig. 10. NAI concentration changes of the green belts over five time intervals

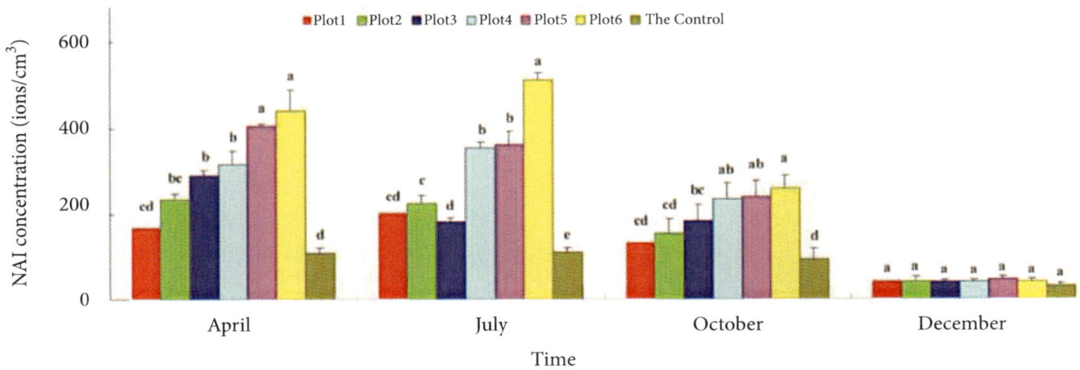

Different letters indicate significant differences among green belts with different widths at 0.05 level.

Fig. 11. The 3-day average of NAI concentration in the green belts

there was no significant difference in the 3-day average NAI concentration among the plots. In July, the NAI concentration benefit was most obvious, and the maximum NAI concentration was 510 ions/cm^3 in plot 6.

For all seasons, the change in the 3-day average NAI concentration grew with the width of the green belt. The width of plots 1–3 had little effect on the increase in the NAI concentration, there was a clear effect in plot 4, and there was a conspicuous, stable effect in plots 5 and 6.

According to the linear correlation analysis of the average NAI concentration (Fig. 12) and the 3-day average NAI concentration (Fig. 11), the NAI benefits in plot 5 were similar to those in plot 6 and higher than those in plot 4, suggesting that the optimal width reference value of an urban green belt for the effect on the increase in NAI was that of plot 5.

2.2.3. Comparison of the air ion assessment coefficient and the air-cleanliness class

The air ion assessment coefficient and the air-cleanliness class were calculated from the data using the formula of CI (air ion assessment coefficient). A higher CI means

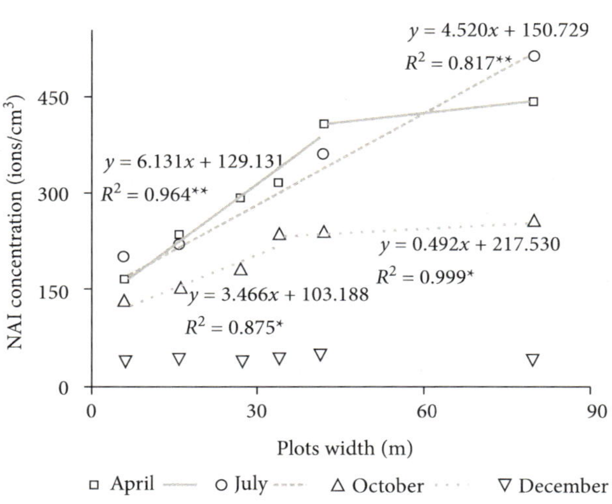

Fig. 12. Linear correlation analysis of average NAI concentrations in green belts

better air quality. When CI > 1.00, the air quality was the best, A degree; when CI was 1.00–0.70, air quality was clean, B degree; when CI was 0.69–0.50, average quality, C degree; when CI was 0.49–0.30, below average quality, D degree; when CI ≤ 0.29, polluted, E degree. In plot 1): CI was 0.29–0.19, lightly polluted, E1 degree; 2): CI was 0.18–0.10, polluted, E2 degree; 3): CI < 0.09, severely polluted, E3 degree.

As seen in Table 4, the order of the air ion assessment coefficient (CI) and the air-cleanliness class was July > April > October > December. In July, plot 6 had a higher air-cleanliness class, whereas there was no obvious difference between plots 4 and 5, which were higher than plots 1–3; plots 1–3 had a lower CI and barely achieved the B degree. In April, air cleanliness was lower and the average air-cleanliness class was of D degree in plots 4–6, whereas the class of plot 3 was higher than plots 1 and 2 and the average air-cleanliness class was E1 degree. The air-cleanliness class of plot 1 was of E2 degree. In October, plot 6 had the highest air ion assessment coefficient, and the air-cleanliness class was of D degree; plots 4 and 5 were significantly higher than plots 1–3, and the air-cleanliness class was of E1 degree. In December, all plots were of E3 degree.

2.3. Effect of different widths of green belts on inhibiting bacteria

2.3.1. Change in bacteria rates over 3 days

The 3-day BR for plots and the control was analysed.

To investigate the air-purification ability of green belts, their effect on decreasing the BR was analysed by atmospheric biological monitoring. Over 3 days, the change in atmosphere BRs showed a similar trend in the six plots. As observed in Figure 13, in July and October, the changes in the BRs in plots 4–6 were stable, and the effect on inhibiting bacteria did not vary with width. In April, the BRs of plots 5 and 6 decreased, whereas in December, there was no obvious variation in BR.

Table 4. Air ion assessment coefficient and air cleanliness class of green belts during seasons

Experi-mental period	Time intervals	Plot 1		Plot 2		Plot 3		Plot 4		Plot 5		Plot 6		The Control	
		CI	Deg-ree	CI	Deg-ree	CI	Deg-ree	CI		CI	Deg-ree	CI	Deg-ree	CI	Deg-ree
April	8: 00–10: 00	0.07	E3	0.24	E1	0.46	D	0.27	E1	0.53	C	0.55	C	0.04	E3
	10: 00–12: 00	0.25	E1	0.19	E1	0.26	E1	0.30	D	0.31	D	0.46	D	0.04	E3
	12: 00–14: 00	0.13	E2	0.12	E2	0.18	E2	0.29	E1	0.25	E1	0.20	E1	0.11	E2
	14: 00–16: 00	0.11	E2	0.16	E2	0.24	E1	0.30	D	0.25	E1	0.23	E1	0.07	E3
	16: 00–18: 00	0.11	E2	0.21	E1	0.33	D	0.58	C	0.37	D	0.25	E1	0.16	E2
	Average	0.13	E2	0.18	E2	0.29	E1	0.35	D	0.34	D	0.34	D	0.09	E3
July	8: 00–10: 00	0.62	C	0.30	D	0.13	E2	2.08	A	2.76	A	3.21	A	0.04	E3
	10: 00–12: 00	0.39	D	0.86	B	0.84	B	1.00	A	0.62	C	1.64	A	0.09	E3
	12: 00–14: 00	0.42	D	0.23	E1	0.19	E1	0.60	C	0.32	D	0.59	C	0.05	E3
	14: 00–16: 00	0.45	D	0.59	C	0.18	E2	0.65	C	0.80	B	0.72	B	0.06	E3
	16: 00–18: 00	0.19	E1	1.51	A	0.58	C	1.08	A	0.86	B	3.56	A	0.37	D
	Average	0.41	D	0.70	B	0.38	D	1.08	A	1.07	A	1.94	A	0.12	E2
October	8: 00–10: 00	0.12	E2	0.04	E3	0.11	E2	0.09	E3	0.35	D	0.31	D	0.27	E1
	10: 00–12: 00	0.18	E2	0.25	E1	0.18	E2	0.35	D	0.18	E1	0.78	B	0.09	E3
	12: 00–14: 00	0.05	E3	0.05	E3	0.33	D	0.20	E1	0.32	D	0.22	E1	0.44	D
	14: 00–16: 00	0.02	E3	0.03	E3	0.21	E1	0.42	D	0.30	D	0.18	E2	0.10	E2
	16: 00–18: 00	0.11	E2	0.16	E2	0.18	E2	0.28	E1	0.14	E2	0.20	E1	0.03	E3
	Average	0.10	E2	0.16	E2	0.17	E2	0.29	E1	0.24	E1	0.38	D	0.10	E2
Decem-ber	8: 00–10: 00	0.06	E3	0.05	E3	0.02	E3	0.03	E3	0.02	E3	0.02	E3	0.02	E3
	10: 00–12: 00	0.09	E3	0.08	E3	0.04	E3	0.07	E3	0.07	E3	0.06	E3	0.04	E3
	12: 00–14: 00	0.07	E3	0.09	E3	0.03	E3	0.07	E3	0.03	E3	0.06	E3	0.02	E3
	14: 00–16: 00	0.13	E2	0.04	E3	0.06	E3	0.04	E3	0.05	E3	0.12	E2	0.01	E3
	16: 00–18: 00	0.01	E3	0.03	E3	0.03	E3	0.05	E3	0.02	E3	0.04	E3	0.14	E2
	Average	0.07	E3	0.06	E3	0.03	E3	0.05	E3	0.04	E3	0.06	E3	0.05	E3

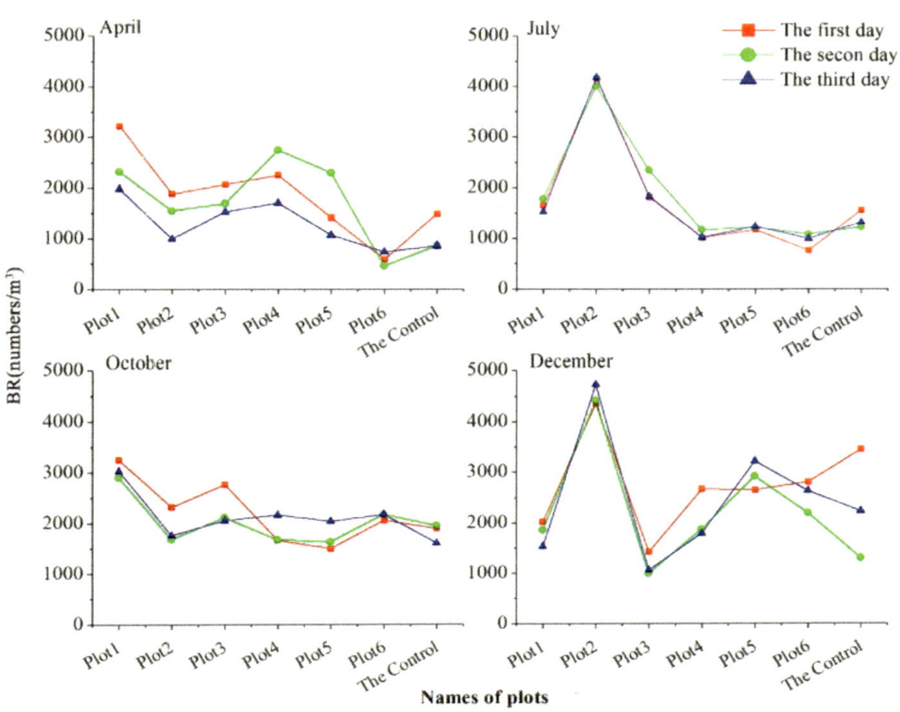

Fig. 13. Air BR changes of green belts over 3 days

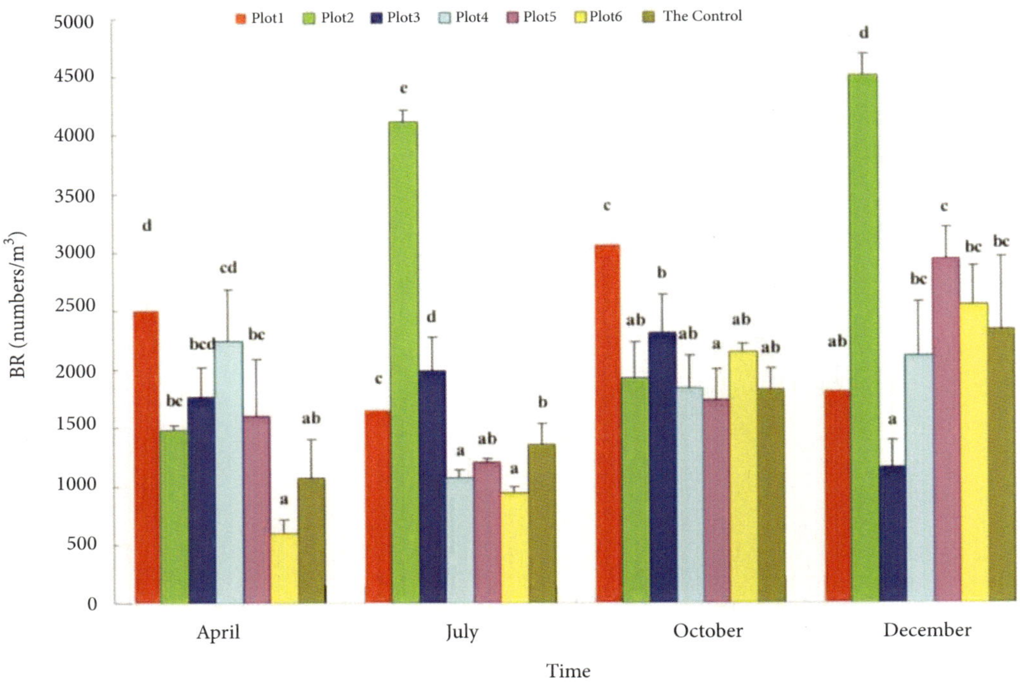

Different letters indicate significant differences among green belts with different widths at 0.05 level.

Fig. 14. The 3-day average BR in the green belts

2.3.2. Change in the average bacteria rate over 3 days

The 3-day average BR was analysed for each plot and the control.

The 3-day average BRs for each season (Fig. 14) showed that the BRs in the atmosphere were higher in October and December than in April and July. Based on Duncan's post hoc tests ($P < 0.05$) and using the data from July as an example, there was no significant difference in the 3-day average BR among plots 4–6, but there was a significant difference between the wider plots and plots 1–3. According to the observations in April, July, October and December, the effect of inhibiting bacteria steadily increased and was regular until the width of green belts reached a certain value.

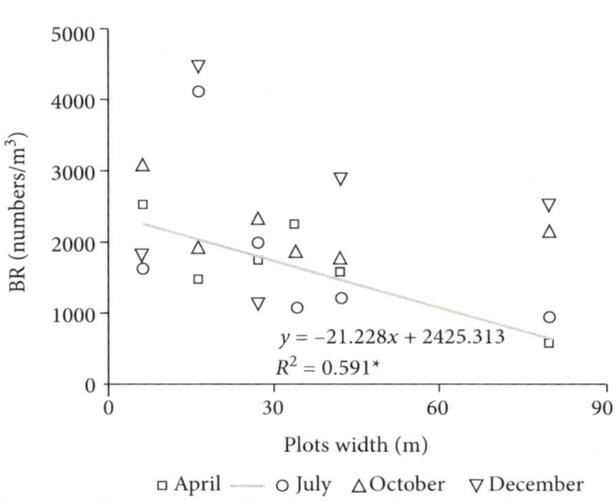

Fig. 15. Linear correlation analysis of average BR in green belts

As observed from the curve in Figures 15, there was a significant linear correlation between the BR and the width of plots ($P < 0.05$) in April and no significant linear correlation in July, October and December.

3. Discussion

3.1. Changes in the air temperature and relative humidity

Previous studies have found that decreases in temperature and increases in humidity caused by urban green space differ according to the vegetation quantity and forest type (Cohen *et al.* 2012; Hamada, Ohta 2010; Onishi *et al.* 2010). Recent research has also found that the environmental temperature may be reduced by area, geometry, location and proportion of built-up near the green space (Chang *et al.* 2007; Cohen *et al.* 2012; Cao *et al.* 2010; Sun, Chen 2012). In accordance with these results, the width of urban green belts was also found to significantly influence local T and RH in Beijing. These results reflected some physical properties of the thermal transmission among green belts.

The T benefit was produced by the crown leaves reducing direct solar insolation as well as by cooling from plant transpiration and the leaves blocking the reflected heat from the ground and adjacent objects to reduce long-wave surface radiation (Georgi, Dimitriou 2010; Oliveira *et al.* 2011; Federer 1976; Berry *et al.* 2013). The belts changed the radiation conditions at the surface and near the surface for atmospheric radiation, which led to the temperature difference between the belts and the bare space. The solar radiation intensity may decrease by more than 80% as a result of crown shadowing (Tang *et al.* 2009), and the belts could consume 60–75% and even 90% of the energy from solar radiation by plant transpiration, decreasing the air temperature (Profous *et al.* 1988). The widths of the green belts significantly affected the decreases in air temperature. To isolate the effect of the width on air temperature from data recorded in the six plots, the internal and external environmental conditions were held constant between plots. In every plot, the canopy was similar; thus, the shading of solar irradiation was essentially the same. The leaf area was different in each plot for the varied widths of the green belt, which meant that the wider green belts had stronger transpiration effects and a greater effect on temperature. In contrast, the narrower plots had lower decreases in T. Plot 1 was narrowest and had the least vegetation; thus, the green belt effect was greatly reduced by the surrounding long-wave surface radiation. Hence, it had less of a T impact on its internal environment. In contrast, plots 5 and 6 were wider and had more vegetation in which transpiration was greater; thus, the internal environment was relatively stable.

There was a very significant negative correlation between RH and solar radiation, surface radiation and air temperature (Tang *et al.* 2009). Narrower plots led to greater variation in the RH of the green belts because of air convection by cold and hot air transmission because of surrounding long-wave surface radiation. Moreover, it was evident that the RH increased with the width of the plots as a result of water evaporation over the wider area. The wider green belts had relatively quiet air movement, which prevented moisture diffusion from plant and soil transpiration; thus, the water vapour content increased in the local air. In addition, the saturation vapour pressure was lower because of decreased temperature, increasing the relative humidity of the air. Thus, plots 4–6 were superior to plots 1–3 at increasing the humidity.

3.2. Changes in the negative air ions concentration

The effect of the green belts on the air-cleanliness class varied by width. Narrower green belts had less vegetation, which produced fewer anions, and they had greater air convection by cold and hot air transmission from surrounding long-wave surface radiation, which further reduced the amount of anions. The green belt plots had a significant edge feature, and the NAI concentration at the edge of the green belt was easily affected by the surrounding environment. Therefore, the narrower plots were also largely affected by automobile exhaust and traffic dust, and the pollutants promoted anion collisions and charged neutralization (Krueger 1985; Retalis, Nastos 2009), which led to the unstable NAI concentrations. The wider plots (plots 4–6) were less affected internally by the surrounding environment; hence, the changes in the NAI concentration were stable, and the maximum NAI concentration was constant across time intervals. In contrast, the NAI concentration was significantly positive correlated with the leaf area because of the leaf tips having a photoelectric effect during photosynthesis in the green belts, which are able to promote electrolysis and produce large quantities of NAI (Liu *et al.* 2011; Tikhonov *et al.* 2004); in addition, aromatic substances released by forests, such as phytoncide, can also promote air ionization and increase NAI concentrations; the wider plots had a larger leaf area. These plots had more photosynthesis and more oxygen, providing the physical conditions necessary for NAI production (Wang, Li 2009). In addition, there was more transpiration in wider plots. The water vapour could first form charged ions, and the oxygen and water molecules had a better electrophilicity than nitrogen or other gases (Krueger 1962; Wu, Lee 2004), making it easier to form NAI. Therefore, all the analyses indicated that the wider green belts had a higher NAI concentration.

The NAI concentrations varied by seasons. From April to July, the air temperature and precipitation increased gradually, enhancing plant photosynthesis and increasing the absorption and purification of pollutants, which was the main reason for the increase in NAI in July. The change in plot 6 was the most obvious because of its greater width. In October, the temperature was lower, and there was a higher frequency of calm conditions when the atmospheric structure was stable. Temperature inversions appeared easily; therefore, the pollutants were not easily diffused, which led to a gradual decrease in NAI. In December, the deciduous plants were dormant, reducing photosynthesis and the effect of air purification, thus lowering the NAI concentration. There was a similar trend in plots 1–6. Previous studies have found that the NAI differed greatly during the four seasons, the highest and the lowest NAI concentrations occurred in the summer and winter, respectively (Wu *et al.* 2011; Liang *et al.* 2014).

Many studies have indicated that the daily changes in NAI are not consistent across seasons, regions, and weather conditions because the variation in the NAI concentration is affected by meteorological conditions such as temperature, humidity, wind direction, wind speed, solar radiation intensity, and atmospheric pressure (Wang, Li 2009; Singh *et al.* 2009; Vana *et al.* 2008). The most important meteorological indexes, air temperature and humidity (Wu *et al.* 2011; Liang *et al.* 2014), were analysed in this study. The results indicated that the NAI changes were affected by air temperature and humidity in April and July, and there was a significantly negative correlation between NAI concentration and air temperature and a positive correlation between NAI concentration and humidity ($P < 0.05$). During the time intervals, as the temperature increased, pollutants diffused and large amounts of NAIs were absorbed; thus, the NAI concentration was reduced. In contrast, as the humidity increased, the pollutants were absorbed by water vapour, increasing the NAI concentration. In October, the NAI concentration was not significantly correlated with air temperature or humidity during daily time intervals. Liu reported that the NAI concentration of plants was "U" or "V" correlated with the solar radiation intensity in the interior of green spaces during daily time intervals (Liu *et al.* 2011). Therefore, in October, plant photosynthesis was weaker, and during five time intervals, the temperature rose and the humidity fell as the solar radiation increased, reducing the NAI concentration. Moreover, the photoelectric reaction of the plants increased with the increase in the solar radiation intensity, causing a maximum NAI concentration at 12:00–14:00 in October.

Overall, the NAI varied across different seasons. In April and July, temperature and humidity were the main indexes affecting the NAI of the green belts; in October, the NAI was affected by temperature and humidity at 8:00–12:00, and it was significantly affected by the photoelectric reaction of plants at 12:00–14:00. For the above reasons, the changes in the NAI were affected by some indexes. In this study, wider plots (plots 4–6) caused the NAI to be more stable than in the narrower plots, which were easily affected by the above indexes. Therefore, when the plot attained a certain value, the width was a dominant factor. In this study, the consistent external environment and sunny, windless weather conditions were set in all plots; thus, the wind speed, atmospheric pressure, and population/vehicle flow are not discussed separately.

3.3. Changes in the bacteria rate

The effect of green belts on inhibiting bacteria was influenced by several factors. Atmospheric BRs were decreased by plant leaves through adsorption, filtration, and the secretion of antibacterial compounds (Zhang *et al.* 1997). The dust-retention functions of tree leaves play an important role in decreasing BRs. Nowak (Nowak 1994) found that the dust retained on the tree could be adsorbed at rates of 82–86% for broadleaved trees and 27–52% for conifers. The BRs were also affected by the traffic on either side of the belts. Traffic caused a significantly greater amount of dust to be retained on the leaf surface, and this hindered the plants' ability to produce antibacterial substances. Furthermore, under certain conditions of temperature and humidity, bacteria propagated more easily (Mesut, Bayram 2010). In the summer, the T of the green space was lower than that of the controls, whereas the RH was higher. If a green belt had poor airflow, it would be easier for bacteria to propagate. Therefore, under certain environmental conditions, the dust retained in wider green belts could be greater than that of narrower belts, but this study found that beyond a certain threshold, wider belt with higher humidity had a lower BR, whereas the plants had a stronger effect on the inhibiting bacteria. Therefore, the atmospheric BRs were higher. Further research may be needed to analyze the role of additional variables (such as species and leaf surface roughness).

In this study, the ordering of the BR was December > October > July > April. Compared with April, plant photosynthesis is stronger in July. There is more bactericidal capacity, but the air relative humidity is also higher; thus, it is easier for the bacteria to breed, leading to a higher BR in July than in April. In April, the plants were at a vigorous growth stage and the bacteriostatic ability was strong, which reduced BR compared with other seasons. In December, the plants were deciduous and dormant; thus, the bacteriostatic ability became weaker and the bacteria rate was higher.

The bacteriostatic ability of the six plots varied by seasons. In April, there was a higher BR in plots 1–5 than in the control because the air temperature and humidity were higher in the mornings, which allowed the bacteria to breed. In plot 1, the bacteria rate was higher. Because of its narrower width, plot 1 was more affected by the dust from traffic and pollen from the flowering "*Prunus triloba*", increasing the BR. As the width gradually increased, the bacteria rates in plots 3–4 increased, showing that the BR was mainly influenced by the temperature of internal plots. There was a significant positive correlation between the BR and air temperature ($P < 0.05$) in the mornings in April. However, the BR was lower in plots 5–6, which shows that the width plays a major role. Compared with the narrower plots, the wider plots were less affected by traffic and by dust deposited onto the leaves of the plants or the ground, reducing the bacterial load and causing a decrease in the bacteria rates in the plots. In July, the narrower plots 1–3 were affected by the surrounding air flow and its suspended dust and other particles, resulting in increased microbial settlement. Thus, there was no obvious effect of inhibition of bacteria in plots 1–3. Plot 2 was dominated by "*Lonicera maackii*", which has rough leaves. The amount of dust trapped was higher than for other species, and the bacteria rates were higher. The wider plots 4–6 had lower BRs, and the changes in the BRs of plots 4–6 were stable; the effect on inhibiting bacteria did not increase with increasing width. In October, all of the plots and the control had a similar BR, and the BR was not significantly correlated with air temperature or humidity ($P < 0.05$). In the mornings in the autumn, the atmospheric structure was stable and temperature inversions were frequent; thus, pollutants were not easily diffused. In addition, solar radiation and plant photosynthesis were weaker, which led to a peak in the microbe concentration. In December, the regular variation of the BRs was not obvious. The plants had a lower bactericidal capability because they were dormant and without leaves. In plots 4–6, there were higher BRs, and the BR was significantly positively correlated with air temperature ($P < 0.05$). In plot 2, the BR was higher, perhaps due to the dust that had accumulated on the "*Lonicera maackii*" returning to the air. Further research and the examination of variables such as solar radiation, wind, and atmospheric pressure may be needed to account for the unexplained variation.

3.4. Weather factors affecting the negative air ions concentration and bacteria rate

The NAI concentration and BR were affected by seasonal weather factors. In this study, from the relationship between T and RH and between the NAI concentration and BRs at 9:00–9:05 (Fig. 16), the correlation between the NAI concentration and the T and RH was affected by the season. There was a significant negative correlation between the NAI concentration and T and a positive correlation between the NAI concentration and the RH ($P < 0.05$) in April, July and October; the NAI concentration increased as the width of green belts increased for a higher RH. There was no significant correlation between the NAI concentration and T and RH ($P > 0.05$) in December. There was a significant positive correlation between the BRs and the T and no significant correlation to RH ($P < 0.05$) in April and December. There was no significant correlation between the BRs and the T and RH ($P > 0.05$) in July and October.

As the effect of green belts could be influenced by the distance in reference to a city centre, it is important to analyse the correlation between green belt positions and the environmental conditions including air T, RH, NAI and BR. Sun *et al.* (2012) investigated the urban cooling island (UCI) intensity of 187 water bodies outside the Fourth Ring Road and 10 water bodies inside the Fourth Ring Road. The results indicated that the UCI intensity was negatively correlated with location in reference to the defined city centre (Sun *et al.* 2012). However, the results reported by Li *et al.* (2004) showed that no significant correlation existed between the underlying surface temperature and the sample point area (inside the Fourth Ring Road) (Li *et al.* 2004). These results may indicate that temperature has no correlation with location inside the Fourth Ring Road in Beijing. However, few results were reported about the green belt position and other environmental conditions. In this research, different correlations were found between air T and other conditions including RH, NAI and BR on six widths of green belts inside the Fourth Ring Road of Beijing. Thus, further studies are needed to uncover the correlation between green belt positions and environmental conditions such as RH, NAI and BR.

Conclusions

Green belts, distributed heterogeneously in urban areas, have significant effects on urban ecological attributes and healthy dwelling environments. Meanwhile, the effective width of urban green belts for better ecological effectiveness is an issue of great concern to many designers and builders, especially in commercial areas and densely populated urban areas. In the present study, an obvious effect of the green belt was observed on the increase of RH and NAI, the decrease in air T and the inhibition of BR, especially in April and July. Compared with the bare and hard ground, the air T of green belts declined by 3.5–10.7% in July; the RH and NAI increased by 9.5–28.3% and 84–364%, respectively; and the effect of green belts

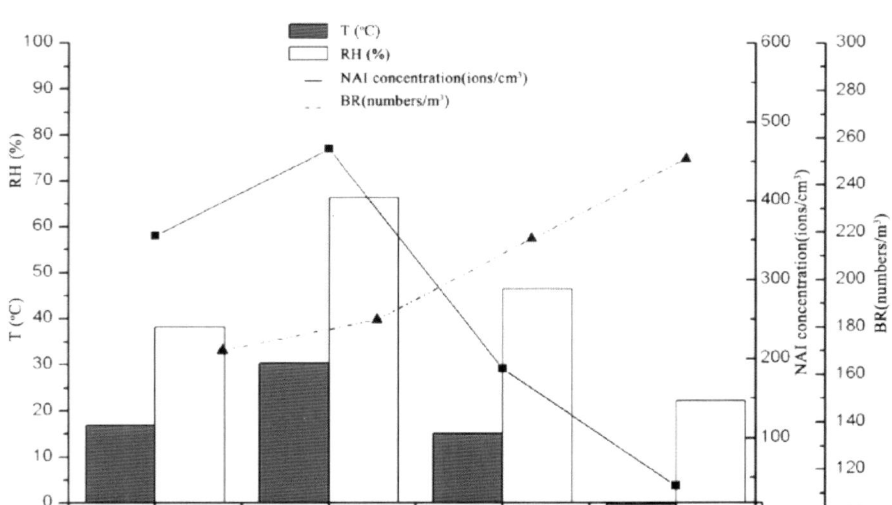

Fig. 16. Correlation between T, RH and the NAI concentration, BR at 9:00–9:05

on BR inhibition was not significant. These results indicated that the width of the green belt played a vital role on the environmental indicators such as air T, RH, NAI etc. As the width of the green belt increased, the effect on air T, RH and NAI was enhanced. However, the effects were stable when the width of green belt was over 42 m, suggesting that the optimal width reference value of an urban green belt for a measureable effect is approximately 42 m. In this case, the green belts will suffice to stabilize their internal space conditions of the ecological environment and contribute largely to the environment.

Effective utilization of the air T-RH and the air quality effects of green belts through appropriate landscape design and planning is becoming increasingly critical given the worsening city environment and continued rapid urbanization. This study provides planners and designers with useful information for achieving optimal landscape design in urban planning projects. Meanwhile, the assessment of urban ecological attributes and the healthy living environment will enable us to understand the urban green belts and their relationship with human beings in a new perspective and further establish their ecological status. In addition, our results will provide new insights into the development of urban green belt construction and the understanding of the sustainable development of urban ecology.

Acknowledgements

This work was supported by the "National Science Foundation of China" under Grant [31500576]; "National Science Foundation of China" under Grant [30972416].

References

Berry, R.; Livesley, S. J.; Aye, L. 2013. Tree canopy shade impacts on solar irradiance received by building walls and their surface temperature, *Building and Environment* 69: 91–100. https://doi.org/10.1016/j.buildenv.2013.07.009

Cao, X.; Onishi, A.; Chen, J.; Imura, H. 2010. Quantifying the cool island intensity of urban parks using ASTER and IKONOS data, *Landscape and Urban Planning* 96: 224–231. https://doi.org/10.1016/j.landurbplan.2010.03.008

Chang, C. R.; Li, M. H.; Chang, S. D. 2007. A preliminary study on the local cool-island intensity of Taipei city parks, *Landscape and Urban Planning* 80: 386–395. https://doi.org/10.1016/j.landurbplan.2006.09.005

Cohen, P.; Potchter, O.; Matzarakis, A. 2012. Daily and seasonal climatic conditions of green urban open spaces in the Mediterranean climate and their impact on human comfort, *Build Environment* 51: 285–295. https://doi.org/10.1016/j.buildenv.2011.11.020

Federer, C. A. 1976. Trees modify the urban microclimate, *Arboricultural Journal* 2: 121–127.

Feyisa, G. L.; Dons, K.; Meilby, H. 2014. Efficiency of parks in mitigating urban heat island effect: an example from Addis Ababa, *Landscape and Urban Planning* 123: 87–95. https://doi.org/10.1016/j.landurbplan.2013.12.008

Fitzsimons, J.; Pearson, C. J.; Lawson, C.; Hill, M. J. 2012. Evaluation of land-use planning in greenbelts based on intrinsic characteristics and stakeholder values, *Landscape and Urban Planning* 106: 23–34. https://doi.org/10.1016/j.landurbplan.2012.01.012

Georgi, J. N.; Dimitriou, D. 2010. The contribution of urban green spaces to the improvement of environment in cities: case study of Chania, Greece, *Build Environment* 45: 1401–1414. https://doi.org/10.1016/j.buildenv.2009.12.003

Gunn, S. C. 2007. Green belts: A review of the regions' responses to a changing housing agenda, *Journal of Environment Planning and Management* 50: 595–616. https://doi.org/10.1080/09640560701475154

Hamada, S.; Ohta, T. 2010. Seasonal variations in the cooling effect of urban green areas on surrounding urban areas, *Urban Forestry and Urban Greening* 9: 1–24. https://doi.org/10.1016/j.ufug.2009.10.002

Hathway, E. A.; Sharples, S. 2012. The interaction of rivers and urban form in mitigating the Urban Heat Island effect: a UK case study, *Building and Environment* 58: 14–22. https://doi.org/10.1016/j.buildenv.2012.06.013

Islam, M. N.; Rahman, K. S.; Bahar, M. M.; Habib, M. A.; Ando, K.; Hattori, N. 2012. Pollution attenuation by roadside greenbelt in and around urban areas, *Urban Forestry and Urban Greening* 11: 460–464. https://doi.org/10.1016/j.ufug.2012.06.004

Ji, P.; Zhu, C. Y.; Li, S. H. 2013. Effects of greenbelt width on air temperature and humidity in urban river corridors, *Chinese Journal of Plant Ecology* 37(1): 37–44 (in Chinese). https://doi.org/10.3724/SP.J.1258.2013.00004

Krueger, A. P. 1962. Air ions and physiological function, *Journal of General Physiology* 45: 233–241. https://doi.org/10.1085/jgp.45.4.233

Krueger, A. P. 1985. The biological effects of air ions, *International Journal of Biometeorology* 3: 205–206. https://doi.org/10.1007/BF02189651

Li, Y. M.; Zhang, J. H.; Gu, R. Z. 2004. Research on the relationship between urban greening and the effect of urban heat island, *Journal of Chinese Landscape Architecture* 1: 72–75 (in Chinese).

Liang, H.; Chen, X. S.; Yin, J. G.; Da, L. J. 2014. The spatial-temporal pattern and influencing factors of negative air ions in urban forests, Shanghai, China, *Journal of Forestry Research* 25: 847–856. https://doi.org/10.1007/s11676-014-0475-9

Liu, X.; Wu, L. H.; Zhang, H.; Wang, X. R. 2011. Study on the concentration of negative air ions and the influential factors in different urban plant communities, *Journal of Fudan University (Natural Science)* 50: 206–212 (in Chinese).

Mackey, C. W.; Lee, X. H.; Smith, R. B. 2012. Remotely sensing the cooling effects of city scale efforts to reduce urban heat island, *Build Environment* 49: 348–358. https://doi.org/10.1016/j.buildenv.2011.08.004

Maimaitiyiming, M.; Ghulam, A.; Tiyip, T.; Pla, F.; Carmona, P. L.; Halik, Ü.; Sawut, M.; Caetano, M. 2014. Effects of green space spatial pattern on land surface temperature: implications for sustainable urban planning and climate change adaptation, *Journal of Photogrammetry Remote Sensing* 89: 59–66. https://doi.org/10.1016/j.isprsjprs.2013.12.010

Mesut, D.; Bayram, K. 2010. A preliminary assessment of the water and air quality parameters in and around the Lake Yenicaga coast, *Journal of the Black Sea/Mediterranean Environment* 16: 167–194.

Nowak, D. J. 1994. Air pollution removal by Chicago's urban forest USDA forest service gen, *Technology Report* 186: 62–83.

Oliveira, S.; Andrade, H.; Vaz, T. 2011. The cooling effect of green spaces as a contribution to the mitigation of urban heat: A case study in Lisbon, *Build Environment* 46: 2186–2194. https://doi.org/10.1016/j.buildenv.2011.04.034

Onishi, A.; Cao, X.; Ito, T.; Shi, F.; Imura, H. 2010. Evaluating the potential for urban heat-island mitigation by greening parking lots, *Urban Forestry and Urban Greening* 9: 323–332. https://doi.org/10.1016/j.ufug.2010.06.002

Profous, G. V.; Rowntree, R. A.; Loeb, R. E. 1988. The urban forest landscape of Athens, Greece: aspects of structure, planning and management, *Arboricultural Journal* 12: 83–107. https://doi.org/10.1080/03071375.1988.9756380

Retalis, A.; Nastos, P. 2009. Study of small ions concentration in the air above Athens, Greece, *Atmospheric Research* 91: 219–228. https://doi.org/10.1016/j.atmosres.2008.05.011

Rohling, J. 1998. Corridors of green, *N C Wild* 5: 22–27.

Schiller, A. 1997. Wildlife conservation in urban greenways of the mid-southeastern, *United States: Urban Ecosystem* 1: 103–116. https://doi.org/10.1023/A:1018515309254

Singh, K.; Malik, A.; Singh, M. 2009. Spatial variation in air ion concentrations under different indoor environments, *International Journal of BioScience, Psychiatrie and Technology* 1: 36–41.

Smith, D. S. 1993. An overview of greenways: their history, ecological context, and specific functions, in D. S. Smith (Ed.). *Ecology of greenways*. Minneapolis: University of Minnesota Press, 1–21.

Sun, R. H.; Chen, L. D. 2012. How can urban water bodies be designed for climate adaptation?, *Landscape and Urban Planning* 105: 27–33. https://doi.org/10.1016/j.landurbplan.2011.11.018

Tang, L. Z.; Li, Z. Q.; Yan, C. F.; Sun, C. H.; Xu, X.; Xiang, H. R. 2009. Mitigative effects of different vegetations on heat island effect in Nanjing, *Ecology and Environmental Sciences* 18: 23–28 (in Chinese).

Tikhonov, V. P.; Tsvetkov, V. D.; Litvinova, E. G. 2004. Generation of negative air ions by plants upon pulsed electrical stimulation applied to soil, *Journal of Plant Physiology* 51: 414–419. https://doi.org/10.1023/B:RUPP.0000028690.74805.e2

Vana, M.; Ehn, M.; Petaja, T.; Vuollekoski, H.; Aalto, P.; de Leeuw, G.; Ceburnis, D.; O'Dowd, C. D.; Kulmala, M. 2008. Characteristic features of air ions at Mace Head on the west coast of Ireland, *Atmospheric Research* 90: 278–286. https://doi.org/10.1016/j.atmosres.2008.04.007

Wang, J.; Li, S. H. 2009. Changes in negative air ions concentration under different light intensities and development of a model to relate light intensity to directional change, *Journal of Environmental Management* 90: 2746–2754. https://doi.org/10.1016/j.jenvman.2009.03.003

Wang, J.; Lin, Y. D. 2006. Role of green land in urban on the reduction of heat island effect, *Grassland and Turf* 6: 56–59 (in Chinese).

Wu, C. C.; Lee, G. W. M. 2004. Oxidation of volatile organic compounds by negative air ions, *Atmospheric Environment* 38: 6287–6295. https://doi.org/10.1016/j.atmosenv.2004.08.035

Wu, C. F.; Lai, C. H.; Chu, H. J.; Lin, W. H. 2011. Evaluating and mapping of spatial air ion quality patterns in a residential garden using a geostatistic method, *International Journal of Environmental Research and Public Health* 8: 2304–2319. https://doi.org/10.3390/ijerph8062304

Wu, W. T.; Zhao, H. Y.; Ren, Y.; Shen, D.; Zhang, Y. 2012. Study on barriers and countermeasures of technological innovation of ecological service function assessment of urban greenbelt, *Energy Procedia* 16: 115–120. https://doi.org/10.1016/j.egypro.2012.01.021

Zhang, X. X.; Gu, R. Z.; Li, Y. M.; Li, H. 1997. The effect of green space in residential area on the air bacteria rates, *Chinese Landscape Architecture* 13: 57–58 (in Chinese).

Chunyang ZHU. D Agr, Lecturer of the College of Horticulture and Forest, Huazhong Agricultural University, Wuhan 430070, China. Publications: author/co-author of over 20 scientific papers. Research interests: ecological environment benefit and landscape planning.

Peng JI. D Agr, Associate Professor of the College of agronomy, Heilongjiang Bayi Agricultural University, Daqing 163319, China. Publications: author/co-author of over 10 scientific papers. Research interests: ecological environment benefit.

Shuhua LI. D Agr, Professor of the Department of Landscape Architecture, School of Architecture, Tsinghua University, Beijing 100084, China. Publications: author/co-author of over 50 scientific papers. Research interests: ecological environment benefit.

4

ECOLOGICAL IMPACT ASSESSMENT OF ROAD NETWORKS AT LANDSCAPE SCALE USING SPATIAL ROAD DISTURBANCE INDEX (SPROADI)

Shekoufeh NEMATOLLAHI, Sima FAKHERAN, Alireza SOFFIANIAN

Department of Natural Resources, Isfahan University of Technology, Isfahan, 84156-83111, Iran

Abstract. Development of roads can have deleterious effects on natural habitats containing species of conservation concern. Fragmentation of habitat into small, non-contiguous patches may result in dramatic population declinesThus appropriate studies quantifying ecological impacts of roads at landscape scale are essential. The main goal of this study was ecological impact assessment of roads network in Eastern part of Isfahan Province, including Abassabad wildlife refuge and Siahkouh National park, which are among the most important habitats for Asiatic Cheetah (Acinonyx jubatus venaticus) classified as Critically Endangered (CR) on the IUCN Red List. For this purpose, Spatial Road Disturbance Index (SPROADI), as a new landscape index was applied for the ecological impact assessment of the roads. This index uses three sub-indices including traffic intensity, vicinity impact and fragmentation grade to calculate the ecological impacts of road networks. Results obtained through quantifying the landscape Index (SPROADI) showed that the degree of disturbance by roads network is between 0 and 54.53. Our results also revealed that 12 percent of Abassabad wildlife and wide range of suitable habitats for Asiatic Cheetah were affected by roads, which presents a conservation concern for this critically endangered species.

Keywords: ecological impact assessment, landscape index, SPROADI, conservation, habitat fragmentation, roads.

Introduction

Roads are an integral part of the transportation infrastructures, which deliver a wide range of social and economic benefits such as providing connectivity for people and enabling communities to rapidly expand into previously remote or inaccessible areas (Freudenberger *et al.* 2013). However, there is also growing evidence of the negative impacts of roads on adjacent habitats, wildlife populations and ecosystems, indicating that these infrastructures are a driving factor of biodiversity loss (Forman *et al.* 2003; Benítez-López *et al.* 2010; Eigenbrod *et al.* 2009; Lin 2015).

The effects of roads on biodiversity are cumulative, time-lagged, complex and often irreversible (Selva *et al.* 2011). Roads cause mortality due to collision with vehicles (van Langevelde *et al.* 2009; Trombulak, Frissell 2000; Forman *et al.* 2003), alternation in behavior (Laurance, Stouffer 2004; Parris, Schneider 2009; Hoskin, Goosem 2010), habitat loss (Rico *et al.* 2007; Ewers, Didham 2007), reduction of habitat quality for wildlife by increasing distance between remaining habitat patches (Goodwin, Fahrig 2002), isolation of populations, decreasing the overall landscape connectivity and restriction of gene flow (Keller *et al.* 2004; Epps *et al.* 2005; Riley *et al.* 2006; Balkenhol, Waits 2009) and reducing the resilience of populations and ecosystems to climate change (Jump, Peñuelas 2005).

To better understand how roads impact on wildlife populations and their habitats at landscape scale, it is necessary to quantify to which extent and degree they are affected by these infrastructures. So far, most of the methods and indices developed to quantify the influence of roads focused on quantifying physical impacts of roads by calculating the degree of habitat fragmentation including landscape cohesion index (Opdam *et al.* 2003), splitting density, splitting index, mesh size (Jaeger 2000) and landscape connectivity indices (Pascual-Hortal, Saura 2006).

The SpreAD-GIS was used to estimate the emission of sound pollution (Madadi *et al.* 2014), the Habitat Evaluation Procedure (HEP) was used to assess the ecological impacts of roads (Makki *et al.* 2013) and the Ecological Footprint indicators used into assess the environmental impact of road access to tourist activities (Martín-Cejas 2015). However, most of the study consider on roads.

But to have a more comprehensive assessment of roads effects, other parameters associated with disturbances of road networks should also be taken into account.

Corresponding author: Sima Fakheran
E-mail: fakheran@cc.iut.ac.ir

Spatial Road Disturbance Index (SPROADI) as recently developed index by Freudenberger *et al.* (2013) to consider cumulative impacts of road networks on natural habitat. This index uses three sub-indices including traffic intensity, vicinity impact and fragmentation grade to calculate the cumulative effects of roads on biodiversity. Traffic intensity is a measure of traffic volume per time and space; vicinity impact is the measure of edge effect of roads on adjacent habitats; and fragmentation grade, indicating the degree to which the landscape is intersected by roads (Freudenberger *et al.* 2013). By applying this index, this would be possible to incorporate and assess the impacts of factors related to traffic density and distribution besides physical impacts of roads on biodiversity.

The main goals of this study were assessment of the ecological impacts of road networks on important habitats such as Abassabad wildlife refuge and Siahkouh national park located in the eastern part of Isfahan province using of SPROAD index. These habitats are among the most important habitats for some species of conservation concern such as Caracal (Caracal caracal), houbarabustard (chlamydotis undulate), sand cat (Felis margarita) and Asiatic cheetah (Acynonix jubaticus venaticus). Nowadays, the most important risk element for Asiatic Cheetah is road networks located in the study area. For example, the Ardakan-Choopanan as a main road, cross the Abassabad wildlife refuge. In recent years, lots of accidents were happened in this road, which showed the importance of this road to biodiversity loss (personal communication with wildlife guards of Abassabad wild life refuge and local people).

1. Materials and methods

1.1. Study area

Our study area, Naein and Khoor-o-Biabanak Counties is located at a distance of 145 kilometres away from east part of Isfahan province in Iran, with an approximate area of 38 000 Km^2. The elevation in this region ranges from 700 m to 2451 m above the sea level with the average annual rainfall between 100–110 mm (Akbari *et al.* 2011). The dominant climate in study area includes hot and dry climate and also some moderate climate mostly found in south-western areas where the mountain ranges are found. Existence of different topographic features including rocky mountain ranges, deserts and sandy areas has provided suitable habitat conditions for a high variety of plant and animal species in this region (Akbari *et al.* 2011). In the study area, the density of road networks is low, though there is a main concern about this network where roads have been passed through some conservation sites in the county including Abassabad wildlife refuge (category IV IUCN) and Siahkouh national park (category II IUCN) and its effects on wildlife habitats. Abassabad is at south-eastern part of study area, with an area of 3000 ha (Fig. 1). This wildlife refuge is among the most important protected areas with high diversity of plant (400 species) and animal species (about 100 species). Some species of high conservation value have been found in this wildlife refuge including Iranian ground jay (podoces pleskei) an endemic species of Iran, sand cat (Felis margarita), Leopard (panther pardus), Caracal (Caracal caracal) houbarabustard

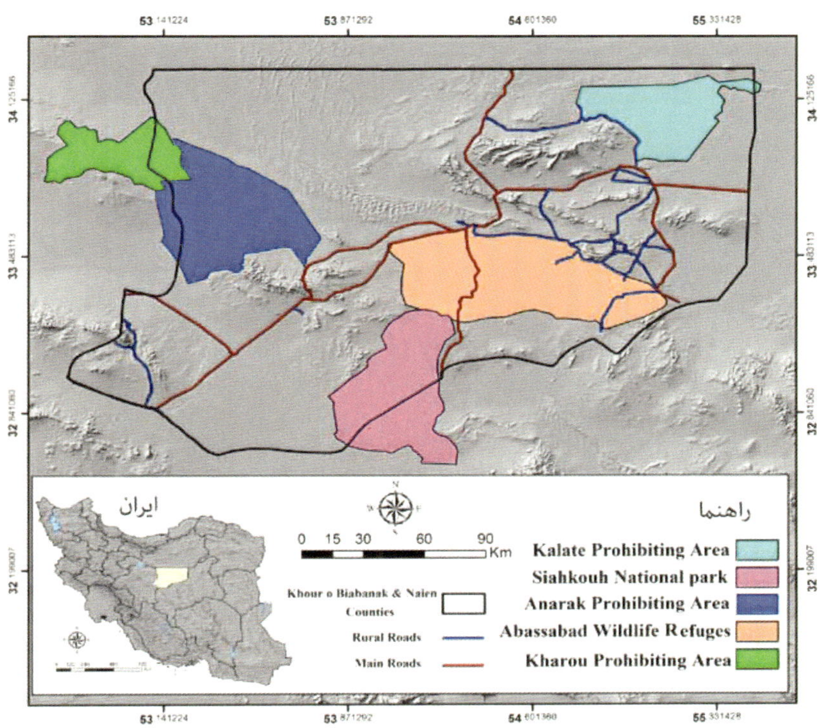

Fig. 1. Location of Study area and roads and protected areas

(chlamydotis undulate), Jebeer gazelle (Gazella bennettii) and Wild goat (Capra aeagagrus). Siahkouh national park covers an area of 80 000 ha and is located at south-west of the study area (Fig. 1) (Akbari *et al.* 2011). Both of these protected areas are well known as important stronghold across Isfahan for a critically endangered member of felidea family with a global distribution range limited only to Iran, Asiatic cheetah (Acynonix jubaticus venaticus). This species especially populations that live in Abassabad wildlife refuge have been seriously affected by these roads through collision with vehicles as a major factor resulted in high rate of the species mortality and destruction of suitable habitats for the species. These roads are also considered as barriers to movement between suitable habitats (Madadi *et al.* 2014).

1.2. Input data

The main input layers required for this study included the road networks and the traffic data maps of the study area. The road networks layer was prepared using topographic map with the scale of 1/250000, Google earth, and converting the final layer to raster format. We excluded the urban area of Naein (some fraction of southwest part) from the study area, because of high density of roads which could bias the results of our study. Figure 1 shows the road networks and the location of protected areas in Eastern part of Isfahan province. As Figure 1 shows, the road density is high around Abassabad wild life refuge and Kalate prohibiting area and can have adverse effects on these protected areas.

1.3. Spatial road disturbance Index development

To calculate the SPROAD index including traffic intensity, vicinity impact and fragmentation grade, traffic data layer with appropriate resolution (1 km^2) (Makki *et al.* 2013; Eigenbrod *et al.* 2009; Forman *et al.* 2003; UNEP 2001) was used as the basic map for producing three sub-indices. To produce the traffic intensity, vicinity impact and fragmentation grade layers, we used the formula developed by Freudenberger *et al.* (2013) as follow: Traffic intensity (T) is expressed as the average number of vehicles per hour (TV) (Table 1) multiplied by road length (R), and summed up for all road sections in a given cell as follow:

$$T_k = \sum_{i=1}^{n_k} R_{k,i} . TV_{k,i} . \quad (1)$$

To achieve the number of vehicle per hour, we used the data from institution of roads transportation (Freudenberger *et al.* 2013) and then, to ensure the accuracy of our calculations, we checked our results with the data gained from Google earth time series in each road section in the study area (Madadi *et al.* 2014).

The vicinity impact (V) for a given cell represents the cumulative effect of all relevant roads as a function of their distance and traffic load. The vicinity impact can be calculated using the attributes of the neighbouring cells (Eq. (2)). In this study, 8 neighbour cells were used to calculate the vicinity impact (Freudenberger *et al.* 2013).

$$V_k = \frac{1}{n_k} . \sum_{v=1}^{n_k} T_{k,v} , \quad (2)$$

where V_K is the vicinity impact on grid k; $T_{k,v}$, traffic intensity in an adjacent (v) grid cell k; $v = \{1, 2, 3,..., n_k\}$, index variable for surrounding cells; n_k, number of surrounding cells of grid cell k. (Freudenberger *et al.* 2013). The effects of adjacent grid cells can be calculated by mentioned index. It means that the adjacent grid cell which roads passed through it can have adverse effects on the grid cell without any roads. To calculate Fragmentation grade, Eq. (3) was used (Freudenberger *et al.* 2013):

$$F_k = \frac{A_k}{\sum_{j=1}^{n_k} A_{j,k} \cdot A_{j,k}^{compl}} , \quad (3)$$

F_k, fragmentation grade of a grid cell k; $j = \{1, 2, 3,..., n_k\}$, index variable for number of polygons; n_k, number of polygons without roads belonging to a grid cell k; A_k, area size of a grid cell k (m^2); $A_{j,k}$, area size of the fraction of a polygon j lying within a grid cell k (m^2); $A_{j,k}^{compl}$, area size of the complete polygon j that $A_{j,k}$ is part of (m^2); A^{compl} total; total area size of all complete polygons intersecting a grid cell k (m^2) (Freudenberger *et al.* 2013).

Because of different range of scales, all three sub-indices were normalized with the Fuzzy membership function in the GIS environment, using a scale ranged between 0 and 100. Then, they were combined to generate the cumulative effects of road networks of the study area as SPROADI using Eq. (4). All sub-indices were weighted equally (w = 0.33) (Freudenberger *et al.* 2013):

$$SPROADI_k = \sum_{i}^{n} w_i \cdot I_{k,norm} = w_T \cdot T_{k,norm} + w_v \cdot V_{k,norm} + w_F \cdot F_{k,norm} . \quad (4)$$

$SPROADI_k$ = spatial road disturbance index of a grid cell k; n = number of sub-indices (in this case: 3); $w_i = \{w_T, w_V, w_F\}$ where w stands for the assigned weight for each of the sub-indices. In this study we assigned equal weights to each, i.e. $w_i = 0.33$, so that $\Sigma w_i = 1$ and $0 \le w_i \le 1$ (Freudenberger *et al.* 2013).

In the second step of our study, we used the developed index for conservation planning target for the Asiatic cheetah. For this purpose, we used maps of the protected area network in the study area and the distribution model of the species along with the *SPROAD* index map to conduct a gap analysis and identifying less impacted unprotected habitats for the Asiatic cheetah in the study area for spatial prioritization of conservation areas.

2. Results and discussion

The total length of road networks across Naien and Khoor-o-Biabanak Counties was approximately 1051 km. Table 1, shows the average number of vehicles per hours calculated for the main roads included in our study area.

Based on the field observation and discussion with the head of Environmental Protection Agency, more than 30 accidents per year (with species such as caracal, wild cat, and houbara and Asiatic cheetah) were happened in the Naien-Anarak, Ardakan-Choopanan and Anarak-Choopanan roads.

Table 1. The average number of vehicles per hours in the main roads

The name of road section	Number of vehicles per hour
Jandagh- Damghan	174
Anarak- Khoor- 1	156
Naien- Anarak	135
Ardakan- Choopanan	102
Bayaze- Khoor	50
Anarak- Khoor-46	39

Raster maps of all three calculated sub-indices and the final map of SPROADI are shown in Figure 2. Comparison of the patterns of values among maps of sub-indices, in terms of having significant correlation revealed that there is a positive spatial correlation between values obtained for vicinity impact and traffic intensity. There is high traffic intensity in the central and southern part of the study area, where the Anarak-khoor and Ardakan-Choopanan main roads with high traffic volume are located. The vicinity impact which reflects the impact of adjacent grid is high in the western and northern parts of the study area and the degree of habitat fragmentation is very high in the eastern part of the study area because of high road density in these parts.

Values for SPROADI map ranged from 0 to 54.53. High levels of road disturbances were obtained for areas at the vicinity of the Ardakan-Choopanan main road stretched from north to south of the Abassabad wild life refuge, and areas around the main road at south-western part of the study area. For these parts of the study area, high levels of traffic intensity and vicinity impact were recorded. However, low levels of road disturbances were obtained for areas around the rural road at the eastern part of the study area.

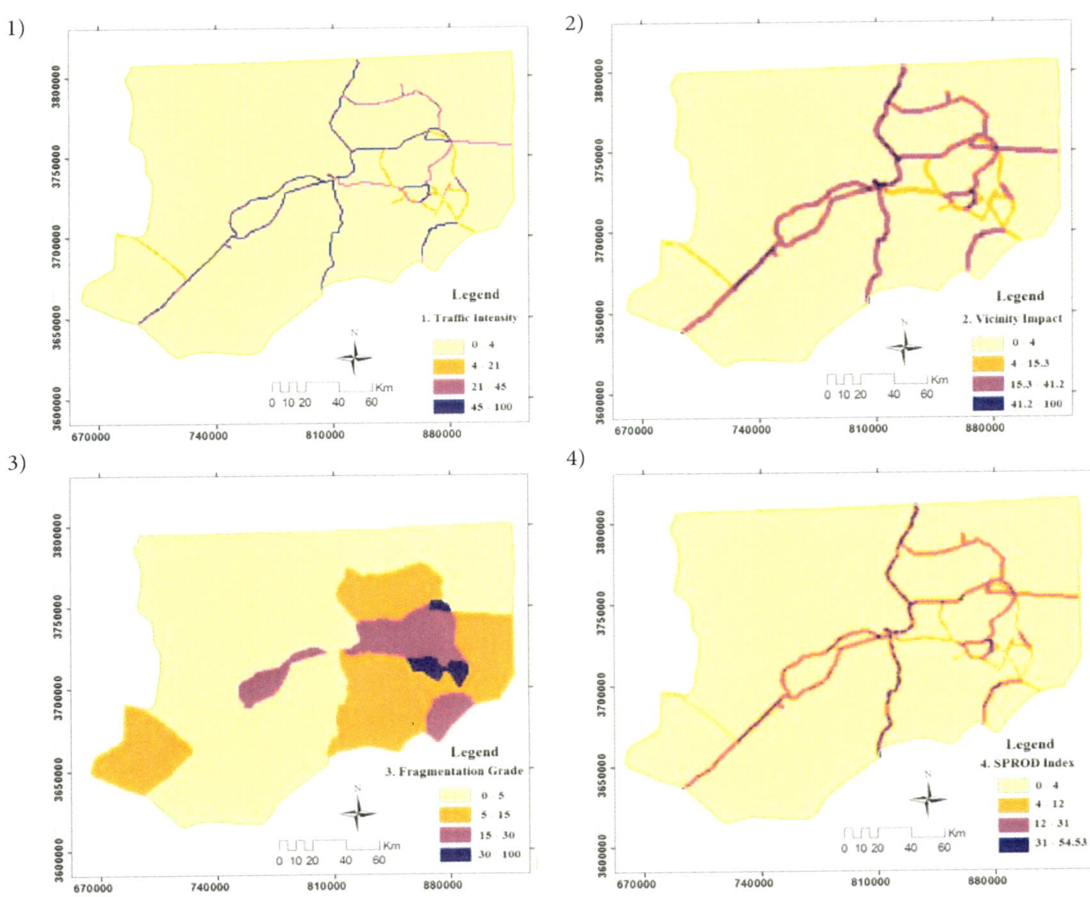

Fig. 2. 1) Traffic intensity (*T*), 2) vicinity impact (*V*), 3) fragmentation grade (*F*) and 4) the combination of the three sub-indices into the spatial road disturbance index (*SPROADI*); high values in red represent high road disturbance, low values in pale pink represent low road disturbance along the different sub-indices or the overall *SPROADI* outcome

The value of SPROAD index, in the study which was done by Freudenberger *et al.* (2013) in Germany, ranged from 0 to 83.67 (Freudenberger *et al.* 2013). Comparison the two study areas, the levels of the road density is higher in Germany. The study area in eastern part of Isfahan is larger comparing the study area in Brandenburg State (38,000 km^2 comparing 29,479 km^2 in Germany), however in this study the volume of traffic in eastern part of Isfahan due to heavy vehicles is much higher in comparison with Germany.

Based on the SPROADI map, approximately 10 percent of the study area was affected by the road networks. The road networks in Naien and Khoor-o-Biabanak counties, caused habitat fragmentation and more accessibility to important conservation areas including Abassabad wild life refuge and Siahkouh national park, increasing the vulnerability of these two areas (Makki *et al.* 2013).

To assess the impacts of the roads network on the habitats of the Asiatic cheetah, we used the distribution map of the species across eastern part of Isfahan produced by Bordkhani (2013). Figure 3, shows the distribution map of suitable habitats for Asiatic cheetah. As the map indicates, the southern and south-eastern parts of the study area are suitable for Asiatic cheetah. The results revealed that 6 percent of the study area consists of the most suitable habitats. The Abassabad wild life refuge and Siahkouh national park are known as the most suitable areas for Asiatic cheetah in Isfahan Province. Important environmental features that are expected to influence the presence of cheetahs include vegetation cover, geographic features such as rocky outcrops, the presence of predators and the availability of water and prey were found in these areas (Bissett *et al.* 2007). Dense vegetation is believed to provide essential cover as refuge from predators, as well as camouflage when stalking prey. Cheetah's diet mainly consists of Jebeer gazelle, Goitered gazelle, wild sheep and wild goat. Water is important, not only for consumption, but as a feature around which prey species tend to congregate (Pettorelli *et al.* 2008).

The results of overlaying the SPROADI map with protected areas and suitable distribution ranges for the Asiatic cheetah in Eastern part of Isfahan province are shown in Figure 4. The roads network has intersected the suitable habitats of Asiatic cheetah in several parts, both inside and outside the protected areas. Southeastern part of the study area, which is the most suitable areas for Asiatic cheetah, has also been affected by the roads network. An important consequence of this human disturbance is habitat fragmentation which in turn has resulted in cheetah's population to form too small and isolated. Therefore, the Asiatic Cheetahs are more vulnerable to stochastic demographic events. The indirect effects on the species include changes or impacts that result from increased contact with humans and human land use

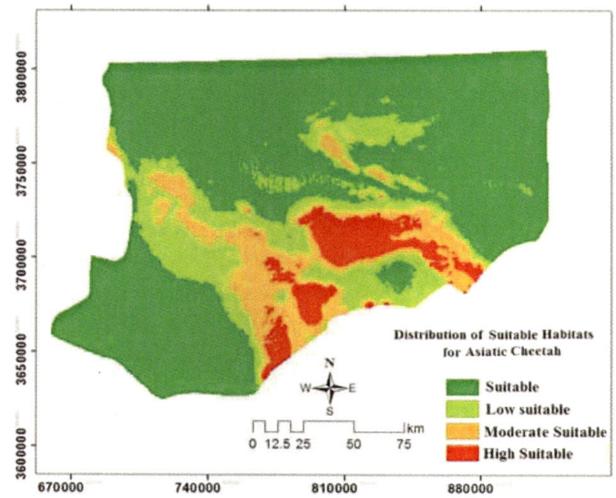

Fig. 3. Distribution and habitat suitability map of Asiatic cheetah at eastern part of Isfahan Province (Bordkhani 2013)

activities. Asiatic cheetah numbers are thought to be limited by the size of the prey populations on which they depend. The road networks increase the probability of prey's collision such as Jebeer gazelle and wild goat. Water is an important environmental variable, not only for consumption, but as a feature around which prey species tend to congregate. Crossing the roads by prey species to access water resources, located on the other side may increase the accidents (Carrillo *et al.* 2000). In addition, illegal hunting has a strongly influenced on the distribution and abundance of Asiatic cheetahs and prey species. Asiatic cheetahs and many prey species that normally use habitats on the edges of their distribution range may be subjected to increased vulnerability to illegal hunting on the habitat boundaries where human density is high (O'Brien *et al.* 2003).

Overlaying the protected areas and SPROADI maps revealed that the Abassabad wildlife refuge undergone the highest degree of roads network disturbances. This level of disturbance is mainly caused by the Ardakan-Choopanan main road passed through the wildlife refuge which caused habitat fragmentation. There are two other main roads in the north and eastern part of Abassabad area. In addition to habitat fragmentation, high number of collisions happened on these road was also responsible for this level of disturbance. One of the reasons of high number of accidents in these roads is that, there are some springs in the other side of the road. Therefore, crossing the roads by wildlife species for accessibility to these water resources caused accidents (Field observation and personal communication with wildlife guards of Abassabad wild life refuge). There are some underpasses, but some species especially mammals did not use them because they escape from human constructions. In addition, these roads can have negative effects on these protected areas

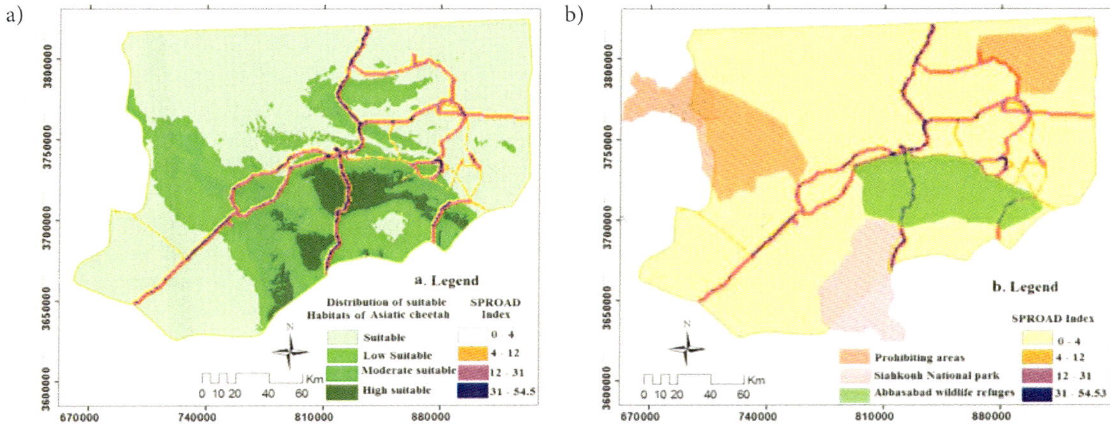

Fig. 4. Schematic representations of the spatial extent of overlaying SPROADI with a) most suitable habitats for the Asiatic cheetah and b) protected areas in study area impacted by the road networks

by increasing the hunter accessibility to these areas and decrease the security in these areas.

Table 2 shows the areas of protected areas, the areas of suitable habitats for Asiatic Cheetah fallen in to different protected areas and the levels of disturbances in different protected areas. Results of this study show that around 12 percent of Abassabad wild life refuge, including 77 percent suitable habitats for Asiatic cheetah, is affected by high level of road disturbances. Also, approximately 2 percent of Siahkouh national park, including 57 percent suitable habitats for Asiatic cheetah is affected by low level of road disturbances.

Table 2. The Area (ha) of Sensitive Habitats and the level of disturbance by roads (SPROADI)

Protected areas	Areas (ha)	Areas of habitat suitability of Cheetah (ha)	Area of *SPROADI* (ha)	
			High	Low
Abassabad wild life refuges	305,433	233,973	14,800	25,400
Siahkouh national park	205,782	116,674	–	400

Conclusions

The present study focused on the ecological impacts of roads network, which passes through Abassabad wildlife refuge and Siahkouh national park in the eastern part of Isfahan province. The key affected species, Asiatic Cheetah (Acinonyx jubatus venaticus) classified as Critically Endangered (CR) on the IUCN Red List was subject to impact analyses.

We used Spatial Road Disturbance Index (SPROADI), as a new landscape index for the ecological impact assessment of the roads network. This index uses three sub-indices including traffic intensity, vicinity impact and fragmentation grade to calculate the ecological impacts of road networks.

Results obtained through quantifying the landscape Index (SPROADI) showed that the degree of disturbance by roads network is between 0 and 54.53. Our results also revealed that 12 percent of Abassabad wildlife and wide range of suitable habitats for Asiatic Cheetah were affected by roads, which presents a conservation concern for this critically endangered species. Our results demonstrate that the roads network not only reduced the area of the original habitat in the refuge but also in addition to the physical occupation of land use and maintenance of the road caused disturbance and thus further reduced the quality of the remaining habitat.

Our study demonstrated that SPROADI method might provide a powerful tool for assessing ecological impact of transportation infrastructures on populations of far-ranging species of conservation concern and for incorporating this index into systematic conservation planning.

Acknowledgements

We express our appreciation to Department of the Environment of Iran (DOE). We are grateful to guards of Abbasabad wildlife refuge, local people, Dr Mahmoud reza Hemami, Dr Hossein Madadi and Dr Hossein Moradi, for their valuable inputs and cooperation to this study.

References

Akbari, H.; Jalalpour, M.; Hojati, V.; Gol Mohamadi, M. 2011. Identification of reptiles and birds fauna of Abassabad wildlife refuge, *Journal of Animal Biology* 3(2): 1–12.

Balkenhol, N.; Waits, L. P. 2009. Molecular road ecology: exploring the potential of genetics for investigating transportation impacts on wildlife, *Molecular Ecology* 18(20): 4151–4164. https://doi.org/10.1111/j.1365-294X.2009.04322.x

Benítez-López, A.; Alkemade, R.; Verweij, P. A. 2010. The impacts of roads and other infrastructure on mammal and bird populations: a meta-analysis, *Biological Conservation* 143(6): 1307–1316. https://doi.org/10.1016/j.biocon.2010.02.009

Bissett, C.; Bernard, R. T. F. 2007. Habitat selection and feeding ecology of the cheetah in thicket vegetation: is the cheetah a savannah specialist?, *Journal of Zoology* 271(3): 310–317.

Bordkhani, M. 2013. *Distribution modelling of Asiatic Cheetah (Acinonyx jubatus venaticus)*: Master's thesis. Isfahan University of Technology.

Carrillo, E.; Wong, G.; Cuarón, A. D. 2000. Monitoring mammal populations in Costa Rican protected areas under different hunting restrictions, *Conservation Biology* 14(6): 1580–1591. https://doi.org/10.1046/j.1523-1739.2000.99103.x

Eigenbrod, F.; Hecnar, S. J.; Fahrig, L. 2009. Quantifying the road effect zone: threshold effects of a motorway on anuran populations in Ontario, *Ecology & Society* 14(1): 24. https://doi.org/10.5751/ES-02691-140124

Epps, C. W.; Palsbøll, P. J.; Wehausen, J. D.; Roderick, G. K.; Ramey, R. R.; McCullough, D. R. 2005. Highways block gene flow and cause a rapid decline in genetic diversity of desert bighorn sheep, *Ecology letters* 8(10): 1029–1038. https://doi.org/10.1111/j.1461-0248.2005.00804.x

Ewers, R. M.; Didham, R. K. 2007. The effect of fragment shape and specie's sensitivity to habitat edges on animal population size, *Conservation Biology* 21(4): 926–936. https://doi.org/10.1111/j.1523-1739.2007.00720.x

Forman, R. T. T.; Sperling, D.; Bissonette, J. A.; Clevenger, A. P.; Cutshall, C. D.; Dale, V. H.; Fahrig, L.; France, R. L.; Goldman, Ch. R.; Heanue, K.; Jones, J.; Swanson, F.; Turrentine, Th.; Winter, Th. C. 2003. *Road Ecology: Science and Solutions*. Washington, DC: Island Press. 504 p.

Freudenberger, L. R.; Hobson, P.; Rupic, S.; Pe'er, G.; Schluck, M.; Sauermann, J.; Kreft, S.; Selva, N.; L. Ibisch, P. 2013. Spatial road disturbance index (SPROADI) for conservation planning: a novel landscape index, demonstrated for the State of Brandenburg, Germany, *Landscape Ecology* 28(7): 1353–1369. https://doi.org/10.1007/s10980-013-9887-8

Goodwin, B. J.; Fahrig, L. 2002. Effect of landscape structure on the movement behaviour of a specialized goldenrod beetle, Trirhabda borealis, *Canadian Journal of Zoology* 80(1): 24–35. https://doi.org/10.1139/z01-196

Hoskin, C. J.; Goosem, M. W. 2010. Road impacts on abundance, call traits, and body size of rainforest frogs in northeast Australia, *Ecology & Society* 15(3): 15.

Jaeger, J. A. G. 2000. Landscape division, splitting index, and effective mesh size: new measures of landscape fragmentation, *Landscape Ecology* 15(2): 115–130. https://doi.org/10.1023/A:1008129329289

Jump, A. S.; Peñuelas, J. 2005. Running to stand still: adaptation and the response of plants to rapid climate change, *Ecology Letters* 8(9): 1010–1020. https://doi.org/10.1111/j.1461-0248.2005.00796.x

Keller, I.; Nentwig, W.; Largiader, C. R. 2004. Recent habitat fragmentation due to roads can lead to significant genetic differentiation in an abundant flightless ground beetle, *Molecular Ecology* 13(10): 2983–2994. https://doi.org/10.1111/j.1365-294X.2004.02310.x

Laurance, S. G. W.; Stouffer, P. C. 2004. Effects of road clearings on movement patterns of understory rainforest birds in central Amazonia, *Conservation Biology* 18(4): 1099–1109.

Lin, S. C. 2015. The width of edge effects of road construction on fauna and ecologically critical road density, *Journal of Environmental Engineering and Landscape Management* 23(4): 241–250. https://doi.org/10.3846/16486897.2015.1054290

Madadi, H.; Moradi, H.; Fakheran, S.; Jokar, M.; Makki, T. 2014. Modelling of emission of the sound pollution of Isfahan's west freeway in the Ghamishloo wild life refuge using the SpreAD-GIS, *Iranian Journal of Applied Ecology* 3(9): 43–55.

Makki, T.; Fakheran, S.; Moradi, H.; Iravani, M.; Senn, J. 2013. Landscape-scale impacts of transportation infrastructure on spatial dynamics of two vulnerable ungulate species in Ghamishloo Wildlife Refuge, Iran, *Ecological Indicators* 31: 6–14. https://doi.org/10.1016/j.ecolind.2013.03.001

Martín-Cejas, R. R. 2015. The environmental impact caused by road access to Timanfaya Natural Park on Lanzarote Island, *Transportation Research Part D: Transport and Environment* 41: 457–466. https://doi.org/10.1016/j.trd.2015.09.027

O'Brien, T. G.; Kinnaird, M. F.; Wibisono, H. T. 2003. Crouching tigers, hidden prey: Sumatran tiger and prey populations in a tropical forest landscape, *Animal Conservation* 6(2): 131–139. https://doi.org/10.1017/S1367943003003172

Opdam, P.; Verboom, J.; Pouwels, R. 2003. Landscape cohesion: an index for the conservation potential of landscapes for biodiversity, *Landscape Ecology* 18: 113–126. https://doi.org/10.1023/A:1024429715253

Parris, K. M.; Schneider, A. 2009. Impacts of traffic noise and traffic volume on birds of roadside habitats, *Ecology & Society* 14(1): 29. https://doi.org/10.5751/ES-02761-140129

Pascual-Hortal, L.; Saura, S. 2006. Comparison and development of new graph-based landscape connectivity indices: towards the prioritization of habitat patches and corridors for conservation, *Landscape Ecology* 21: 959–967. https://doi.org/10.1007/s10980-006-0013-z

Pettorelli, N.; Hilborn, A.; Broekhuis, F.; Durant, S. M. 2008. Exploring habitat use by cheetahs using ecological niche factor analysis, *Journal of Zoology* 277(2): 141–148. https://doi.org/10.1111/j.1469-7998.2008.00522.x

Rico, A.; Kindelmann, P.; Sedlacek, F. 2007. Barrier effects of roads on movements of small mammals, *Folia Zoology* 56(1): 1.

Riley, S. P. D.; Pollinger, J. P.; Sauvajot, R. M.; York, E. C.; Bromley, C.; Fuller, T. K.; Wayne, R. K. 2006. A southern California freeway is a physical and social barrier to gene flow in carnivores, *Molecular Ecology* 15(7): 1733–1741. https://doi.org/10.1111/j.1365-294X.2006.02907.x

Selva, N.; Kreft, S.; Kati, V.; Schluck, M.; Jonsson, B.; Mihok, B.; Okarma, H.; Ibisch, P. L. 2011. Road less and low-traffic areas as conservation targets in Europe, *Environmental Management* 48(5): 865–877. https://doi.org/10.1007/s00267-011-9751-z

Trombulak, S. C.; Frissell, C. A. 2000. Review of ecological effects of roads on terrestrial and aquatic communities, *Conservation Biology* 14(1): 18–30. https://doi.org/10.1046/j.1523-1739.2000.99084.x

UNEP. 2001. *Globio: global methodology for mapping human impacts on the biosphere*. UNEP/DEWA/TR.01–3. Environment Information and Assessment Technical Report. UNEP. Nairobi, UNEP: 55.

Van Langevelde, F.; Van Dooremalen, C.; Jaarsma, C. F. 2009. Traffic mortality and the role of minor roads, *Journal of Environmental Management* 90(1): 660–667.

Shekoufeh NEMATOLLAHI. PhD Student of Environmental Sciences, Department of Natural Resources, Isfahan University of Technology. She obtained MSc and BSc in Environmental Sciences, Isfahan University of Technology. Her research interests include: Ecological Impacts of Transportation Infrastructure, Spatial analysis, Landscape Pattern Changes, Land use Planning. She has published more than 10 refereed and conference papers.

Sima FAKHERAN. Assistant Professor of Environmental Sciences, Department of Natural Resources, Isfahan University of Technology. President of Iranian Regional Chapter of International Association for Landscape Ecology (IALE-Iran). She got PhD in Environmental Sciences, Institute of Evolutionary Biology and Environmental Studies (IEU), University of Zurich, Switzerland. MSc of Environmental Science, Faculty of Natural Resources, University of Tehran, Iran, and BSc of Environmental science, from Faculty of Natural Resources, Isfahan University of Technology, Iran. Her research interests include: Landscape Ecology, Conservation Planning, Wildlife corridor design, Ecological Impacts of Climate Change. She has published more than 30 refereed scientific papers.

Alireza SOFFIANIAN. Associate Professor of Environmental Sciences, and Dean of Department of Natural Resources, Isfahan University of Technology. He got PhD of Geography (GIS), from Sorbonne University, Master of Environmental science, from University of Tehran, Iran, and BSc of Fisheries and Environmental Science, University of Tehran, Iran. His research interests: Spatial analysis, Land use Planning, Change detection. He has published more than 40 refereed scientific papers.

5

IMPACT OF AGRO-ENVIRONMENTAL SYSTEMS ON SOIL EROSION PROCESSES AND SOIL PROPERTIES ON HILLY LANDSCAPE IN WESTERN LITHUANIA

Gintaras JARAŠIŪNAS, Irena KINDERIENĖ

Vėžaičiai Branch of the Lithuanian Research Centre for Agriculture and Forestry, Gargždų st. 29, 96216 Vėžaičiai, Klaipėda distr., Lithuania

Abstract. The objective of this study was to evaluate the impact of different land use systems on soil erosion rates, surface evolution processes and physico-chemical properties on a moraine hilly topography in Lithuania. The soil of the experimental site is Bathihypogleyi – Eutric Albeluvisols (ABe–gld–w) whose texture is a sandy loam. After a 27-year use of different land conservation systems, three critical slope segments (slightly eroded, active erosion and accumulation) were formed. Soil physical properties of the soil texture and particle sizes distribution were examined. Chemical properties analysed for were soil pH, available phosphorus (P) and potassium (K), soil organic carbon (SOC) and total nitrogen (N). We estimated the variation in thickness of the soil Ap horizon and soil physico-chemical properties prone to a sustained erosion process. During the study period (2010–2012) water erosion occurred under the grain–grass and grass–grain crop rotations, at rates of 1.38 and 0.11 m^3 ha^{-1} yr^{-1}, respectively. Soil exhumed due to erosion from elevated positions accumulated in the slope bottom. As a result, topographic transfiguration of hills and changes in soil properties occurred. However, the accumulation segments of slopes had significantly higher silt/clay ratios and SOC content. In the active erosion segments a lighter soil texture and lower soil pH were recorded. Only long-term grassland completely stopped soil erosion effects; therefore geomorphologic change and degradation of hills was estimated there as minimal.

Keywords: erosion processes, hilly landscape, long–term land use, crop rotations, transects, slope segments, soil losses.

Introduction

Soil erosion is a natural process, but its rate has been increased by human activities. Soil erosion is one of the most relevant agro-environmental processes in slope areas. In the EU, over 150 million hectares of soil are affected by erosion (Bucur *et al.* 2011) and implementation of appropriate measures at the European scale has been estimated to cost ~€9.3 billion per year (Kuhlmann *et al.* 2010). Unfortunately, until recently soil conservation in Europe has generally not received sufficient attention (Boardman, Evans 2006).

Numerous studies have shown that many soil properties are related to slope gradient and to the particular position of soil on slopes (Birkeland 1984; Zirlewagen, von Wilpert 2010). Conacher and Dalrymple (1977) discussed soil-slope relations in considerable detail and proposed a nine-unit landsurface model based on process and response. They found strong relationships between the spatial distribution of soils and topography. They also assumed that summits are a point of maximum erosion. However, it is sufficient to use a simple five-unit (i.e. summit, shoulder, backslope, toeslope and footslope) model based on slope form, because in reality not all hillslope elements necessarily exist along any given hillslope (Birkeland 1984; Thompson *et al.* 2012).

Olson *et al.* (2002) and Jankauskas (2012) indicated that soil erosion causes irreversible processes in chemical properties of soil along hillslopes by thinning the Ap thickness. The major expenses related to soil erosion in a farm are from the soil nutrients, productivity and quality restoration (Bucur *et al.* 2011; Lal 2001). Erosion has a marked impact on soil physico-chemical properties (Maetens *et al.* 2012; Cerdan *et al.* 2010; Šurda *et al.* 2007) and increase a cost of soil conservation measures in agricultural areas characterised by hilly topography (Leyva *et al.* 2007).

Corresponding author: Gintaras Jarašiūnas
E-mail: gintaras.jarasiunas@gmail.com

Vegetation cover reduces runoff and nutrient losses (Morgan 2007). In addition, generic soil quality under vegetative cover is improved with a larger amount of plant roots reinforcing the soil body to resist to erosion and nutrients leaching (Kinderienė *et al.* 2013). Several studies have indicated that application of suitable cropping systems (crop rotations) significant mitigates land degradation in sloping areas (Jankauskas 2012; Cerdan *et al.* 2010).

Soil erosion on the hilly agricultural landscape is a complex phenomenon, involving land degradation in Lithuania (Jankauskas 2012). Soil erosion is the most intensive factor of Lithuanian relief transformation, given that the estimated volume of all deposits re-deposited by erosion reaches 47.6 million m^3 sediments per year (Česnulevičius 2011). The depth of denudation reaches ≤0.5 mm^{-1} yr^{-1} (Česnulevičius 2011), thus in conformity with Dėnas *et al.* (2006) varies from 0.12–3.5 mm^{-1} yr^{-1}, with maximum values in uplands and river valleys, respectively. The main problems requiring agro-environmental land management measures in Lithuanian uplands are the degradation degree of fields by erosion, deterioration of soil structure and compaction. Seventeen per cent of Lithuania's agricultural land is eroded with levels of ≤43–58% in hilly regions (Lietuvos dirvožemiai 2001). Surface runoff generation and soil erosion in the Žemaičiai Uplands mostly depend on the erodible glacial parent material, precipitation intensity, land use and cover condition.

The aim of investigations is to study the long-term land use impact of soil erosion in several slope positions, surface evolution processes and soil properties in different agro-environmental systems located on hilly landscape in Western Lithuania.

1. Material and methods

1.1. Study site

The erodible ground moraine material, abundance and intensity of precipitation have created favourable natural conditions for erosion of parent materials by water in the Žemaičiai Uplands (Kudaba 1983). Sandy clay loams and sandy loams Gleyic Albeluvisols and Luvisols prevail in the upland (Lietuvos dirvožemiai 2001). The mean and maximum elevations in the Žemaičiai Uplands are 119.0 and 234.6 m a.s.l., respectively. The topography of Kaltinėnai catchment area consists of large and steeply dissected hills (≥8–12°) where the average slope length is ≥100 m (Lietuvos dirvožemiai 2001). Major crops cultivated in the catchments are wheat, barley and oats with mean annual yields of 3.75, 4.27 and 2.0 t^{-1} ha^{-1} y^{-1}, respectively.

The Lithuanian climate is conducive to the occurrence of water erosion (Jankauskas 2012). Based on 52-year cumulative precipitation data from the nearest meteorological station based in Laukuva (55°61′N, 22°23′E, 166 m a.s.l.), annual precipitation is estimated to be 837 mm, with summer and autumn dominance. Mean annual precipitation in the area during the study period was 831–966 mm. The mean annual air temperature is estimated to be 6.5 °C (−2.5 °C in January and 17.3 °C in July) (Meteorological bulletin of Laukuva meteorological station 1960–2012).

The study was conducted in the fields of a 27-year long experiment on moraine hilly terrain of the southern-central Žemaičiai Uplands (55°34′N, 22°29′E) in Western Lithuania. Research data were obtained in 2010–2012 on a sandy loam Bathihypogleyi-Eutric Albeluvisol (ABe-gld-w) at the Kaltinėnai Research Station of the Lithuanian Research Centre for Agriculture and Forestry. The catchment area has a maximum elevation of 174.2 m a.s.l. The field experiments investigating water erosion rates under different agro-environmental systems in the sloping area have been carried out since 1983. The field experiments (environmental monitoring) are part of the Core Research Programme of the Global Change and Terrestrial Ecosystem (GCTE) Project, a component of the International Geosphere Biosphere Programme (IGBP).

1.2. General methodological framework

The formed critical slope segments (slightly eroded, active erosion and accumulation) and sustained erosion processes were investigated on sloping agricultural lands. Indicators used to estimate formed slope segments included estimated transects of Ap horizon thickness and conformity to topographic attributes. In most cases, differences in thickness of the Ap horizon in hilly areas may be due to soil erosion (Lal 1998). The use of transects is well known in field research (Gillison, Brewer 1985). Nine downslope transects (three replicates within each study plot) 98 m in length were set up along the northern slope with a width of 3.60 m; the relative elevation of slope transects ranged between 13.64–15.02 m.

Based on the spatial heterogeneity of Ap horizon thickness, the studied transects were divided into three slope segments on each study field. Relative flat summit form-unit of slope is called slightly eroded segment and it could serve as a control site due to very weak soil erosion. Meanwhile, the shoulder (under the long-term grassland) and backslope form-units (under the grain-grass and grass-grain crop rotations) served as a strongly eroding site and was entitled as active erosion segments. Finally, the footslope form-unit acted as accumulation segment because all sediment, induced down by the throughflow of water, is trapped in this topographic low.

1.3. Agro-environmental systems and field measurements

The erosion preventive agro-environmental systems (crop rotations) were developed for a 6-course (years). In this study we represent results obtained during 3-course (years). More detailed descriptions and explanations of crop rotations are given in Jankauskas *et al.* (2009). The three crop rotations of the long-term field experiment were compared:

The grain-grass crop rotation (containing 33% grasses and 67% cereal grains): winter rye (*Secale cereale* L.) (in 2010); spring barley (*Hordeum vulgare* L.) (during 2011–2012).

The long-term grassland: multi–species mixture of perennial grasses which contribute to soil humus formation. This grass mixture consisted of 20% each of common timothy (*Phelum pratense* L.), red fescue (*Festuca rubra* L.), white clover (*Trifolium repens* L.), Kentucky bluegrass (*Poa pratensis* L.) and birdsfoot trefoil (*Lotus corniculatus* L.) (during the period 2010–2012). Multi-species mixtures of perennial grasses for long-term use as sod–forming grasses.

The grass-grain crop rotation (containing 67% grasses and 33% cereal grains): winter rye (*Secale cereale* L.) (in 2010); spring barley (*Hordeum vulgare* L.), undercrop of cocksfoot and red fescue mixture (*Dactylis glomerata* L. – *Festuca rubra* L.) (in 2011), and one year of red clover and timothy mixture (*Trifolium pratense* L. – *Phleum pratense* L.) (in 2012).

The topographic attributes (elevations, distances and angles) of study fields (Fig. 1) were measured using precision geometrical land levelling methods (equipment Zeiss Ni002®), providing measurement accuracy of ±0.03 m, according to Kriaučiūnaitė–Neklejonovienė *et al.* (2008).

The topographic measurements showed the slope inclination of each experimental field to be 8.0° (17.8%) (the grain-grass crop rotation), 8.9° (19.8%) (the long-term grassland) and 8.7° (19.3%) (the grass-grain crop rotation) (Table 1). Data of soil erosion rate for the period 1983–2009 was obtained by Jankauskas (2012).

Table 1. Experimental agro-environmental plots description

Variables	Grain-grass crop rotation	Long-term grassland	Grass-grain crop rotation
Altitude range (m)	170.54–156.91	171.05–156.02	171.73–157.06
Slope gradient (°)	8.0	8.9	8.7
Slope length (m)	98	98	98
Slope width (m)	3.60	3.60	3.60
Aspect	north	north	north
Textural class	Sandy loam	Sandy loam	Sandy loam
Soil erosion rate (m^3 ha^{-1} yr^{-1})			
2010–2012	1.38	0	0.11
1983–2009	15.88	0.01	4.69

1.4. Soil laboratory analyses

Soil samples (~500 g) were taken annually from topsoil (200 mm) using a sterile soil corer. The samples were taken randomly from the studied slope segments across the slope. Three soil samples per slope segment were collected considering the entire slope. A total of 27 composite soil samples were collected for determination of soil properties. Soil samples were examined for textural class, particle size (sand 2000–63 μm; silt 63–2 μm and clay <2 μm) distribution and chemical properties (soil pH_{KCl}, available phosphorus (P) and potassium (K), soil organic carbon (SOC) and total nitrogen (N)). Soil particle size distribution was established at the Agrochemical Research Laboratory of the LRCAF using the particle-size analysis method (ISO 11277: 2009). Soil pH

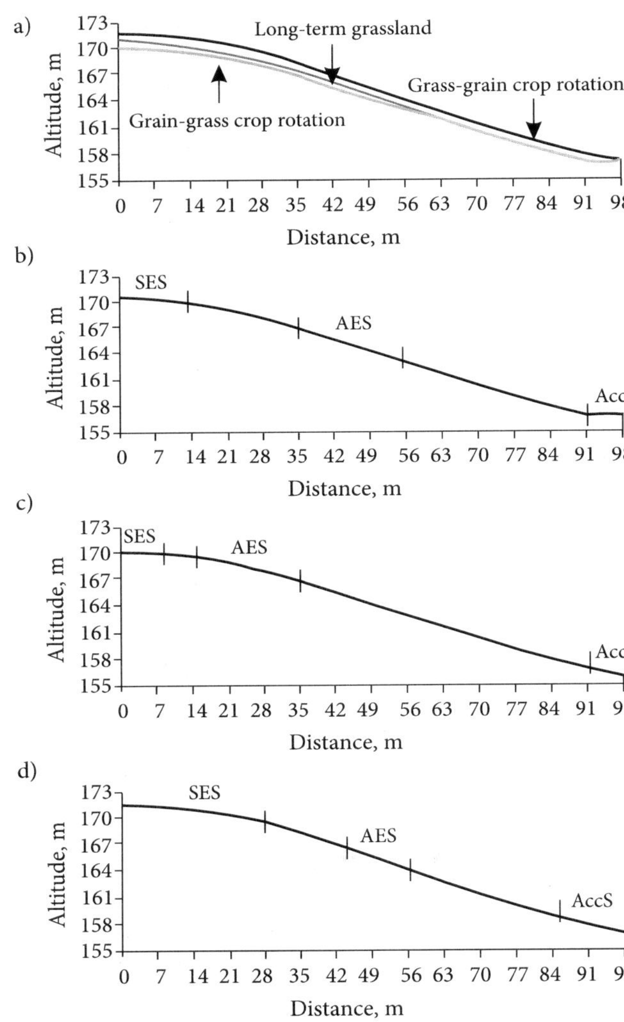

Fig. 1. The localization and topographical attributes of the slope segments: transects in the experimental sites under: a) all agro-environmental systems, b) the grain-grass crop rotation, c) long–term grassland, and d) grass-grain crop rotation. Abbreviations: SES – Slightly eroded segment; AES – Active erosion segment; AccS – Accumulation segment.

was determined in 1 M KCl soil sample extracts using a calibrated digital pH meter (ISO 10390: 1994), mobile P_2O_5 and K_2O extracted using the A–L method (available P and K were extracted with ammonium acetate–lactate solution, pH 3.7, ratio 1:20) (GOST 26208-91: 1993). SOC content was determined by the Tyurin dichromate-oxidation method (Nikitin 1999; ISO 14235: 1998) and total N by the Kjeldahl method (ISO 11261: 1995). The silt/clay ratio, an index of soil weathering, was calculating according to FAO (1988).

1.5. Assessment of water erosion

Water erosion rates on environmental monitoring site were assessed by measuring the length and cross-sectional area of rills, to calculate soil loss volume (Chambers *et al.* 2000; Jankauskas *et al.* 2008). Lost soil volume was calculated using the formula (1):

$$x = [(\Sigma l_1 p_1 + \Sigma l_2 p_2 + \ldots \Sigma l_n p_n) : n] : y, \qquad (1)$$

where: x – volume of erosion rills (m^{-3} ha^{-1}); l, l_1, ... l_n – rill depth (cm^{-1}); p, p_1, ... p_n – rill width (cm^{-1}, accuracy ± 0.1 cm^{-1}; n – number of rills on the measured plot width; y – measured plot width (m^{-1}), and Σ – sum of performed measurements from selected 1 m length segments located at equal distances on the experimental plot.

Methods of rill volume measurements may be performed relatively well in areas where rill erosion represents the dominant sediment source (Govers, Poesen 1988; Vanmaercke *et al.* 2012). Plot runoff and erosion were measured on a regular basis, up to weekly after rainfall and sowing. Measurements were taken from spring sowing (typically late April or early May) to mid-June for cereals. During the research period 72 rill volume measurements were performed.

1.6. Statistical analysis

One-way analysis of variance (ANOVA) was used to estimate significance of differences at $P \leq 0.05$ between treatment means using Fisher's LSD test and correlations among soil properties were determined using Pearson correlation coefficients on a significance levels of $p \leq 0.05$ using the statistical programme STAT ENG from the software package *Selekcija* (Tarakanovas, Raudonius 2003).

2. Results

2.1. Plot topography change and soil texture

Variations in the soil Ap layer thickness and soil properties prone to long-term erosion process are potentially indicative of the relative effects of both accelerated (human induced) and natural erosion. In experimental fields, measurements of Ap layer thickness showed a great variation under various erosion preventive agro-environmental systems by slope segments (Table 2).

Under the influence of erosive processes on cultivated study plots higher soil weathering and thinner Ap horizon at a higher altitude was observed, therefore as after-effect the thicker accumulation segments were determinate as well. The silt/clay ratios were highest in the accumulation segments under all crop rotations.

Soil texture was classified as sandy loamy in all agro-environmental systems and slope segments. The results showed a strong correlation between sand and silt contents ($r = -0.95$, $p \leq 0.05$, $n = 27$) under the grain–grass crop rotation. However, sand content under the long-term grassland strongly correlated with clay and silt contents ($r = -0.85$–0.86, $p \leq 0.05$, $n = 27$). Sand and clay particles under the grass–grain crop rotation were significantly correlated ($r = -0.96$, $p \leq 0.05$, $n = 27$).

Table 2. Topography and soil properties of the slope segments in the different agro-environmental systems after 27-years of study

Variables	Grain-grass crop rotation			Long-term grassland			Grass-grain crop rotation		
	SES	AES	AccS	SES	AES	AccS	SES	AES	AccS
Topography									
Gradient (°)	3.3	10.9	0.8	3.5	13.8	8.5	5.5	7.4	7.3
Length (m)	14	21	7	8.9	17.3	7	28	13	12.8
Area (m^{-2})	50.40	75.60	25.20	32.04	62.28	25.20	100.80	46.80	46.08
Soil properties									
Ap horizon thickness (cm)	18.3±4.26	14.6±0.84	75.7±4.72	25.3±2.17	23.3±0.68	46.3±2.26	21.4±2.53	18.8±0.59	63.7±2.11
Sand (%)	69.6±1.11	64.5±0.39	62.7±1.23	70.3±2.85	60.9±0.35	65.2±0.29	66.2±0.59	65.2±2.18	69.5±0.36
Silt (%)	16.0±0.44	19.7±0.50	22.6±0.73	16.1±1.50	20.8±0.46	21.2±0.09	18.5±0.49	18.8±0.67	19.1±0.23
Clay (%)	14.2±0.82	15.8±0.13	14.7±0.55	13.6±1.43	18.3±0.81	13.6±0.27	15.3±0.09	16.0±1.51	11.4±0.23
Silt/clay ratio (%)	1.15	1.25	1.55	1.20	1.14	1.60	1.21	1.21	1.70

Abbreviations: SES – Slightly eroded segment; AES – Active erosion segment; AccS – Accumulation segment.
Mean ± standard error.

The silt/clay ratio varied from 1.14–1.25% on slightly eroded and active erosion segments to 1.55–1.70% in accumulation segments (Fig. 2).

The correlation between silt/clay ratios and Ap thickness values showed that under all crop rotations the active erosion segments have lighter soil textures than the other segments ($r = 0.93$, $p \leq 0.05$, $n = 27$).

2.2. Soil chemical properties

The results showed that under all crop rotations the active erosion segments had significantly ($P \leq 0.05$) lower soil pH compared with other segments. Slope gradient values were correlated negatively with soil pH values ($r = -0.68$, $p \leq 0.05$, $n = 27$). Mean soil pH ranged from 5.8–6.5. The lowest soil pH was established under the grain-grass crop rotation and long-term grassland in active erosion segments, where relatively acidic soils prevailed.

It was found that under all crop rotations accumulation segments had significantly ($P \leq 0.05$) higher SOC content compared with the other slope segments. The SOC under the grain–grass crop rotation varied from 0.85% in the slightly eroded segment to 0.93 and 1.25% in the active erosion and accumulation segments, respectively. The SOC under the long-term grassland and grass-grain crop rotations was higher in the slightly eroded segment and lower under the grain-grass crop rotation (decrease 38%

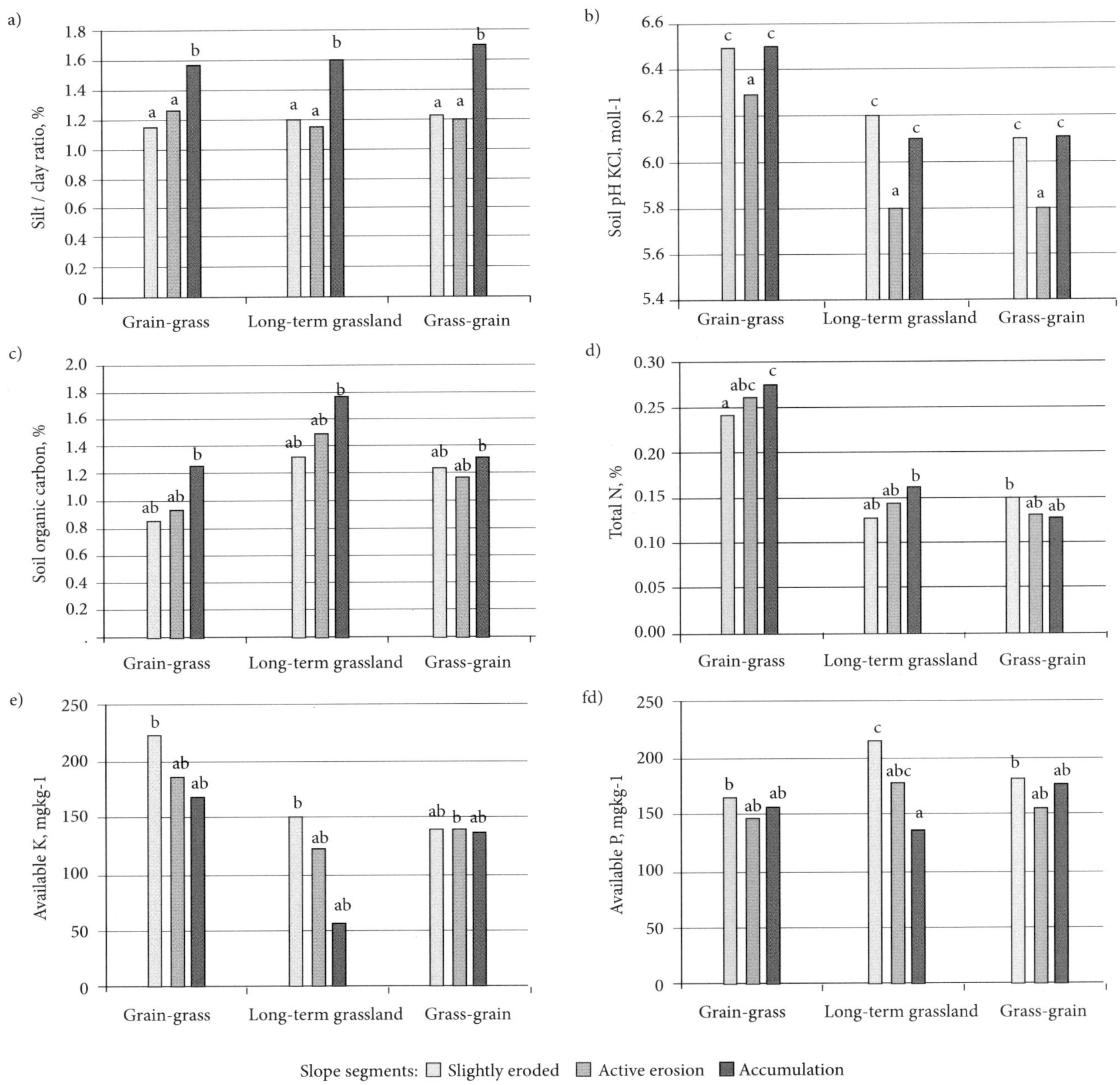

Fig. 2. The comparison of mean values by slope segments under separate crop rotations for: a) Silt/clay ratio, b) Soil pH, c) SOC, d) Total N, e) Available K, and f) Available P. Different letters represent significant differences among treatments at $p \leq 0.05$; $n = 27$

in relative value), respectively. The SOC was generally high under long-term grassland compared to the other crop rotations. Under the grain-grass crop rotation and long-term grassland the slightly eroded segments had less SOC, only under the grass-grain crop rotation there was recorded lower SOC in the active erosion segment.

Analysis of total *N* showed that the upper mean (0.28%) was obtained under the grain-grass crop rotation in the accumulation segment, the lower (0.13%) under the grass-grain crop rotation in the accumulation segment. Under the grass-grain crop rotation there was identified significantly ($P \leq 0.05$) lower total N content than under other crop rotations.

Under all crop rotations was observed significant ($P \leq 0.05$) differences between slope segments with higher available *P* in slightly eroded segments. Under the long-term grassland the accumulation and active erosion segments had 55 and 29% lower means of available P versus slightly eroded segment. The highest concentration of available *K* (139–224 mg kg^{-1}) under all crop rotations was in slightly eroded segments; the least mean (57 mg kg^{-1}) was under the long-term grassland in the accumulation segment. The higher ($P \leq 0.05$) available K was estimated under the grain-grass crop rotation and long-term grassland in the slightly eroded segments. Significantly ($P \leq 0.05$) lower available K was determined under the grass-grain crop rotation in slightly eroded and accumulation segments.

2.3. Soil erosion rate under the crop rotations

Rill erosion prevailed in the study fields. The streams of water runoff developed small and large rills on slopes. In this form of erosion, soil and particles displaced by water were intensively separated and sorted. The soil erosion rate during three-year experimental period was estimated under the grain-grass and grass-grain crop rotations of 1.38 and 0.11 m^3 ha^{-1} yr^{-1}, respectively. Tillage erosion was a dominant process of soil redistribution. However, even the grass-grain crop rotation could not completely prevent soil erosion. Only long-term grassland prevented water erosion where there was no evidence of rilling.

3. Discussion

The erosion-preventive agro-environmental systems developed over a 27-year period, tillage practice and influence of annual rainfall indicate important impacts on land degradation processes on the moraine hilly terrain in Western Lithuania. Soil was differentially eroded along the slopes. Soil erosion was mainly caused by tillage and water under continuous intensive cropping. Variations in the Ap thickness and soil properties prone to erosion are potentially indicative of the relative **effects** of both accelerated and natural erosion. Thinner Ap horizons not only reflect soil erosion rates. Soil compaction by agricultural machinery may also play a significant role. The results from this study suggested that transects of Ap thickness and the same agro-chemical soil properties can be successfully adapted determining critical slope segments.

The silt content was a very meaningful indicator of the extent of soil erosion along the slope. Eroded soil occurred in accumulation segments on footslope considering to increased silt quantity. Bronick and Lal (2004) observed that on sloping soils erosion tends to preferentially remove light particles, including silts. Grasses exhibit high resistance to soil erosion and sediment accumulation, while cereals perform poorly in this respect (Račinskas 1990). Actually, when soil erosion occurs, silt particles are suspended in surface water and are transported downslope, thus leaving coarser material at the top slope positions (Halim *et al.* 2007). Previous studies (Morgan 2007; Aksoy, Kavvas 2005) have also reported that eroded materials are enriched by silt sized particles relative to the original soil where the erosion event commenced. Bronick and Lal (2004) reported that elevation influences the rate of weathering in soils. Sand content under the grain-grass crop rotation decreased with decreasing elevation, while the silt increased with decreasing elevation. Clay content was significantly higher in active erosion segments. Silt content under the long-term grassland increased with decreasing elevation. It was found, that more weathered soils (e.g. in active erosion segments) contained more clay than less weathered ones. Where the rills eroded through the topsoil, there was a marked reduction in SOC, from 2.4–3.9 to <0.9%. Clay content was reduced from 7–9 to 3–4%. The fraction 0.002–0.1 mm (silt+very fine sand) increased from 84–90 to about 92%. Especially the reduction in SOC makes the soil more erodible and gives a sudden decrease in the resistance to erosion when concentrated water flows into the layer with a high silt and sand content.

Higher silt/clay ratios in accumulation segments compared with others can be explained by their lower locality on slopes, where silt particles can be washed out by runoff and might be deposited in the topographic low (Zachar 1982; Øygarden 2003). Thus, silts, as SOC, were probably eroded, which occurred at more severe rates on the higher elevation segments. Similar results were obtained in Nordic countries on silt loam and clay loam soil. Silt is very vulnerable to water erosion, since it is small and has low resistance to both cohesion and friction (Ulen *et al.* 2012; Bechmann 2012). Combined with a high silt/sand content, more severe rilling and ephemeral gullies develop (Øygarden 2003).

Nutrient dynamics are closely linked to soil movement by tillage (Ni, Zhang 2007). Soil pH could become one of very sensitive and suitable criteria for identifying critical slope segments on moraine hilly terrain in agricultural areas affected by erosion. This could be because eroding soils are exhuming calcareous subsoils. Variations of soil pH and SOC changes prone to the erosion process and are potentially indicative of the relative extent of erosion along the slopes. Soil pH and SOC strongly correlated ($r = 0.81$, $p \leq 0.05$, $n = 27$). We also estimated a positive correlation ($r = 0.80$, $p \leq 0.05$, $n = 27$) between soil pH and total N within each slope segment.

Apparently, the accumulation segments in all agro-environmental systems showed higher SOC contents than in other vulnerable erosion headland segments. In general, the SOC of slightly eroded and active erosion segments were redistributed and deposited in the accumulation segments. These differences may be due to finest soil particles and SOC translocation caused by tillage and water erosion from upper to lower parts of the slope. As confirmation, significant correlation ($r = 0.50$, $p \leq 0.05$, $n = 27$) between the SOC and silt contents was indicated. Fullen *et al.* (2006) and Jankauskas *et al.* (2009) also reported that silt can play an important role in influencing SOC storage. The higher SOC content under long–term grassland compared to the grain–grass and grass–grain crop rotations are impacted by differences in soil erosion, vegetation cover, biomass return and it provides evidence of effective SOC sequestration in soil.

Results regarding total N under the grain–grass crop rotation revealed an increasing trend from slightly eroded slope segment, which might be due to their downward movement with surface runoff from top slopes and accumulation on basal slopes. Birkinshaw and Ewen (2000) and Aksoy and Kavvas (2005) have confirmed that on upland areas total *N* is transported with sediments by surface runoff to topographic lows. The difference found in total *N* content by slope segments can be also justified by their difference in SOC content (Dercon *et al.* 2006; Pennock *et al.* 1994). Our investigations established significant correlation bettwen SOC and total *N* under the grain-grass crop rotation ($r = 0.86$, $p \leq 0.05$, $n = 9$) and the long-term grassland ($r = 0.79$, $p \leq 0.05$, $n = 9$). Only under the grass-grain crop rotation these relationship was not observed. It could be so because under the grass-grain crop rotation, where grasses predominate, the fertilizer of nitrogen on the soil surface is spreading and as a result nitrogen became more mobile and relation with SOC is incoherent. Though under the grain-grass crop rotation the nitrogen in the soil are inserted as well. In general, different soil cover (vegetation), water runoff volume and methods of fertilization have influence to agrochemical properties of soil on eroded and steep slopes. The significant correlative occurrence of total *N* and SOC was confirmed in eroded agricultural fields by Haag and Kaupenjohann (2001).

The soil under the long-term grassland was richest in available *P*. The influence of the SOC explains available *P* retention and supply under this highly vegetated cover. It was estimated that only under the grass-grain crop rotation all slope segments contained sufficient available *P*.

Previous researchers (Brubaker *et al.* 1993; Kinderiene, Karcauskiene 2012) argued that erosion hazard on eroded lands is the main factor of controlling potassium (K) and other soil fertility parameters. Our study data showed an decreasing trend in soil available K down the slopes under grain-grass crop rotation and long-term grassland. Thus, contents of available K under grass-grain crop rotation in all slope segments were similar. According to Jankauskas (2003) due to water erosion on a sandy loam Bathihypogleyi – Eutric Albeluvisols losses of K with soil are 20 kg t^{-1}.

In Europe, soil erosion caused mainly by water and rill erosion affects the largest area (Gobin *et al.* 2004). The upland area is usually considered to have rill erosion (Aksoy, Kavvas 2005). Soil loss due to rill erosion is important processes of soil degradation that cause significant on-site and off-site problems (Montgomery 2007).

In Lithuania vegetation have a great influence on soil erodibility. In the study the anti-erosion agro-environmental system (grass-grain crop rotation) decreased soil losses due to erosion by ninety-two per cent compared with those of the grain-grass crop rotation. This shows the importance of applying soil conservation measures on hilly terrain and agrees with Van Rompaey *et al.* (2001; 2002) idea that the potential for surface runoff and soil erosion is mostly affected by land use and cultivation. The surface of the long-term grassland was not subject to rill erosion. Kinderienė *et al.* (2013) observed that the slopes occupied with permanent grasses, where slightly and moderately eroded Eutric Albeluvisol (AB–e) prevailed, were also resistant to soil erosion. It should be mentioned, that according to data collected in 1983–2000 in the long-term experiments at the Kaltinėnai Experimental Station, under the perennial grasses, winter rye and spring barley the annual soil losses amounted to 0.15, 34.44 and 108.35 $m^3\ ha^{-1}$, respectively (Jankauskas, Jankauskienė 2003). The risk of erosion of agricultural land is an important factor for land abandonment (Koulouri, Giourga 2007; Pointereau *et al.* 2008).

Conclusions

The findings of the current study suggested that after 27-years of use of different erosion preventive agro-environmental systems on moraine hilly landscape, sustained soil erosion resulted in micro-topographic landscape changes. Erosion processes can affect spatial variability of soil particle distribution, thus resulting in soil texture becoming lighter in active erosion and increasing the silt/clay ratios in accumulation segments.

Soil pH was significantly negatively correlated with slope gradient. Thus, soil pH was the lowest in active erosion segments. The highest concentration of available P and K was in slightly eroded segments. Higher SOC content was in the accumulation segments compared with others. The SOC content was generally higher under the long-term grassland.

During the experimental period water erosion occurred under the grain-grass and grass-grain crop rotation fields. The grass-grain crop rotation decreased rill erosion and soil erosion rate by ninety-two per cent compared to grain-grass crop rotation. Only long-term grassland completely stopped soil erosion.

Therefore, our research data relating to slope segments as the critical parts of sustained erosion process and experience of progressive soil conservation practices on the moraine hilly terrain are important for sustainable agricultural development in the temperate climate zone. The observed topographic changes and alterations in soil properties, resulting from land use management, would lead to further land degradation, with negative implications on slope use for agricultural needs.

Acknowledgments

The authors thank Mrs. Vida Sodienė (Department of Civil Engineering, Klaipėda University, Lithuania) for assistance with levelling experimental fields.

References

Aksoy, H.; Kavvas, M. L. 2005. A review of hillslope and watershed scale erosion and sediment transport models, *Catena* 64(2–3): 247–271. http://dx.doi.org/10.1016/j.catena.2005.08.008

Bechmann, M. 2012. Effect of tillage on sediment and phosphorus losses from a field and a catchment in south eastern Norway, Acta Agriculturae Scandinavica, Section B–Soil & Plant Science. Special Issue. *Soil Erosion in the Nordic Countries* 2: 206–216. http://dx.doi.org/10.1080/09064710.2012.715183

Birkeland, P. W. 1984. *Soils and geomorphology*. New York, Oxford: Oxford University Press.

Birkinshaw, S. J.; Ewen, J. 2000. Nitrogen transformation component for SHETRAN catchment nitrate transport modelling, *Journal of Hydrology* 230(1–2): 1–17. http://dx.doi.org/10.1016/S0022-1694(00)00174-8

Boardman, J.; Evans, R. 2006. Britain, in J. Boardman, J. Poesen (eds.). *Soil erosion in Europe*. Chicester: Wiley, 439–453.

Bronick, C. J.; Lal, R. 2004. Soil structure and management: a review, *Geoderma* 124(1–2): 3–22. http://dx.doi.org/10.1016/j.geoderma.2004.03.005

Brubaker, S. C.; Jones, A. J.; Lewis, D. T.; Frank, K. 1993. Soil properties associated with landscape positions and management, *Soil Science Society of America Journal* 57: 235–239.

Bucur, D.; Jitareanu, G.; Ailincai, C. 2011. Effects of long-term soil and crop management on the yield and on the fertility of eroded soil, *Journal of Food, Agriculture & Environment* 9(2): 207–209.

Cerdan, O.; Govers, G.; Le Bissonnais, Y.; Van Oost, K.; Poesen, J.; Saby, N.; Gobin, A.; Vacca, A.; Quinton, J.; Auerswald, K.; Klik, A.; Kwaad, F. J. P. M.; Raclot, D.; Nonita, I.; Rejman, J.; Rousseva, S.; Muxart, T.; Roxo, M. J.; Dostal, T. 2010. Rates and spatial variations of soil erosion in Europe: a study based on erosion plot data, *Geomorphology* 122(1–2): 167–177. http://dx.doi.org/10.1016/j.geomorph.2010.06.011

Chambers, B. J.; Garwood, T. W. D.; Unwin, R. J. 2000. Controlling soil water erosion and phosphorus losses from arable land in England and Wales, *Journal of Environmental Quality* 29(1): 145–150. http://dx.doi.org/10.2134/jeq2000.00472425002900010018x

Conacher, A. J.; Dalrymple, J. B. 1977. The nine-unit land surface model: and approach to pedogeomorphic research, *Geoderma* 18: 1–154.

Česnulevičius, A. 2011. The morphometric structure of Lithuanian relief and its influence on erosion processes, *Baltica, Special Issue//Geosciences in Lithuania: challenges and perspectives* 24: 137–142.

Dėnas, Ž.; Kumetaitis, A.; Šliaupa, S.; Zakarevičius, A.; Šliaupienė, R. 2006. GIS model of the erosion potential of Lithuania, *Geodesy and Cartography* 32(3): 57–61 (in Lithuanian).

Dercon, G.; Clymans, E.; Diels, J.; Merckx, R.; Deckers, J. 2006. Differential ^{13}C isotopic discrimination in maize at varying water stress and at low to high nitrogen availability, *Plant and Soil* 282(1–2): 313–326. http://dx.doi.org/10.1007/s11104-006-0001-8

FAO. 1988. *Soil map of the world*. Revised Legend, World Soil Resources. Report 60. Rome. 119 p.

Fullen, M. A.; Arnalds, A.; Bazzoffi, P.; Booth, C. A.; Castillo, V.; Kertész, Á.; Martin, P.; Ritsema, C.; Benet, A. S.; Souchére, V.; Vandekerckhove, L.; Verstraeten, G. 2006. Government and agency response to soil erosion risk in Europe, in J. Boardman, J. Poesen (Eds.). *Soil erosion in Europe*. Chicester: John Wiley, 805–828.

Gillison, A. N.; Brewer, K. R. W. 1985. The use of gradient directed transects or gradsects in natural resource surveys, *Journal of Environmental Management* 20: 103–127.

Gobin, A.; Jones, R.; Kirkby, M.; Campling, P.; Govers, G.; Kosmas, C.; Gentile, A. R. 2004. Indicators for pan-European assessment and monitoring of soil erosion by water, *Environmental Science & Policy* 7(1): 25–38. http://dx.doi.org/10.1016/j.envsci.2003.09.004

GOST 26208-91: 1993. Soils. Determination of mobile compounds of phosphorus and potassium by Egner-Riem-Domingo method (AL-method). Moscow, Russia.

Govers, G.; Poesen, J. 1988. Assessment of the interrill and rill contributions to total soil loss from an upland field plot, *Geomorphology* 1(4): 343–354. http://dx.doi.org/10.1016/0169-555X(88)90006-2

Haag, D.; Kaupenjohann, M. 2001. Landscape fate of nitrate fluxes emissions in Central Europe: a critical review of concepts, data, and models for transport and retention, *Agriculture, Ecosystems and Environment* 86(1): 1–21.

Halim, R.; Clemente, R. S.; Routray, J. K.; Shrestha, R. P. 2007. Integration of biophysical and socio-economic factors to assess soil erosion hazard in the Upper Kaligarang watershed, Indonesia, *Land Degradation & Development* 18(4): 453–469. http://dx.doi.org/10.1002/ldr.774

ISO 11277: 2009. Soil quality – Determination of particle size distribution in mineral soil material – Method by sieving and sedimentation. International Organization for Standardization. Geneva, Switzerland. 46 p.

ISO 10390: 1994. Soil quality. Determination of pH. TC 190/SC 3.

ISO 14235: 1998. Soil quality. Determination of organic carbon by sulfochromic oxidation. TC 190/SC 3.

ISO 11261: 1995. Soil quality. Determination of total nitrogen – Modified Kjeldahl method. TC 190/SC 3.

Jankauskas, B. 2012. Soil erosion: case study, Lithuania, in C. Jakobsson (Ed.). *Sustainable agriculture. Ecosystem health and sustainable agriculture*. Uppsala: Baltic University Press, 231–238.

Jankauskas, B.; Jankauskienė, G. 2003. Erosion-preventive crop rotations for landscape ecological stability in upland regions of Lithuania, *Agriculture, Ecosystems and Environment* 95: 129–142.

Jankauskas, B.; Jankauskienė, G.; Fullen, M. A.; Booth, C. A.; Šlepetienė, A. 2009. Soil organic matter changes in Lithuanian soils: experiences and results, *Zemdirbyste–Agriculture* 96(1): 85–97.

Jankauskas, B.; Jankauskienė, G.; Tiknius, A. 2008. The contribution of quality assessment of eroded agricultural soil on hilly-undulating landscapes to sustainable community development, in I. Petrosillo *et al.* (Eds.). *Use of landscape sciences for the assessment of environmental security*. Netherlands: Springer, 431–451.

Kinderienė, I.; Jarašiūnas, G.; Karčauskienė, D. 2013. Augalų maisto medžiagų (N, P, K) nuo kalvų šlaitų praradimas su dirvožemiu ir vandeniu [Loss of plant nutrients (N, P, K) with soil loss and water runoff from hill slopes], *Žemės ūkio mokslai* 20(1): 10–19 (in Lithuanian). http://dx.doi.org/10.6001/zemesukiomokslai.v20i1.2634

Kinderiene, I.; Karcauskiene, D. 2012. Effects of different crop rotations on soil erosion and nutrient losses under natural rainfall conditions in Western Lithuania, *Acta Agriculturae Scandinavica, Section B – Soil & Plant Science*, 62(sup2): 199–205. http://dx.doi.org/10.1080/09064710.2012.714400

Koulouri, M.; Giourga, C. 2007. Land abandonment and slope gradient as key factors of soil erosion in Mediterranean terraced lands, *Catena* 69(3): 274–281. http://dx.doi.org/10.1016/j.catena.2006.07.001

Kriaučiūnaitė–Neklejonovienė, V.; Šližienė, G.; Šližys, J.; Stravinskienė, V.; Venckus, S.; Zivatkauskas, A. 2008. *Geodezija* [Geodesy]. Vilnius: Vaistų žinios. 329 p. (in Lithuanian).

Kudaba, Č. 1983. *Lietuvos aukštumos.* [Lithuanian Uplands]. Vilnius: Mokslas. 186 p. (in Lithuanian).

Kuhlmann, T.; Reinhard, S.; Gaaff, A. 2010. Estimating the costs and benefits of soil conservation in Europe, *Land Use Policy* 27(1): 22–32. http://dx.doi.org/10.1016/j.landusepol.2008.08.002

Lal, R. 1998. Soil erosion impact on agronomic productivity and environment quality, *Critical Reviews in Plant Sciences*, 17(4): 319–464. http://dx.doi.org/10.1080/07352689891304249

Lal, R. 2001. Soil degradation by erosion, *Land Degradation & Development* 12(6): 519–539. http://dx.doi.org/10.1002/ldr.472

Leyva, J. C.; Martínez, J. A. F.; Roa, M. C. G. 2007. Analysis of the adoption of soil conservation practices in olive groves: the case of mountainous areas in southern Spain, *Spanish Journal of Agricultural Research* 5(3): 249–258.

Lietuvos dirvožemiai. 2001. Vilnius: Lietuvos mokslas. 1244 p. (in Lithuanian).

Maetens, W.; Vanmaercke, M.; Poesen, J.; Jankauskas, B.; Jankauskienė, G.; Ionita, I. 2012. Effects of land use on annual runoff and soil loss in Europe and the Mediterranean: A meta-analysis of plot data, *Progress in Physical Geography* 36(5): 12–55. http://.doi.org/10.1177/0309133312451303

Meteorological bulletin of Laukuva meteorological station. 1960–2012. Vilnius: Lietuvos hidrometeorologijos tarnyba.

Montgomery, D. R. 2007. Soil erosion and agricultural sustainability, *Proceedings of the National Academy of Sciences of the United States of America* 104(33): 13268–13272.

Morgan, R. P. C. 2007. Vegetative-based technologies for erosion control, in A. Stokes, I. Spanos, J. E. Norris, E. Cammeraat (Eds.). Eco- and Ground Bio-Engineering: The use of vegetation to improve slope stability, *Developments in Plant and Soil Sciences* 103: 265–272.

Ni, S. J.; Zhang, J. H. 2007. Variation of chemical properties as affected by soil erosion on hillslopes and terraces, *European Journal of Soil Science* 58(6): 1285–1292. http://dx.doi.org/10.1111/j.1365-2389.2007.00921.x

Nikitin, B. A. 1999. Metod opredelenija gumusa počvy, *Agrohimija* 5: 91–93 (in Russian).

Øygarden, L. 2003. Rill and gully development during an extreme winter runoff event in Norway, *Catena* 50: 217–242. http://dx.doi.org/10.1016/S0341-8162(02)00138-8

Olson, K. R.; Gennadiyev, A. N.; Jones, R. L.; Chernyanskii, S. 2002. Erosion patterns on cultivated and reforested hillslopes in the Moscow Region, Russia, *Soil Science Society of America Journal* 66(1): 193–201. http://dx.doi.org/10.2136/sssaj2002.1930

Pennock, D. J.; Anderson, D. W.; de Jong, E. 1994. Landscape-scale changes in indicators of soil quality due to cultivation in Saskatchewan, Canada, *Geoderma* 64(1–2): 1–19. http://dx.doi.org/10.1016/0016-7061(94)90086-8

Pointereau, P.; Coulon, F.; Girard, P.; Lambotte, M.; Stuczynski, T.; Sanchez Ortega, V.; Del Rio, A. 2008. *Analysis of farmland abandonment and the extent and location of agricultural areas that are actually abandoned or are in risk to be abandoned*, in E. Anguiano, C. Bamps, J. M. Terres (Eds.). JRC Scientific and Technical Reports.

Račinskas, A. 1990. *Dirvožemio erozija* [Soil erosion]. Vilnius: Mokslas. 136 p. (in Lithuanian).

Šurda, P.; Šimonides, I.; Antal, J. 2007. A determination of area of potential erosion by geographic information systems, *Journal of Environmental Engineering and Landscape Management* 15(3): 144–152. http://dx.doi.org/10.1080/16486897.2007.9636922

Tarakanovas, P.; Raudonius, S. 2003. *Agronominių tyrimų duomenų statistinė analizė taikant kompiuterines programas ANOVA, STAT, SPLIT-PLOT iš paketo SELEKCIJA ir IRRISTAT* [The statistical analysis of agronomic research data using the software programs Anova, Stat, Split-Plot from package Selekcija and Irristat]. Akademija. 58 p. (in Lithuanian).

Thompson, J. A.; Roecker, S.; Grunwald, S.; Owens, P. R. 2012. Digital soil mapping: interactions with and applications for hydropedology, *Hydropedology*: 665–709. http://dx.doi.org/10.1016/B978-0-12-386941-8.00021-6

Ulen, B.; Bechmann, M.; Øygarden, L.; Kyllmar, K. 2012. Soil erosion in Nordic countries – future challenges and research needs, *Acta Agriculturae Scandinavica, Section B–Soil & Plant Science*. Special Issue. Soil Erosion in the Nordic Countries 2: 176–184.

Van Rompaey, A. J. J.; Govers, G.; Puttemans, C. 2002. Modeling land use changes and their impact on soil erosion and sediment supply to rivers, *Earth Surface Processes and Landforms* 27(5): 481–494. http://dx.doi.org/10.1002/esp.335

Van Rompaey, A. J. J.; Govers, G.; Van Hecke, E.; Jacobs, K. 2001. The impacts of land use policy on the soil erosion risk: a case study in central Belgium, *Agriculture Ecosystems and Environment* 83(1–2): 83–94. http://dx.doi.org/10.1016/S0167-8809(00)00173-0

Vanmaercke, M.; Maetens, W.; Poesen, J.; Jankauskas, B.; Jankauskienė, G.; Verstraeten, G.; de Vente, J. 2012. A comparison of measured catchment sediment yields with measured and predicted hillslope erosion rates in Europe, *Journal of Soils and Sediments* 12(4): 586–602. http://dx.doi.org/10.1007/s11368-012-0479-z

Zachar, D. 1982. *Soil erosion. Developments in soil science.* New York: Elsevier Scientific. 547 p.

Zirlewagen, D.; von Wilpert, K. 2010. Upscaling of environmental information: support of land-use management decisions by spatio-temporal regionalization approaches, *Environmental Management* 46(6): 878–893. http://dx.doi.org/10.1007/s00267-010-9468-4

Gintaras JARAŠIŪNAS. He is PhD student (Agriculture Sciences, Agronomy) at the Lithuanian Research Centre for Agriculture and Forestry. Author of two research articles and two presentations at scientific conference in Lithuania. Research interests include soil erosion processes and natural biophysical handicaps.

Irena KINDERIENĖ. She works as a scientist in Vėžaičiai Branch of the Lithuanian Research Centre for Agriculture and Forestry. She is a Doctor of Biomedical (Agronomy) Sciences. She is the author and co-author of over 50 research papers and author and co-author 2 scientific books. She has attended 8 international and 10 local conferences. Her research interests include soil erosion processes, soil degradation, soil and water pollution, sustainable agriculture and soil properties.

6

USE OF RESTORATION PLANTINGS TO ENHANCE BIRD SEED DISPERSAL AT THE ROADSIDE: FAILURES AND PROSPECTS

Rocío de TORRE[a], María Dolores JIMÉNEZ[a], Álvaro RAMÍREZ[b], Ignacio MOLA[c], Miguel A. CASADO[a], Luis BALAGUER[d]

[a] *Department of Ecology, Complutense University, Madrid 28040, Spain*
[b] *Department of Zoology and Physical Anthropology, Complutense University, Madrid 28040, Spain*
[c] *Research, Development and Innovation (R&D) Service, Obrascón Huarte Lain, S.A. (OHL), Madrid 28046, Spain*
[d] *Department of Plant Biology I, Complutense University, Madrid 28040, Spain*

Abstract. Plantings are commonly used in roadside reclamation for ornamental purposes and for increasing slope stability and road safety. However, the role of these plantings in restoring ecological processes, such as seed dispersal, has received little attention. We carried out a study to assess the potential role of plantings on roadside embankments to attract frugivorous birds and to enhance seed dispersal mediated by birds from the surrounding landscape. We examined: (1) bird species richness and abundance; (2) patterns of avian spatial distribution within embankments and (3) seed dispersal mediated by birds. Bird richness and abundance did not differ between embankments with and without plantings. However, birds were not distributed randomly within embankments, with levels of species richness and abundance for facultative frugivorous between 4.8–8 times higher in areas closer to plantings. An analysis of bird droppings showed that birds only dispersed seeds of the planted species since no seeds of woody plants from matrices were detected. These results suggest that plantings acted as selective bird attractors, providing food and perches for frugivorous species. Nevertheless, the scarcity of seed-dispersing birds in the surrounding agricultural landscape prevented plantings from enhancing seed dispersal and connectivity to adjacent habitat.

Keywords: ecological connectivity, ecological memory, landscape structure, mobile links, ornamental plantings, landfills.

Introduction

Roadsides are a common scenario all over the world, in which earthwork grading and road operations largely transform the geomorphologic, hydrologic, edaphic, atmospheric, and biotic local conditions (Mola *et al.* 2011). In ecological terms, this environmental damage translates into a loss of the ecosystem storage pools of water, organic matter, soil banks, and biomass, among other impacts (i.e. reserves *sensu* Ludwig *et al.* 1997). Such environmental loss compromises ecosystem recovery and restoration, since these reserves are key components of the local 'ecological memory' (Bengtsson *et al.* 2003). As a consequence, vegetation re-establishment in roadsides does not usually rely on germination of a remnant soil seed bank, but on two other main sources: (1) Natural immigration of propagules from the surrounding matrix (Bochet *et al.* 2007; de la Riva *et al.* 2011), and (2) Introduction of plant species by humans through revegetation practices such as hydroseeding or planting. The former leads to an over-representation of wind-dispersed plant species on roadslopes and those lacking specific traits for seed dispersal (Bochet *et al.* 2007), while the latter only includes a few cultivated species.

Disentangling the role of revegetation practices in activating ecological processes is critical for restoring ecosystems damaged by road construction (Coffin 2007).

Plantings along roadsides are usually designed for ornamental purposes, and for increasing road safety and slope stability. However, plantings might also have important effects on ecological processes, for example, by activating seed dispersal by birds and therefore enhancing the connectivity of the roadsides with its surrounding landscape.

Corresponding author: Rocío de Torre
E-mail: rtorrece@ucm.es

While the relevance of such processes is not well known, the potential of birds in helping to promote vegetation recovery has been shown in other types of degraded lands (Bonet 2004; Méndez *et al.* 2008).

Bird dispersal is considered a major driver of plant community dynamics over time (Howe, Miriti 2004). Specifically, frugivorous birds constitute the most effective mobile links for connecting habitats, acting as dispersers of seeds of fleshy fruit plants (Howe, Smallwood 1982; Sekercioglu 2006). In fact, these mobile organisms can provide connectivity between isolated areas in fragmented environments (Wilkinson 1997; Nathan *et al.* 2008). Roadslopes usually lack an internal memory, and birds are considered to be the main mobile links providing an external ecological memory, acting as vectors for seeds and propagules from the matrix (García *et al.* 2010). Thus, inclusion of plants that provide seeds for birds in the design of roadside restoration strategies may favour the integration of the damaged ecosystem with its surrounding matrix through flows and exchanges. These strategies would alleviate the deficit in bird-dispersed species in roadsides. Previous studies have shown that road infrastructures may create corridors connecting habitat patches (Meunier *et al.* 1999; Huijser, Clevenger 2006; Reijnen, Foppen 2006, Kociolek *et al.* 2011) and may provide scavenger bird species with new food sources (Lambertucci *et al.* 2009). However, it is not known whether this is the case in other guilds or trophic groups.

The main objective of this study is to test whether the standard revegetation practices on roadsides designed for ornamental, safety and slope stability purposes, may activate ecological processes such as seed dispersal. Thus, two ornamental plant species commonly used in roadside revegetation were selected to assess the effect of current planting practices on seed-dispersal by birds in human-altered landscapes. We examined plantings on roadsides of a new-constructed motorway section to address the following questions: (1) Do plantings on roadslopes act as "attractors" for frugivorous birds? We hypothesized that plantings provide food, shelter, perches, roosting or nesting sites for birds, as observed in other ecosystems (Holl *et al.* 2000; Wydhayagarn *et al.* 2009), and (2) Is there an active process of seed dispersal mediated by birds on roadslopes? Our hypothesis was that plantings enhance bird-dispersed seed arrival from the surrounding matrices to embankments.

Our ultimate aim is to stimulate reflections on effective strategies to foster bird-mediated connectivity between the roadslope and its surrounding matrix, with positive effects on seed arrival, as reported in other settings (e.g., Zamora *et al.* 2010), to enhance restoration efforts at the roadside.

1. Material and methods

1.1. Study area

The study was conducted on embankments along the A1 highway, in El Molar (Madrid, central Spain; 40° 44′ N, 3° 35′ W, 800 m asl), which was opened to traffic in June 2009. The regional climate is continental Mediterranean, characterized by cold winters and dry hot summers.

Average annual rainfall is less than 500 mm. Agriculture and livestock farming have generated a highly deforested surrounding landscape, dominated by cereal crops, xerophytic grasslands and scrublands. The dominant woody species are *Cytisus scoparius* (L.) Link., *Thymus zygis* Loefl. ex L. and *Rubus ulmifolius* Schott. Small woodland remnants consisting of *Quercus ilex* subsp. *ballota* (Desf.) Samp., *Fraxinus angustifolia* Vahl, *Juniperus oxycedrus* subsp. *badia* (H.Gay) Debeaux and *Pinus pinea* L. are also sparsely scattered across the surrounding landscape.

1.2. Experimental design

Six different embankments (roadslopes constructed by compacting successive layers of earth) located along 5 km of the A1 Highway were selected. We chose top-soiled embankments because of their favourable conditions for vegetation establishment (Tormo *et al.* 2007), but also because bird traffic kills are less likely to occur on raised sections of road than on flat verges or buried roadcuts (Clevenger *et al.* 2003). Three embankments remained untreated as controls (C1, C2, and C3). On the other three embankments (P1, P2, and P3), we located experimental plots of 2.25 m^2 with clump plantings of fleshy-fruited woody species (hereafter "planted islands") every 45 m (Fig. 1). The total number of planted islands of woody species was 8, distributed among roadslopes depending on the embankment area, according to a stratified sampling scheme (Table 1).

The planted island layout consisted of one individual of the deciduous tree *Prunus cerasifera* Ehrh. var. *pissardii*

Table 1. Characteristics of the six embankments used in this study

Embankment	N° planted islands	Surface (m^2)	Width (m)	Maximum height (m)
P1	3	4400	220	20
P2	2	1600	160	10
P3	3	4800	160	38
C1	–	1920	120	28
C2	–	3600	180	42
C3	–	11400	340	79

Note: P1 to P3 correspond to embankments planted with woody species and C1 to C3 are control sites.

(Carrière) C.K. Schneid (cherry plum) (2.75–3 m tall and 10–12 cm diameter at breast height), surrounded by four individual, thorny, evergreen shrubs of the species *Pyracantha coccinea* M. Roem (firethorn) (50–60 cm tall) (Fig. 1b). We expected these structures to be attractive to birds as perches, shelter and food resources. For example, the fleshy berries of *P. coccinea* have been found in the diet of some frugivorous birds in other Mediterranean areas (Debussche, Isenmann 1994). Therefore, only individual firethorns with ripe, conspicuous, fleshy fruits over at least 30% of the shrub crown were selected. Fleshy-fruited individuals of *P. coccinea* were planted in February 2009 and those that died were replaced in October 2009. These two ornamental species were selected because they are commonly used in current revegetation practices. Neither of these two exotic, non-invasive species (Sanz-Elorza *et al.* 2004) was present in the surrounding vegetation, so we could assume that any propagules of these species came from our plantings.

1.3. Bird surveys

Bird communities and their spatial distribution within roadslopes were sampled by means of two methods. First, line transect surveys were conducted to identify species composition and to estimate bird densities at the roadsides. We surveyed each embankment on foot along its entire width in the middle of the embankment, recording all birds that were using the embankment. Since the distance from the walking line to the edge of the embankment was always less than 40 m (Table 1), and vegetation was mostly herbaceous, we assumed that virtually all birds were detected during censuses. Surveys were conducted between 8:00 and 13:00 h, over two consecutive years every two-three weeks from February to April 2009 and from October 2009 to March 2010 (eleven visits in total). In each census, we calculated bird species richness (number of species/ha) and density (number of individuals/ha), for both the whole set of birds and frugivorous species (birds were classified according to diet based on Cramp *et al.* 1977–1994; SEO BirdLife 2008). Second, the spatial distribution of birds on experimental embankments with plantings was estimated by point counts. Each embankment was surveyed for 30 minutes from a distance that was sufficiently far from the embankment to avoid interferences with birds (Verner 1985). For each bird using the embankment, we recorded the species identity and its spatial position (X, Y coordinates). We performed these counts twice per week between 8:00 and 13:00 h, randomizing the order of embankment surveys to minimize possible time effects on bird activity patterns. A total of 25 point count samples per embankment were taken from October 2009 to mid-March 2010. No survey was conducted in conditions of strong wind or rain.

1.4. Seed dispersal by birds

The effectiveness of birds as seed dispersers on embankments with plantings was assessed using dropping traps (Fujita, Koike 2009). We placed 32 traps (wooden frames of 50×50 cm with a piece of tulle) on the three embankments with plantings. We placed 16 dropping traps within the planted islands (two per island in a diagonal under the cherry plum, then sampling 22% of the area of the 2.25 m^2 experimental plots). The other 16 traps were placed outside the planted islands with a similar layout, halfway (22.5 m) between two consecutive planted islands (Fig. 1 b). Bird droppings from the traps were collected every two weeks from February 2009 to April 2009 and from October 2009 to March 2010. The seed content of

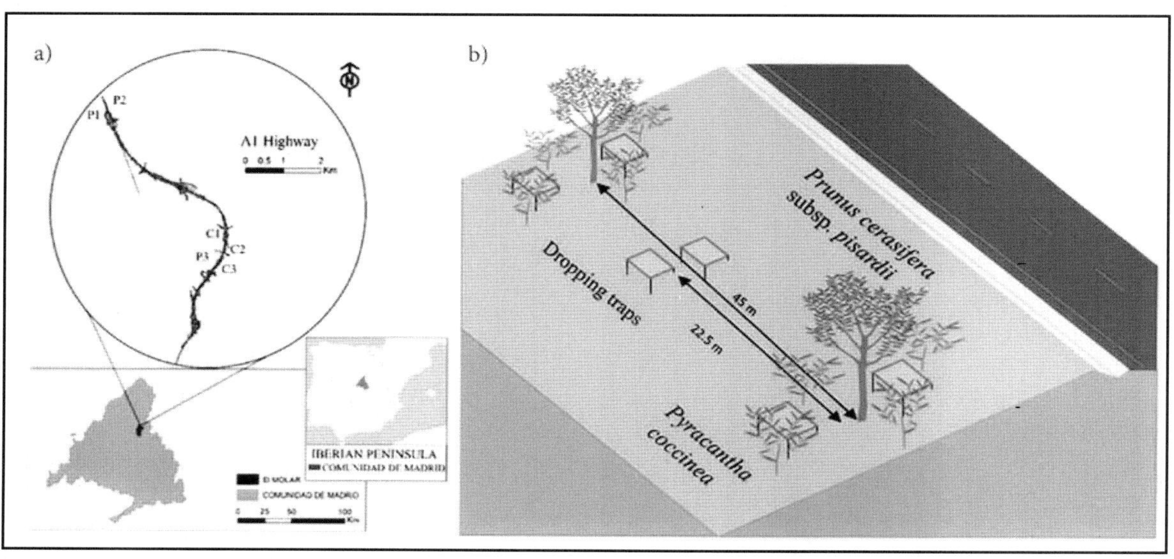

Fig. 1. Maps showing the location of the control embankments (C1 to C3) and the embankments with plantings (P1 to P3) (a). Experimental layout showing two "planted islands" and dropping traps (b)

the droppings was examined under low power magnification (10x). All seeds were identified using an *ad hoc* catalogue of seeds collected from the embankments and their surrounding matrices. The viability of seeds was estimated by means of germination tests in peat at 15% RH and 22 °C, under a 12 h light photoperiod for 12 months (Chamorro *et al.* 2013). We also measured the abundance of plants with fleshy fruits in surrounding matrices to test the effect of fruit availability on frugivorous birds. We established circular plots with a radius of 150 m located at the bottom and in the middle of each embankment. This distance is considered suitable for seed dispersal by small- and medium-sized birds (Jordano *et al.* 2007). Each circle surface was segmented into four cover layers by means of aerial photography: tree layer, shrub layer, herbaceous layer, and that occupied by infrastructures. We sampled 50 percent of the area of each of the first three layers, using one or two randomly distributed plots for the vegetation sampling. During 2009–2010 we also quantified the density of woody species providing fleshy fruits (number of individuals/ha) within the circles.

1.5. Data analysis

Statistical differences in richness and density of birds between the embankments with plantings and controls were analysed by means of repeated measures ANOVA. We analysed both the whole bird community and only frugivorous birds. The same analysis was used to compare the dropping rate in the traps placed inside and outside the planted islands on the embankments with plantings. In all analyses, sampling date was treated as a repeated measures factor. Bird density and the dropping rate were logarithmically transformed. We assessed Pearson correlations between the density of plants with fleshy fruits in the surrounding matrices and the density of birds (total and facultative frugivorous) on the embankments. All statistical analyses were performed with the software STATISTICA 6.0 (StatSoft, Inc. 2001).

Spatial distribution of birds was analysed by two methods. Due to the low number of observations, we examined the cumulative distribution pattern of the 25 censuses in both methods. First, differences in the occurrence of birds in plots of 4x4 m inside and outside the planted islands were analysed using chi-square tests. Richness (number of bird species) and abundance (number of birds) inside 4x4 m plots centred in the planted islands were used as the observed frequencies. To obtain the expected frequencies associated with each planted island, we randomly distributed 20 4x4 m plots per island in the corresponding embankments (160 4x4 m random plots in total). We then used the average richness and abundance of each of these 20 random plots as expected frequencies associated with each island. We repeated chi-square analyses for both the whole set of species and only facultative frugivorous. Second, we used Ripley's K function to study the exact location of frugivorous birds. Point patterns of bird occurrence were analysed with the use of the pair-correlation function g (r). The function g is the expected density of points at a given distance divided by the intensity of the pattern, making it more suitable for pattern analyses (Stoyan, D., Stoyan H. 1994; Wiegand, Moloney 2004). We used a cell size of 1x1 m and the ring width was one cell unit. The functions were calculated to a radius of 5 m and a maximum value of r equal to the maximum 40 m (which was the maximum distance between the centre and the edge of the largest embankment). Monte Carlo simulation techniques were used to construct confidence intervals. Envelopes for the 95% confidence level were determined by the 25th-lowest and the 25th-highest value of 999 Monte Carlo simulations of heterogeneous Poisson process, alternative to the null model Complete Spatial Randomness (CSR) for point patterns in irregularly shaped areas. Under conditions of CSR, when g (r) values fall between the envelopes, the spatial pattern is not significantly different from that expected by chance. When g (r) values fall outside the envelopes, values < 1 indicate regular distribution and values > 1 indicate aggregated patterns. We performed bivariate spatial pattern analysis, which considers both the pattern of the frugivorous birds at random as well as the position of the planted islands as a pattern of fixed points on the treated embankments. Ripley's K-function analyses were performed with Programita software (Wiegand, Moloney 2004).

2. Results

2.1. Seed dispersal limitation by bird availability

A total of 3746 observations of 27 bird species on the embankments was made using both line transects and point counts (Table 2). *Fringillidae* was the family most commonly observed (76% of all observations), followed by *Motacillidae* (10%) and *Embericidae* (7%). The most frequently recorded bird species were linnet (*Carduelis cannabina*), meadow pipit (*Anthus pratensis* L.), goldfinch (*Carduelis carduelis* L.) and Eurasian chaffinch (*Fringilla coelebs* L.). Nearly 70% of the bird species were obligate or facultative granivorous, representing 87% of the observations. 13.4% of the observations were of species that can feed on fleshy fruits, and only 5 species were facultative frugivorous but not granivorous, accounting for just 2% of the observations (Table 2).

After controlling for embankment area, no differences were found in species richness (S) between embankments with plantings (Mean ± S.E., S = 5.0 ± 0.5 species/ha per visit) and without them (S = 5.4 ± 3.5 species/ha per visit) (Treatment: $F_{1,4} = 0.05$, $p = 0.84$; Treatment x Date:

$F_{10, 40} = 0.46$, $p = 0.91$). Likewise, bird densities (D) were no different in embankments with (Mean ± S.E., D = 29.4 ± 21.5 birds/ha per visit) and without plantings (D = 47.9 ± 49.9 birds/ha per visit), (Treatment: $F_{1, 4} = 0.35$, $p = 0.59$; Treatment x Date: $F_{10, 40} = 0.70$, $p = 0.72$). When considering only those bird species that can feed on fleshy fruits no differences were found between embankments with plantings (Mean ± S.E., S = 1.16 ± 0.48 species/ha per visit; D = 5.41 ± 6.64 birds/ha per visit) and without plantings (S = 1.11 ± 1.11 species/ha per visit; D = 3.32 ± 4.53 birds/ha per visit), either in species richness or in bird density (Repeated measures ANOVA; $p > 0.05$ in both cases).

The analyses of spatial distribution of birds within embankments with plantings showed a preference for planted islands. First, chi-square analyses showed a positive selection for planted islands. The total observed species richness and abundance were 4.2 and 1.9 times higher inside than outside the islands (Richness: $\chi2 = 136.41$, df = 7, $p < 0.001$; Abundance: $\chi2 = 149.77$, df = 7, $p < 0.001$). Richness and abundance of facultative frugivorous species were between 8 and 4.8 times higher inside the islands (Richness: $\chi2 = 108.75$, df = 7, $p < 0.001$ (Fig. 2 a); Abundance: $\chi2 = 681.83$, df = 7, $p < 0.001$ (Fig. 2 b)).

Second, bivariate spatial pattern analyses indicated that frugivorous birds aggregated at the planted islands (points outside of the envelopes at distances of 1 m in embankments P1 and P3, Fig. 3) but showed a random distribution across most of the embankment surface.

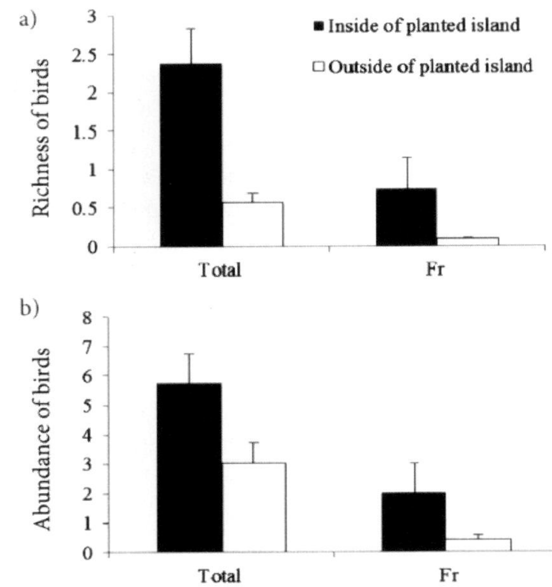

Note: Bars represent standard error.

Fig. 2. Bird species richness (a) and abundance (b) outside and inside of planted islands for all trophic groups (Total), and frugivorous birds (Fr)

2.2. Seed dispersal limitation by roadside resource availability

A total of 207 droppings were collected from all traps, with higher dropping rates found in midwinter (from December 1st to February 15th) with 52.2% of bird droppings (Fig. 4). There were significant differences in the dropping rate both among embankments and over time (Table 3), considering the period of arrival of wintering birds (from October to January) (Herrera 1995). Traps located inside the planted islands contained 1.7 times more feces than traps outside the islands. Embankment P3 contained between 2.4 and 2.9 times more feces than the other two embankments. Only 9% of feces contained seeds, of which 30.6% germinated. The most abundant plant species in the feces was *P. coccinea* (57.9% of the seeds), followed by the herbaceous *Solanum nigrum* L. and *Vicia dasycarpa* Ten. No seeds of woody plant species from the surrounding matrix were found in the traps.

Rosa spp., *Rubus ulmifolius* Schott and *Crataegus monogyna* Jacq. were the only three species of plants with fleshy fruits found in the matrices samples. The average density of plants with fleshy fruits ranged from 12 (P1) to 93 (P3) individuals/ha. Only the density of facultative frugivorous birds was positively correlated with that

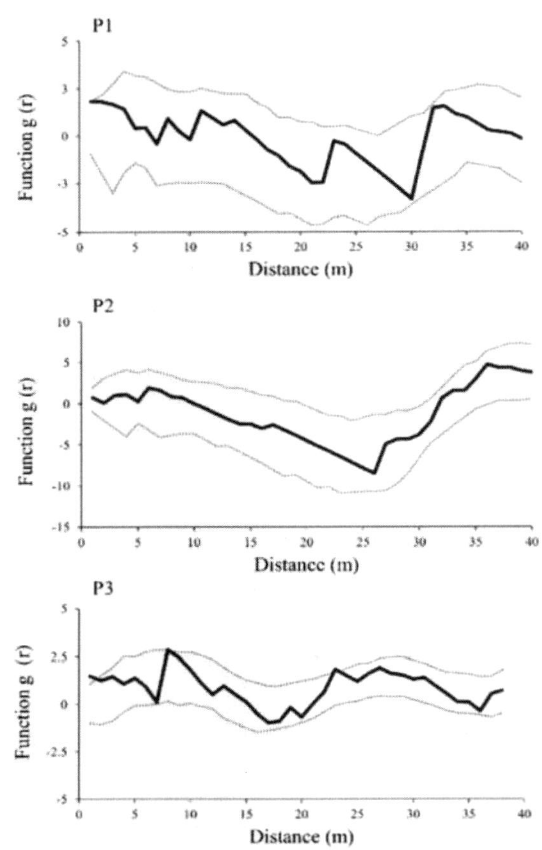

Fig. 3. Bivariate Ripley's K-function analysis of the spatial distribution of frugivorous birds at random and the position of the planted island as a pattern of fixed points for embankments with plantings (P1 to P3)

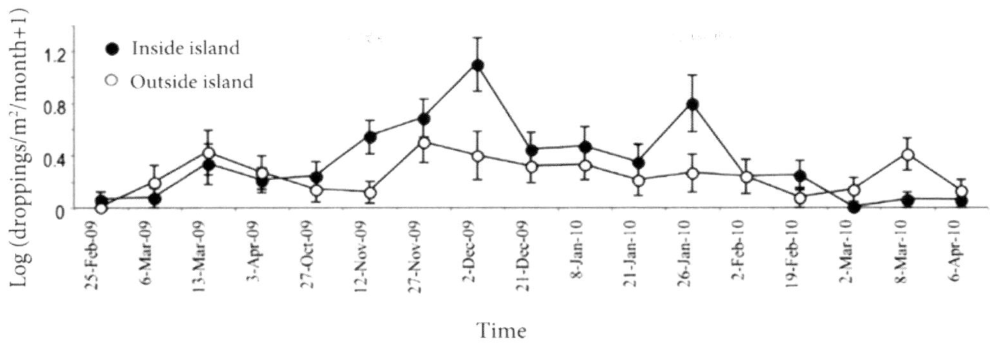

Fig. 4. Temporal variation of the dropping rate (mean ± SE) in the traps located inside and outside the planted islands

Table 2. Bird species recorded in the embankments by two sampling methods

Scientific name	Common name	Family	Trophic group	Line transects	Point counts	Total appearance (%)
Alectoris rufa	Red-legged partridge	Phasianidae	Gr	8	2	0.27
Anthus pratensis	Meadow pipit	Motacillidae	Gr-In	200	118	8.49
Bubulcus ibis	Cattle egret	Ardeidae	In	1	10	0.29
Carduelis cannabina	Linnet	Fringillidae	Gr-In	612	131	19.83
Carduelis carduelis	Goldfinch	Fringillidae	Gr	186	740	24.72
Carduelis chloris	Greenfinch	Fringillidae	Gr-In	3	0	0.08
Columba palumbus	Common woodpigeon	Columbidae	Gr	0	16	0.43
Emberiza cia	Rock bunting	Emberizidae	Gr-In	10	4	0.37
Emberiza cirlus	Cirl bunting	Emberizidae	Gr-In	4	0	0.11
Erithacus rubecula	Robin	Turdidae	In-Fr	0	6	0.16
Fringilla coelebs	Eurasian Chaffinch	Fringillidae	Gr-In-Fr	121	234	9.48
Galerida cristata	Crested lark	Alaudidae	Gr-In	25	19	1.17
Lanius meridionalis	Woodchat shrike	Laniidae	In-Ca	0	1	0.03
Miliaria calandra	Corn bunting	Emberizidae	Gr-In	93	138	6.17
Motacilla alba	White wagtail	Motacillidae	In	20	31	1.36
Parus caeruleus	Blue tit	Paridae	Gr-In-Fr	0	1	0.03
Parus major	Great tit	Paridae	In-Fr	0	4	0.11
Passer domesticus	House sparrow	Passeridae	Om	13	5	0.48
Passer montanus	Eurasian tree sparrow	Passeridae	Om	4	0	0.11
Phoenicurus ochruros	Black redstart	Turdidae	Gr-In-Fr	3	28	0.83
Phylloscopus collybita	Common chiffchaff	Sylviidae	In-Fr	9	54	1.68
Pica pica	Common magpie	Corvidae	Om	11	8	0.51
Saxicola torquata	Common stonechat	Turdidae	Gr-In-Fr	13	26	1.04
Serinus serinus	Serin	Fringillidae	Gr-In	129	83	5.66
Sturnus unicolor	Spotless starling	Sturnidae	Gr-In-Fr	0	1	0.03
Sylvia atricapilla	Blackcap	Sylviidae	In-Fr	0	2	0.05
Sylvia melanocephala	Sardinian warbler	Sylviidae	In-Fr	0	1	0.03
Troglodytes troglodytes	Winter wren	Troglodytidae	In	1	0	0.03
Not identified	–	Fringilidae	-	0	610	16.28
Not identified	–	-	-	0	7	0.19

Note: line transects (n = 1466 observations) and point counts (n = 2280). The trophic group is indicated for each species: granivorous (Gr), insectivorous (In), frugivorous (Fr), carnivores (Ca) and omnivores (Om) according to Cramp *et al.* 1977–1994; SEO BirdLife 2008. The number of individuals detected per species is shown for each sampling method. The percentage of bird appearance was calculated across the two sampling methods.

of woody plants with fleshy fruits ($r = 0.976$, $p < 0.001$, $n = 6$). Embankment P3 presented the highest density of both facultative frugivorous birds and plants with fleshy fruits in its immediate vicinity. On the contrary, embankment P1 had the lowest values for both variables.

Table 3. Repeated measures ANOVAs considering the effect of site (embankment) and trap location (inside/outside island) on dropping rate with time as repeated factor

	February 09–April 10			October 09–January 10		
	df	F	p	df	F	p
Embankment	2	5.95	0.007	2	13.57	<0.001
Inside/Outside island	1	3.78	0.063	1	3.65	0.04
Embankment x Inside/Outside island	2	0.09	0.913	2	0.14	0.871
Time	16	5.91	<0.001	7	3.51	<0.001
Time x Embankment	32	2.85	<0.001	14	1.84	0.036
Time x Inside/Outside island	16	2.63	<0.001	7	1.43	0.195
Time x Embankment x Inside/Outside island	32	1.76	0.007	14	2.15	0.011

Note: the analysis was performed over the whole sampling period (17 surveys: from February 2009 to April 2010) and the most active period for birds (8 surveys: from October 2009 to January 2010; see also Fig. 4).

3. Discussion

In ecological restoration, local actions such as plantings, are expected to deliver outcomes at larger scales, for instance, development of dispersal linkages (see for instance Wydhayagarn *et al.* 2009; Rey Benayas *et al.* 2008). However, most restoration actions in roadslopes are designed either for ornamental purposes, increasing safety or reducing slope instability, without taking into account their potential effect to restore ecological processes. Our results show that plantings designed specifically as bird attractors implemented in roadside revegetation are successful in attracting bird but fail to consistently foster bird-mediated seed dispersal from the agricultural and peri-urban surrounding matrices. This failure is not due to a low bird density at the roadside but rather to an avian community shaped by the history and structure of the surrounding landscape.

3.1. Seed dispersal limitation by bird availability

Despite being a human-altered habitat, we find high bird densities in the embankments. High abundances in agricultural landscapes are often found in strip-cover habitats such as railways or roadsides in winter (Best *et al.* 1995; Delgado, Moreira 2000). During this season, many bird species aggregate in flocks due to less selective habitat use or higher avian sociality (Morse 1982; Wiens 1989). The roadsides mainly include grasses and other herbaceous species, which are not usually mowed in winter. Granivorous birds may find suitable habitats in the roadsides during winter, when the availability of food (seeds, insects, plants, etc.) and shelter is very low in most agricultural stands in the surrounding area. However, these high densities in roadsides are paralleled by relatively low species richness, which is consistent with previous reports on road effects on bird community composition through habitat loss (Laurance 2004), traffic noise disturbance (Bautista *et al.* 2004), predation (Bergin *et al.* 2000), or collisions with vehicles (Erickson *et al.* 2005). Nevertheless, for our case, the dominance of granivorous can better explain the high density of birds but the low richness of species.

The avian community of the study embankments is dominated by granivorous species, in consonance with the surrounding agricultural landscape. Granivorous species sporadically use planted stands of woody plants, apart from not being efficient seed dispersers. Thus, the exozoochorous contribution of this type of birds appears to be bound to particular environmental conditions (Twigg *et al.* 2009). Conversely, the low abundance of frugivorous birds find in the embankments is not related to bird density, which was ten times higher if considering all birds present at the roadside than what had been reported for the surrounding area (Carrascal *et al.* 2002) and for similar scenarios (Díaz, Tellería 1994). Considering that the abundance of frugivorous birds is a good predictor for seed dispersal capacity in human-altered landscapes (García *et al.* 2010), the low number of frugivorous birds found in our study severely hampers the potential seed dispersal.

3.2. Seed dispersal limitation by roadside resource availability

Woody islands act as selective attractors for birds and several frugivorous species use the planted stands as food resources and perches. Consumption of fleshy fruits is evidenced by the presence of viable seeds from the planted *P. coccinea* shrubs in bird feces. The highest values of bird droppings were found between October and January, when the influx of winter migrants in the Mediterranean region is higher (Herrera 1995). These findings support the potential role of plantings in activating or facilitating ecological processes such as seed dispersal. In fact, persistence or establishment of islands of woody vegetation in extensive agricultural landscapes has often been recommended with regard to fostering biota restoration to respond to land use changes (Rey Benayas *et al.* 2008; Herrera, García 2009; Manning *et al.* 2009). Furthermore, previous reports have suggested that bird perching may even offset frugivorous bird scarcity in human-altered environments (Herrera, García 2009). Recent studies revealed that in agricultural landscapes, isolated trees are able to disperse to nearby roadside environments via frugivore activity (Coulson *et al.* 2014). In the present study, the attraction of frugivorous birds by plantings was highly skewed among roadslopes. This site (embankment) dependency is linked to the high correlation observed between the density of frugivorous birds at the roadside and the density of fleshy-fruited plants in the surrounding matrix. Nevertheless, these results should be regarded with caution due to the small sample size. More years of monitoring may be required due to the time needed by birds to gather new resources.

We surmise that a mismatch between the local revegetation actions and the landscape structure of the matrix may explain why the seeds from the fleshy-fruited species (*Rosa* spp., *Rubus ulmifolius*, or *Crataegus monogyna*) in the adjacent matrices never reached the planted stands. In practical terms, the connection of the planted stands would have required restoration action in the surrounding matrix, such as the establishment of stepping-stone corridors (*sensu* Gillies, St. Clair 2010), through planting of clumps of fleshy-fruited shrubs to channel movements of seed-dispersing birds. In summary, current landscape heterogeneity provides the structural frame into which disturbed sites, such as roadslopes, should be 'plugged' in order to restore ecological connectivity within the ecosystem mosaic.

Conclusions

1. Plantings of ornamental woody species with fleshy fruits acted as selective attractors for frugivorous birds on embankments. These experimental woody islands provided perches and food to some bird species. These results suggest that revegetation practices can be designed to restore ecological processes, not just for slope stability and ornamental purposes.
2. Effective seed dispersal by birds was not detected probably because the low abundance of frugivorous birds (the bird community in the area was dominated by granivorous species, as a result of a long history of urbanization and agricultural land use).
3. Local revegetation practices must consider the surrounding landscape. The lack of awareness of the surrounding landscape structure may jeopardize the success of ecological restoration actions at the roadside.
4. Our findings support top-down approaches to design restoration actions that first consider broad-scale drivers and then restoration protocols and local actions.

Acknowledgments

Sadly, our dear colleague and friend Luis Balaguer passed away on the 19th of March 2014. This publication is dedicated in grateful tribute to him. We are indebted to Manuel Villén and Luís García-Linares at the R and D Department of Obrascon Huarte Lain, S.A. (OHL) for their support, permission to publish this paper and, in particular, for their commitment to the environmental integration of transport infrastructures. We thank Ana Vázquez Benítez and Silvia Murillo Burgaña for their assistance in the laboratory and field work. This research was funded by the Spanish Ministry of Economy and Competitiveness (ECONECT project: CDTI IDI-20120317). We also wish to thank the Regional Government of Madrid for financial support through the network REMEDINAL3-CM (S2013/MAE-271).

References

Bautista, L. M.; García, J. T.; Calmaestra, R. G.; Palacín, C.; Martín, C. A.; Morales, M. B.; Bonal, R.; Viñuela, J. 2004. Effect of weekend road traffic on the use of space by raptors, *Conservation Biology* 18: 726–732. http://dx.doi.org/10.1111/j.1523-1739.2004.00499.x

Bengtsson, J.; Angelstam, P.; Elmqvist, T.; Emanuelsson, U.; Folke, C.; Ihse, M.; Moberg, F.; Nyström, M. 2003. Reserves, resilience and dynamic landscapes, *Ambio* 32: 389–396. http://dx.doi.org/10.1579/0044-7447-32.6.389

Bergin, T. M.; Best, L. B.; Freemark, K. E.; Koehler, K. J. 2000. Effects of landscape structure on nest predation in roadsides of a midwestern agroecosystem: a multiscale analysis, *Landscape Ecology* 15: 131–143.

Best, L. B.; Freemark, K. E.; Dinsmore, J. J.; Camp, M. 1995. A review and synthesis of habitat use by breeding birds in agricultural landscapes of Iowa, *American Midland Naturalist* 134: 1–29. http://dx.doi.org/10.2307/2426479

Bochet, E.; García-Fayos, P.; Tormo, J. 2007. Road slope revegetation in semiarid Mediterranean environments. Part I: Seed dispersal and spontaneous colonization, *Restoration Ecology* 15: 88–96. http://dx.doi.org/10.1111/j.1526-100X.2006.00193.x

Bonet, A. 2004. Secondary succession of semi-arid Mediterranean old-fields in south-eastern Spain: insights for conservation and restoration of degraded lands, *Journal of Arid Environments* 56: 213–233. http://dx.doi.org/10.1016/S0140-1963(03)00048-X

Carrascal, L. M.; Palomino, D.; Lobo, J. M. 2002. Patrones de preferencias de hábitat y de distribución y abundancia invernal de aves en el centro de España. Análisis y predicción del efecto de factores ecológicos, *Animal Biodiversity and Conservation* 1: 7–40 (in Spanish).

Chamorro, D.; Luna, B.; Moreno, J. M. 2013. Germination response to various temperature regimes of four Mediterranean seeder shrubs across a range of altitudes, *Plant ecology* 214(12): 1431–1441. http://dx.doi.org/10.1007/s11258-013-0264-0

Clevenger, A. P.; Chruszcz, B.; Gunson, K. E. 2003. Spatial patterns and factors influencing small vertebrate fauna road-kill aggregations, *Biology Conservation* 109: 15–26. http://dx.doi.org/10.1016/S0006-3207(02)00127-1

Coffin, A. W. 2007. From roadkill to road ecology: a review of the ecological effects of roads, *Journal of Transport Geography* 15(5) : 396–406.
http://dx.doi.org/10.1016/j.jtrangeo.2006.11.006

Coulson, C.; Spooner, P. G.; Lunt, I. D.; Watson, S. J. 2014. From the matrix to roadsides and beyond: the role of isolated paddock trees as dispersal points for invasion, *Diversity and Distribution* 20: 137–148.
http://dx.doi.org/10.1111/ddi.12135

Cramp, S.; Simmons, K. E. L.; Perrins, C. 1977–1994. *Handbook of the Birds of Europe, the Middle East and North Africa.* 9 Volumes. Oxford: Oxford University Press.

de la Riva, E. G.; Casado, M. A.; Jiménez, M. D.; Mola, I.; Costa-Tenorio, M.; Balaguer, L. 2011. Rates of local colonization and extinction reveal different plant community assembly mechanisms on road verges in central Spain, *Journal of Vegetation Science* 22: 292–302.
http://dx.doi.org/10.1111/j.1654-1103.2010.01248.x

Debussche, M.; Isenmann, P. 1994. Bird-dispersed seed rain and seedling establishment in patchy Mediterranean vegetation, *Oikos* 69: 414–426. http://dx.doi.org/10.2307/3545854

Delgado, A.; Moreira, F. 2000. Bird assemblages of an Iberian cereal steppe, *Agriculture, Ecosystems and Environment* 78: 65–76. http://dx.doi.org/10.1016/S0167-8809(99)00114-0

Díaz, M.; Tellería, J. 1994. Predicting the effects of agricultural changes in central Spanish croplands on seed-eating overwintering birds, *Agriculture, Ecosystems and Environment* 49: 289–298. http://dx.doi.org/10.1016/0167-8809(94)90058-2

Erickson, W. P.; Johnson, G. D.; Young Jr., D. P. 2005. A summary and comparison of bird mortality from anthropogenic causes with an emphasis on collisions, *US Forest Service General Technical Report*, PSW 191: 1029–1042.
http://dx.doi.org/10.1111/j.0022-0477.2004.00882.x

Fujita, M.; Koike, F. 2009. Landscape effects on ecosystems: birds as active vectors of nutrient transport to fragmented urban forests versus forest-dominated landscapes, *Ecosystems* 12: 391–400. http://dx.doi.org/10.1007/s10021-009-9230-z

García, D.; Zamora, R.; Amico, G. C. 2010. Birds as suppliers of seed dispersal in temperate ecosystems: conservation guidelines from real-world landscapes, *Conservation Biology* 24: 1070–1079.
http://dx.doi.org/10.1111/j.1523-1739.2009.01440.x

Gillies, C. S.; St. Clair, C. C. 2010. Functional responses in habitat selection by tropical birds moving through fragmented forest, *Journal Applied Ecology* 47: 182–190.
http://dx.doi.org/10.1111/j.1365-2664.2009.01756.x

Herrera, C. M. 1995. Plant-vertebrate seed dispersal systems in the mediterranean: ecological, evolutionary, and historical determinants, *Annual Review of Ecology and Systematics* 26: 705–727. http://dx.doi.org/10.1146/annurev.ecolsys.26.1.705

Herrera, J. M.; García, D. 2009. The role of remnant trees in seed dispersal through the matrix: being alone is not always so sad, *Biological Conservation* 142: 149–158.
http://dx.doi.org/10.1016/j.biocon.2008.10.008

Holl, K. D.; Loik, M. E.; Lin, E. H. V.; Samuels, I. A. 2000. Tropical montane forest restoration in Costa Rica: overcoming barriers to dispersal and establishment, *Restoration Ecology* 8: 339–349. http://dx.doi.org/10.1046/j.1526-100x.2000.80049.x

Howe, H. F.; Miriti, M. N. 2004. When seed dispersal matters, *Bioscience* 54: 651–660. http://dx.doi.org/10.1641/0006-3568(2004)054%5B0651:WSDM%5D2.0.CO;2

Howe, H. F.; Smallwood, J. 1982. Ecology of seed dispersal, *Annual Review of Ecology and Systematics* 13: 201–228.
http://dx.doi.org/10.1146/annurev.es.13.110182.001221

Huijser, M. P.; Clevenger A. P. 2006. Habitat and corridor function of rights-of-way, in J. Davenport, J. L. Davenport (Eds.). *The ecology of transportation: managing mobility for the environment.* London: Springer, 233–254.
http://dx.doi.org/10.1007/1-4020-4504-2_11

Jordano, P.; García, C.; Godoy, J. A.; García-Castaño, J. L. 2007. Differential contribution of frugivores to complex seed dispersal patterns, *Proceedings of the National Academy of Sciences USA* 104: 3278–3282.
http://dx.doi.org/10.1073/pnas.0606793104

Kociolek, A. V.; Clevenger, A. P.; St Clair, C. C.; Proppe, D. S. 2011. Effects of road networks on bird populations, *Conservation Biology* 25: 241–249.
http://dx.doi.org/10.1111/j.1523-1739.2010.01635.x

Lambertucci, S. A.; Speziale, K. L.; Rogers, T. E.; Morales, J. M. 2009. How do roads affect the habitat use of an assemblage of scavenging raptors?, *Biodiversity Conservation* 18: 2063–2074. http://dx.doi.org/10.1007/s10531-008-9573-3

Laurance, S. G. W. 2004. Responses of understory rain forest birds to road edges in Central Amazonia, *Ecological Applications* 14: 1344–1357. http://dx.doi.org/10.1890/03-5194

Ludwig, A. J.; Tongway, D. J.; Freudenberger, D.; Noble, J.; Hodgkinson, K. (Eds.). 1997. *Landscape ecology, function and management: principles from Australia's rangelands.* Melbourne: CSIRO Publishing.

Manning, A. D.; Gibbons, P.; Lindenmayer, D. B. 2009. Scattered trees: a complementary strategy for facilitating adaptive responses to climate change in modified landscapes?, *Journal Applied Ecology* 46: 915–919.
http://dx.doi.org/10.1111/j.1365-2664.2009.01657.x

Méndez, M.; García, D.; Maestre, F. T.; Escudero, A. 2008. More ecology is needed to restore mediterranean ecosystems: a reply to Valladares and Gianoli, *Restoration Ecology* 16: 210–216. http://dx.doi.org/10.1111/j.1526-100X.2008.00390.x

Meunier, F.; Verheyden, C.; Jouventin, P. 1999. Bird communities of highway verges: influence of adjacent habitat and roadside management, *Acta Oecologica* 20: 1–13.
http://dx.doi.org/10.1016/S1146-609X(99)80010-1

Mola, I.; Jiménez, M. D.; López-Jiménez, N.; Casado, M. A.; Balaguer, L. 2011. Roadside reclamation outside the revegetation season: management options under schedule pressure, *Restoration Ecology* 19: 83–92.
http://dx.doi.org/10.1111/j.1526-100X.2009.00547.x

Morse, D. H. 1982. *Behavioral mechanisms in ecology.* Cambridge: Harvard University Press.

Nathan, R.; Schurr, F. M.; Spiegel, O.; Steinitz, O.; Trakhtenbrot, A.; Tsoar, A. 2008. Mechanisms of long-distance seed dispersal, *Trends in Ecology and Evolution* 23: 638–647. http://dx.doi.org/10.1016/j.tree.2008.08.003

Reijnen, R.; Foppen, R. 2006. Impact of road traffic on breeding bird populations, in J. Davenport, J. L. Davenport (Eds.). *The ecology of transportation: managing mobility for the environment.* London: Springer, 255–274.
http://dx.doi.org/10.1007/1-4020-4504-2_12

Rey Benayas, J. M.; Bullock, J. M.; Newton, A. C. 2008. Creating woodland islets to reconcile ecological restoration, conservation, and agricultural land use, *Frontiers in Ecology and the Environment* 6: 329–336. http://dx.doi.org/10.1890/070057

Sanz-Elorza, M.; Dana, E. D.; Sobrino, E. 2004. *Atlas de las plantas alóctonas invasoras en España*. Madrid: Ministerio de Medio Ambiente, Organismo Autónomo Parque Nacionales (in Spanish).

Sekercioglu, C. H. 2006. Increasing awareness of avian ecological function, *Trends Ecology and Evolution* 21: 464–471. http://dx.doi.org/10.1016/j.tree.2006.05.007

SEO BirdLife. 2008. *La enciclopedia de las aves de España* [on line], [cited 16 October 2014] (in Spanish). Available from Internet: http://www.enciclopediadelasaves.es/originales/Enciclopedia_Aves.htm

StatSoft Inc. 2001. *Statistica for Windows 6.0*. Tulsa, Oklahoma: Statsoft Inc.

Stoyan, D.; Stoyan, H. 1994. *Fractal, random shapes and point field*. Chichester: Wiley, 389.

Tormo, J.; Bochet, E.; García-Fayos, P. 2007. Roadfill revegetation in semiarid Mediterranean environments. Part II: topsoiling, species selection, and hydroseeding, *Restoration Ecology* 15: 97–102. http://dx.doi.org/10.1111/j.1526-100X.2006.00194.x

Twigg, L. E.; Lowe, T. J.; Taylor, C. M.; Calver, M. C.; Martin, G. R.; Stevenson, C.; How, R. 2009. The potential of seed-eating birds to spread viable seeds of weeds and other undesirable plants, *Austral Ecology* 34: 805–820. http://dx.doi.org/10.1111/j.1442-9993.2009.01992.x

Verner, J. 1985. Assessment on counting techniques, in R. F. Johnson (Ed.). *Current ornithology*. Vol. 2. New York: Plenum Press, 247–302.

Wiegand, T.; Moloney, K. A. 2004. Rings, circles, and null-models for point pattern analysis in ecology, *Oikos* 104: 209–229. http://dx.doi.org/10.1111/j.0030-1299.2004.12497.x

Wiens, J. A. 1989. Spatial scaling in ecology, *Functional Ecology* 3: 385–397.

Wydhayagarn, C.; Elliott, S.; Wangpakapattanawong, P. 2009. Bird communities and seedling recruitment in restoring seasonally dry forest using the framework species method in Northern Thailand, *New Forests* 38: 81–97. http://dx.doi.org/10.1007/s11056-009-9133-z

Wilkinson, D. M. 1997. Plant colonization: are wind dispersed seeds really dispersed by birds at large spatial and temporal scales?, *Journal of Biogeography* 24: 61–65. http://dx.doi.org/10.1111/j.1365-2699.1997.tb00050.x

Zamora, R.; Hódar, J. A.; Matías, L.; Mendoza, I. 2010. Positive adjacency effects mediated by seed disperser birds in pine plantations, *Ecological Applications* 20: 1053–60. http://dx.doi.org/10.1890/09-0055.1

Rocío de TORRE. Doctor of Ecology, Conservation and Restoration of Ecosystems of the University of Alcalá, Spain, 2014. Author of 2 scientific papers under review. Scientific interests: plant ecology and plant-animal and plant-environment interactions, across different levels of organisation as individuals, communities, ecosystems and landscapes.

María Dolores JIMÉNEZ. Assistant Professor of Ecology at the Complutense University, Madrid, Spain. Doctor of Biology of the Politécnica University, Madrid, Spain, 2007. Author of 18 scientific papers. Scientific interests: urban ecology, forest ecology and restoration ecology, in order to take into account the anthropogenic factor on ecosystem and communities functioning.

Álvaro RAMÍREZ. Lecturer at the Department of Zoology and Physical Anthropology, at the Complutense University, Madrid, Spain. Doctor of Biology of the Complutense University, Madrid, Spain, 2004. Author of 14 scientific papers. Scientific interests: animal ecology, biogeography and conservation biology, using birds as a study model.

Ignacio MOLA. Technical Advisor of the Research, Development and Innovation (R&D) Service of the Construction Company OHL (Obrascón Huarte Laín S. A.), Madrid, Spain. Bachelor in Biology from Autónoma University, Madrid, Spain, 2002. Author of 6 scientific papers and co-editor of one book (co-author of 4 chapters). Scientific interests: factors which are driving the vegetation establishments at the roadside under Mediterranean conditions.

Miguel A. CASADO. Associate Professor of Ecology at the Complutense University, Madrid, Spain. Doctor of Biology of Autónoma University, Madrid, Spain, 1987. Author of 65 scientific publications, co-editor of 4 books and 24 book chapters. Scientific interests: ecology of agrosilvopastoral systems, as well as restoration ecology.

Luis BALAGUER. Associate Professor of Plant Biology at the University Complutense, Madrid, Spain. Unfortunately, our dear friend and colleague Luis Balaguer passed away on March 19th, 2014. He published 48 scientific papers and several book chapters. His research and teaching interest was focussed on restoration ecology and evolutionary ecology from a holistic point of view.

THE RESEARCH ON THE IMPACT ASSESSMENT OF VISUAL LANDSCAPE OF COUNTRY PARKS IN BEIJING

Chuanan LIU, Tong QI , Xiao MA

College of Resource Environment and Tourism, Capital Normal University, 105 West Third Ring Road North, Haidian District, Beijing, 100048 China

Abstract. The attribute of a country park and its visual environment are not only influenced by the distance from the park to the downtown area, but also the degree of urbanization. Based on the landscape geography, landscape ecology, landscape aesthetics theories and the Guideline for Technical Review of Environmental Impact Assessment of Construction Projects (HJ 616-2011) and aiming to evaluate the visual quality of the country parks, we put forward three first-grade indicators (visual sensitive capacity, visual space capacity and visual absorption capacity) and seven second grade indicators (sighting distance, visual disturbance frequency, green looking ratio, inter-visibility, openness, relative angle and the relative area of artificial matrix) in our research. By virtue of the established entropy weight method model, combined with field data, the quantitative research of the visual environment of landscape in country parks was carried out so that the visual quality of the park can be improved. At same time, this can also provide a scientific basis for spatial planning of urban fringe area and repositioning of the park attribute.

Keywords: country park, urban-rural area, visual landscape, visual environment quality, impact assessment, entropy weight method.

Introduction

With the acceleration of urbanization, urban fringe areas where originally belong to green space have been replaced by many high-rise buildings, which seriously interferes with the edge of the city green-land. This especially affects the visual environment of the country park and destroys the integrity, openness and continuity of urban green space, resulting in lacking of green landscape preservation in the urban-rural area. In 2011, Ministry of Environmental Protection of the People's Republic of China promulgated the Guideline for Technical Review of Environmental Impact Assessment of Construction Projects (HJ 616-2011), which was mainly aimed at roads, railways, mines, quarries, scenic areas, water conservancy and hydropower projects, and large construction projects in urban areas, in particular, reckoning in the impact assessment of landscape aesthetics (Ministry of Environmental... 2011). Country park as a typical representative of green space in the urban-rural area, its impact assessment of visual landscape is imperative. Based on previous studies, using a new method of the impact assessment of visual environment evaluates the quality of country parks. As a result, the visual environment of urban-rural green space can be quantified, which provides a scientific basis for spatial planning of urban fringe area and repositions the park attribute.

Visual landscape, in short, is a visual resource. For the concept of the visual landscape, scholars have not reached on consensus and each sticks to his argument. When researching on the visual landscape quality, Daniel and Vining (1983) summed up the evaluation models, and those were ecological model, formal aesthetic model, psychophysical model, psychological and phenomenogical model. Through the analysis of landscape structure and image, Krause (2001) wanted to protect and develop high-quality natural and cultural beauty. Dramstad *et al.* (2006) studied the visual landscape preference of the different groups for different landscape structures. For the problems of accessing data of traditional landscape perception and preference in the visual landscape assessment, Roth (2006) used the network to obtain data in order to solve the above problems. Roads dominated our daily surroundings and were essential conditions of the access to the visual landscape, Garré explored the dual role of road in the

Corresponding author: Tong Qi
E-mail: qitongdi@vip.sina.com

visual landscape (Garré *et al.* 2009). Along with the pace of urbanization, the countryside transformed itself from a rural area to an urban nation. In reaction to this phenomenon, Spanish scholars studied the rural landscape changed by human, which corresponded to the changes of visual perception (Cloquell-Ballester *et al.* 2012). In order to better understand the future landscape as the aesthetic subject, Nohl (2001) established a conceptual framework. Unlike other visual landscape research, Kaplan *et al.* (2006) studied the visual landscape preference of the urban-rural fringe. Nakamae *et al.* (2001) applied the rendering technology generated by computer graphics and video sequences to evaluate the quality of visual landscape. Researches on the visual impact of other countries involved in the wind power plants, greenhouse parks, mines and other aspects. Spanish researchers had developed a predictive model which could be used to predict the visual impact of the wind plant and as a query tool to analyze and evaluate the proposed wind plant project (Hurtado *et al.* 2004; Molina-Ruiz *et al.* 2011). Along with the advance of modern geographic research techniques, GIS and 3D simulation technology had been widely used in the impact assessment of visual landscape (Ayad 2005). Rogge *et al.* (2008) utilized the GIS to quantify and value the visual impact brought by the development of greenhouse cluster, putting forward the measures to reduce the visual impact of greenhouse park. During the process of mining and restoration, Menegaki *et al.* (2012) took LETOPID (Landscape Evaluation Tool for Open Pit Mine Design) method to reduce the visual impact. The impact assessment of visual landscape, often involving in the occlusion and avoidance, usually used GIS to establish the 3D visual model, then made sight and viewshed analysis (Hernández *et al.* 2004a, 2004b). Through the questionnaire survey, Hernández *et al.* (2004b) determined the influence of building on the surrounding landscape in order to verify the effectiveness of the GIS for site selection.

In China, Kongjian Yu as one of the researchers who studied the visual landscape, proposed the concept of landscape threshold (Kongjian 1991). Binyi Liu introduced American visual resources management into China, and deepened its development (Binyi 1990). However, the visual landscape impact research started relatively late in the last century, some scholars researched on the concept, evaluation process, content and methods of visual impact of landscape, and conducted a quantitative evaluation (Renhai 1999; Yang, Xiaolin 1999). In the 21st century, this field really began to be concerned about, but only made some theoretical researches initially (Tan *et al.* 2000; Yuhu 2001). With the maturity of some theories and methods, visual landscape impact research had gradually applied to different evaluation objects. On the basis of literature review research, Tong *et al.* (2013) emphasized that visual landscape should give the viewer a strong visual perception and impression in a particular area. However, most of Chinese researchers were limited to the internal area for visual landscape environment impact, lacking in the research on the negative impact arising from the external disturbances (Chuanfa, Jianhua 2009; Youbo, Yue 2007; Hui *et al.* 2009; Xüling *et al.* 2012; Jianhua 2008).

Whether it is the natural landscape or cultural landscape, environmental problems lead to specific requirements needed in various sectors for visual impact assessment. But, evaluation factors involved are too subjective, and less factors that can be objective and quantitative, which makes the impact assessment of visual landscape is too subjective (Mortberg *et al.* 2007). In this article, objective quantitative evaluation of interference factors (especially the artificial building) outside the park has an effect on the visual environment, which is a good reference for solving the above problems.

1. Materials and methods

1.1. Study area

In this article, 14 parks are selected in Beijing (12 country parks, two city parks) to carry on the case study. 12 country parks selected are very representative (there are three in the west and the east, two in the south and four in the north), and the two city parks are the typical parks which are used to compare with the country parks (Fig. 1). By conducting field data collected, combined with the indoor processing, the evaluation indicators of the initial data are obtained.

1.2. The index system of the impact assessment of visual landscape

There is a close relationship among the interference of high-rise buildings in the urban-rural area and their spatial position, physical attribute. Korean scholar put forward the visual threshold carrying capacity (VTCC) in the study of city landscape management. Kyushik (1998) considered that the influence of external disturbance factors on the visual landscape was a cumulative effect, and not a single effect. The study of visual environment should examine its internal constitution and surroundings, including some sensitive landscapes, spatial attribute and the coordination between different landscapes. We find many problems in Beijing Limited Construction Area Planning (BLCAP): the construction limited area is not clear. The boundaries between construction areas and green space are uncertain, and construction guidelines only make demands on building height (Ying *et al.* 2012). There are also other problems, such as lacking in the consideration of architectural appearance, visual control, visual corridor, coordination and other aspects. Based on these, three first-grade indicators are proposed for the impact assessment of visual landscape: visual sensitive capacity (VSEC),

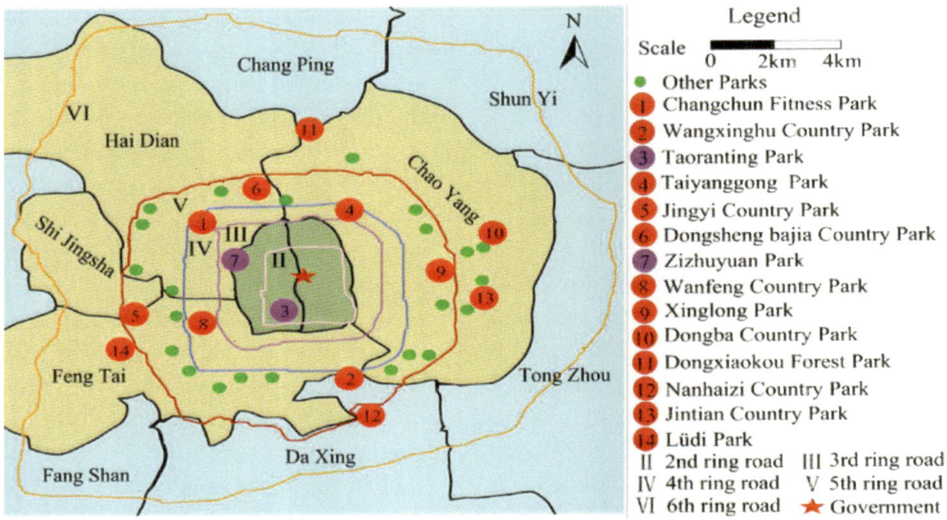

Fig. 1. The location map of the 14 sample parks

visual space capacity (VSPC), visual absorption capacity (VABC).

1.2.1. Visual sensitive capacity

"Sensitive" comes from the psychology, referring to responding to the external things quickly. Kongjian Yu put forward the "landscape sensitivity" in the study of conservation planning of South Taihang Mountain Canyon landscape, introducing the term "sensitive" to landscape geography. He deemed that the landscape sensitivity was a measure of the degree of landscape to be noticed, the comprehensive reflection of relative slope, the distance between landscape and the observer, the probability of landscape appearance in the field of vision and striking degree of landscape (Kongjian 1991). This was the evaluation of the natural heritage landscape, meanwhile, it also had a reference for city landscape evaluation. Aoki Yoji in 1987 presented the notion of "Green Looking Ratio", defining it as the share within the vision of green plants (Xiaojun, Honggang 2002). Visual sensitive capacity is mainly in the light of the impact assessment of physical properties of artificial architecture and the natural landscape, including sighting distance, visual disturbance frequency and green looking ratio.

Sighting distance (D): From the point of view to outside the building is D, can be measured by a laser range finder, if no building outside the area, then $D = 0$.

Visual disturbance frequency (V): According to the green space size, at 100 m (small scale) or 200 m (large scale) interval separately setting a viewpoint, between adjacent viewpoints, the visible distance of outside the building is L. Then $V = L/100$ or $V = L/200$, if never see outside structure between adjacent viewpoints, then $V = 0$; if the building has always been seen outside the region between adjacent viewpoints, then $V = 1$.

Green looking ratio (G): The share within the vision of green plants, importing photos to Photoshop software, calculating the area of green vegetation in the photograph and photo area, then obtains green looking ratio (green vegetation area/photo area).

1.2.2. Visual space capacity

Space not only belongs to the geography, but also the aesthetic category. It could be divided into three-dimensional physical space – objective space, psychological space – subjective space, behavior space – social space and symbolic space. Space discussed in this article only refers to physical space which is composed of the base surface and the three-dimensional boundary. When it comes to visual space, we should consider the openness of viewshed (the horizontal direction) and the continuity of the line of sight (the vertical direction). Visual space capacity mainly focus on visual control range and visual corridor of the human eye, referring to the degree of vision openness, spatial category and characteristics of the horizon constitution, which depends on the scale of landscape element and visual space openness surrounded by the skyline. To evaluate the visual space capacity, two second grade indicators are proposed in this article: inter-visibility and openness.

Inter-visibility (I): By investigating the continuity of the line of sight, we use a laser range finder to measure the distance between observer and the object that blocks line of sight.

Openness (O): To examine the extent of open of viewshed, photographs are imported into Photoshop software for processing and computed the skyline share in the field of vision (the length of skyline in the photograph/the width of photograph).

1.2.3. Visual absorption capacity

In 1960s–1970s, due to the improvement of public's concernment for the visual environment quality, as well as the introduction of legal documents relating to the environment, America's environmental impact assessment emphasized on the visual and aesthetic resource value. In the United States, there were three important scenic resources management systems: Visual Management System (VMS), Visual Resources Management (VRM) and Landscape Resources Management (LRM) (Xiaojun 1993). Scenic resources management system was mainly used in landscape evaluation, and it was based on landscape quality assessment and visual impact assessment to divide different level of regions. Visual absorption capacit (VABC) evaluation was put forward in VRM assessment. VABC had the ability to assimilate outside environment or visual impact, maintaining the inherent characteristics of the landscape (Anderson 1979). In landscape Ecology, landscape structure is divided into patch (punctiform), corridor (threadiness) and matrix (planar). This is a large-scale landscape structure, zooming to the city landscape, artificial construction can be regarded as a kind of matrix. The height and distance of artificial construction outside the park determine the observer's angle, which will affect the coordination of the entire park and its ability to assimilate visual environment. Therefore, the visual absorption capacity evaluates the coordination between artificial architecture outside and the natural landscape, including the relative angle and the relative area of artificial matrix.

Relative angle (A): Measure the angle of elevation of sight between artificial architecture of outside and viewer through a laser range finder. The angle of elevation of observing landscape is α, its sine value is relative angle. $A = \sin\alpha$ ($0^o \leq \alpha < 90^o$). When $\alpha = 0$, $A = 0$.

Relative area of artificial matrix (C): We import photographs into Photoshop software to calculate the area of artificial construction and photo area, and calculate the visual share of artificial construction (artificial construction area/photo area).

2. The process of the impact assessment of visual landscape

The impact assessment of visual landscape covers three first-grade indicators and seven second grade indicators. This article intends to build a mathematical model of the impact assessment of visual landscape by entropy weight method to achieve quantitative evaluation of the visual quality.

2.1. Division of factor subset

The entire visual environment quality (Z) is divided into three subsets: visual sensitive capacity (Y_1), visual space capacity (Y_2) and visual absorption capacity (Y_3). Then $Z = \{Y_1, Y_2, Y_3\}$, three subsets are: $Y_1 = \{X_1, X_2, X_3\}$, $Y_2 = \{X_4, X_5\}$, $Y_3 = \{X_6, X_7\}$. X_i represents the seven second grade indicators.

2.2. Quantitative classification indicators

Sighting distance: visual effects observed from the object is closely related to sighting distance. Distance is 300 m, which can be observed the form and details of the single structure; from 300 m to 1500 m, building groups can be observed; from 1500 m to 3000 m, only can see the outline of building groups. This principle is related to the division of close view, medium view and distant view, and also affects the scope of each landscape control point. Based on this, the sighting distance is divided into four grades: 0 m, more than 1000 m, (300 m, 1000 m], (0 m, 300 m].

Visual disturbance frequency: If the landscape can been seen between two points, then l = L, and sensitivity is the highest; l = L/2, medium sensitivity; l = L/7, the lowest sensitivity. Visual disturbance frequency is divided into four grades: 0, (0, 15%], (15%, 50%], (50%, 100%].

Green looking ratio: When the green looking ratio is less than 15%, the artificial traces are increased. The green looking ratio is more than 15%, and natural feeling will increase. When the green looking ratio reaches 25%, people will feel most comfortable, and it's beneficial to human health. Chunfa and Jianhua (2009) evaluated the visual environment of Yuntai mountain, dividing the herb coverage into four grades which based on the three thresholds: 70%, 80% and 90%. Reference to the above research results, this article will divide green looking ratio into four grades: [90%, 100%], [70%, 90%), (25%, 70%), [0, 25%].

Inter-visibility: Road traffic had strict requirements for the inter-visibility, in general, the visual distance of road's longitudinal slope design must reach 400–600 m (Jishuang, Yong 2001). According to the degree of occlusion, this article will divide inter-visibility into four grades: 600 m, [400 m, 600 m], [200 m, 400 m), (0 m, 200 m).

Openness: "Golden section" has a close inner link mechanism with human vision. Based on this, openness is divided into the following four grades: 100%, (61.8%, 100%), (0, 61.8%], 0.

Relative angle: Based on the 50 m contour interval, Kongjian Yu (1991) calculated the sensitivity of relative slope during the evaluation of canyon landscape of Wangxiangyan. The relative angle is divided into the following four grades: 1, [1/2, 1), [1/4, 1/2), [0, 1/4).

Relative area of artificial matrix: Based on the share of artificial landscape outside the park, we divide the relative area of artificial matrix into four grades in this paper (Guxin, Feng 2009): [0, 4%], (4%, 10%], (10%, 20%], (20%, 100%].

Table 1. The indicators classification of the impact evaluation of visual landscape

Assignment Indicators	7 points	5 points	3 points	1 points
Sighting distance (m)	0	D > 1000	300 < D ≤ 1000	0 < D ≤ 300
Visual disturbance frequency (%)	0	0 < V ≤ 15	15 < V ≤ 50	50 < V ≤ 100
Green looking ratio (%)	90 ≤ G ≤ 100	70 ≤ G < 90	25 < G < 70	0 ≤ G ≤ 25
Inter-visibility (m)	Unobstructed or I > 600	400 ≤ I ≤ 600	200 ≤ I < 400	0 < I < 200
Openness (%)	1	0.618 < O < 1	0 < O ≤ 0.618	0
Relative angle	0	0 < A ≤ 1/4	1/4 < A ≤ 1/2	1/2 < A ≤ 1
Relative area of artificial matrix (%)	0 ≤ C ≤ 4	4 < C ≤ 10	10 < C ≤ 20	20 < C ≤ 100

2.3. Entropy empowerment

In order to overcome the subjectivity of the traditional methods of empowerment, we use entropy weight method to determine the weight of each indicator factor (Wanhua 2010). The weight depends on the information loaded by each indicator, and the information content can be represented by entropy, that is the entropy of information (Yu, Zongfang 2006). The reduction of entropy means the increase of the information content. According to the view of information theory, indicator variation is usually used to examine the role of each indicator in the evaluation model. The change of indicator is larger, and the information content is greater, then the indicator of "recognition" is stronger (Meihong, Aihua 2006). In short, if the change of a certain indicator X in different evaluations is large, then it has an important role on differentiating the subject of evaluation and should be given greater weight. Entropy model includes many formulas and calculation process. The formula of W_i and H_i are used to respectively calculate the values of weight and the entropy of each indicator. f_{ij} is the proportion of each indicator in standardization value, and k is the correction coefficient.

Entropy model:

$$W_i = \frac{1-H_i}{m-\sum_{i=1}^{m} H_i}, \quad H_i = -k\sum_{j=1}^{n} f_{ij} \ln f_{ij}, \quad f_{ij} = \frac{r_{ij}}{\sum_{j=1}^{n} r_{ij}},$$

$$k = \frac{1}{\ln n}.$$

Where, W is the weight value for each single indicator. H is entropy, m is the number of indicator, k is the correction coefficient, n is the number of evaluation subject, f is normalized indicator matrix, r is the normalized value of indicator.

Before analyzing the data, we usually need to standardize the data. After standardization, the original data will be converted into dimensionless indicator value and the value of each indicator will be in the same level, which could be used to make a comprehensive evaluation analysis. The role of formula r_{ij} is to standardize the data.

Standardized formula:

$$r_{ij} = \frac{r_{ij}' - \min\{r_{ij}'\}}{\max\{r_{ij}'\} - \min\{r_{ij}'\}},$$

r_{ij} is the normalized value of indicator, r_{ij}' is the original value.

2.4. The mathematical model of the impact assessment of visual landscape

On the basis of the weight value of each indicator calculated by entropy weight method and the field investigation data, making the visual environment quality (*VEQ*) of country park as the dependent variable and each evaluation indicator as independent variable, to establish the mathematical model of the impact assessment of visual landscape according to the function relationship between the evaluation indicators and the visual environment quality. In this model, *VEQ* is to calculate the value of visual quality, and the influence of each indicator on the visual environment quality is independent.

$$VEQ = m^{-1}\sum_{i=1}^{n}\sum_{j=1}^{m} w_i x_{ij}.$$

Where, VEQ represents the value of visual quality, x_{ij} is the quantity value of indicator i of point j. w_i is the weight of indicator i. n is the number of indicator, m is the number of viewpoint.

2.5. The model of the impact assessment of visual landscape of country park

The mathematical formula *VEQ* is a multivariate first-order equation of linear regression established by the weight value of each indicator. Through regression analysis, we summarize the degree of significant differences and the collinearity, finding out the main factors which impact the visual landscape environment of country park. In this formula, X_i is the weight value of indicator.

$$VEQ = 0.154\,X_1 + 0.378\,X_2 + 0.026\,X_3 + 0.276\,X_4 + 0.115\,X_5 + 0.027\,X_6 + 0.024\,X_7.$$

2.6. Model validation and sample evaluation

For the Beijing urban green space system, the country park is an important part (Beijing Municipal... 2009). The "country park ring" plan has been launched since 2007 in Beijing, and the main function is to prevent the central area of the city and peripheral groups forming into one, controlling urban disordered extension, improving the efficiency of urban land use, avoiding the development of city "pie" type. As a significant greenbelt, country parks play important roles in the capital's economic and social development, urban ecological pattern, and the living quality of residents. At the same time, this also improves the city's sense of identity and belonging, and changes leisure pattern of Beijing city. Considering the difference of types and distributions, 12 representative country parks are selected among the 81 country parks, and the two typical city parks which are used to compare with the country parks.

Table 2. The weight of each indicator of the impact assessment of visual landscape

The first grade indicators (Y)	The second grade indicators (X)	Weight
Visual sensitive capacity (Y_1)	Sighting distance (X_1)	0.154
	Visual disturbance frequency (X_2)	0.378
	Green looking ratio (X_3)	0.026
Visual space capacity (Y_2)	Inter-visibility (X_4)	0.276
	Openness (X_5)	0.115
Visual absorption capacity (Y_3)	Relative angle (X_6)	0.027
	Relative area of artificial matrix (X_7)	0.024

Table 3. Regression analysis

Models	Factors	P	Tolerance
1	(Constant)	.510	
	Sighting distance	.030	.011
	Visual disturbance frequency	.000	.343
	Green looking ratio	.217	.626
	Inter-visibility	.006	.033
	Openness	.010	.087
	Relative angle	.320	.028
	Relative area of artificial matrix	.437	.043
2	(Constant)	.212	
	Sighting distance	.004	.017
	Visual disturbance frequency	.000	.345
	Green looking ratio	.133	.675
	Inter-visibility	.002	.049
	Openness	.000	.229
	Relative angle	.282	.028
3	(Constant)	.480	
	Sighting distance	.000	.251
	Visual disturbance frequency	.000	.373
	Green looking ratio	.155	.680
	Inter-visibility	.000	.210
	Openness	.000	.368
4	(Constant)	.007	
	Sighting distance	.000	.255
	Visual disturbance frequency	.000	.374
	Inter-visibility	.000	.230
	Openness	.000	.389

In order to verify the validity of the initial model, we use step-by-step backward regression analysis to analyze the significance of difference degree and colinearity of each factor (Table 3). The result shows, P value of relative area of artificial matrix (model 1), relative angle (model 2) and green looking ratio (model 4) are all greater than 0.05. This illustrates significant difference is in low degree, in other words, the degree of recognition of factor is low, and the effect for evaluation is very small. Therefore, excluding the above three factors, and the four factors: sighting distance, visual disturbance frequency, inter-visibility and openness are preserved. Tolerance values are all greater than 0.1 (model 4), meeting the conditions of the colinearity, and the model evaluation by four factors are effective. Then the four factors were assigned weight by the standard deviation method. The final evaluation model is established: $VEQ = 0.272D+0.3V+0.231I+0.197O$.

Urban-rural area, except for urban construction land, is a large area of natural or near-natural state which belongs to urban green space. Along with the spread of urban expansion, the contradiction of the land use of urban-rural fringe is increasing. In 2007, the Beijing municipal government launched the construction of the first green isolating areas into "country park ring". The purpose was to form the space layout of "a ring, six districts, one hundred parks" which was around the center of the city, and to build a "chain cluster" structure of "the whole into a ring, the segmentation into pieces". It becomes a landscape green belt and ecological protection zone, meanwhile, a place for public recreation and exercise.

So far, 81 country parks have been built in Beijing. However, according to our initial investigation, we find the following problems in Beijing country parks: (1) The space layout of country park is not balanced. Large numbers are in the east and south, only a few portions distribute in the north and west. (2) Obvious traces of artificial landscape are seen in somewhere. Golf courses, business club and other illegal constructions occur frequently in the country park. This seriously interferes with the natural landscape. (3) The design of country parks is not a big difference compared with city parks, which lacks in the property of wildness. (4) The boundary is disturbed by high-rise buildings. The features of visual landscape are inconsistent with the attribute of country park.

We calculate the value of visual environment quality of each country park by using the impact assessment of visual landscape model, and visual environment quality of each country park is graded.

3. Discussion

The level of the visual environment quality of the country park reflects the visual impact of peripheral artificial construction on the park and the urbanization level in a certain area. Theoretically, with the increasing of distance from the park to downtown area, the level of urbanization is gradually decreasing, and the visual environment quality of the country park should gradually increase and the park's overall visual environment should gradually improve. However, through our research, we found the visual environment quality of the selected 14 sample parks did not completely comply with the above hypothesis. The main reason is that the direction of urbanization development in Beijing mainly towards the north and west areas, which results in the visual environment quality sharply reducing in Changchun Fitness Park (between west 3rd ring road and west 4th ring road) and Dongsheng bajia Country Park (Between north 4th ring road and north 5th ring road) (Fig. 2). With regards to this, the Beijing City Master Plan (2004–2020) posed that the urban space resources should be rationally utilized and controlled the large-scale population and industrial agglomeration in western and northern mountainous areas of Beijing, and the development direction of city should towards the eastern and southern plain regions (Beijing Municipal People's Government 2005). The results of this research confirm the thought of the Beijing City Master Plan (2004–2020).

The development of Beijing urbanization is not balanced, and in the west and north part areas are significantly higher than other urban areas, which leads to the so-called country park wearing "country park" hat, but dressing in "city park" coat. It is difficult to make distinguish from the attribute of country park. In this article, by calculating the visual environment quality of the country parks, we find that the score of visual environment quality should reach at least 4.2 which could meet the basic requirements of a country park. Besides, according to Beijing Green Space System Planning (BGSSP), country

Table 4. The grade of comprehensive evaluation of the visual environment quality

Visual environment quality	$1 \le VEQ < 2.8$	$2.8 \le VEQ < 4.2$	$4.2 \le VEQ < 5.6$	$5.6 \le VEQ \le 7$
Centesimal system	$14 < VEQ < 40$	$40 \le VEQ < 60$	$60 \le VEQ < 80$	$80 \le VEQ \le 100$
Grade	iv	iii	ii	i
Outcome	Lower visual environment quality	Low visual environment quality	High visual environment quality	Higher visual environment quality

Table 5. Visual environment quality and grade of the 14 sample parks

Sample parks	Location	Distance to the city center (Tiananmen Square) (km)	Visual environment quality/ Centesimal system (%)	Grade
Changchun Fitness Park	Between north 3rd ring road and north 4th ring road	10.8	1.14/16.3	iv
Wangxing Lake Country Park	Between south 4th ring road and south 5th ring road	10.9	2.7/38.6	iv
Taoranting Park	Within south 2nd ring road	3.9	2.79/39.9	iv
Taiyanggong Park	Between north 3rd ring road and north 4th ring road	8.6	3/42.9	iii
Jingyi Country Park	Between west 4th ring road and west 5th ring road	15.1	3.08/44	iii
Dongsheng bajia Country Park	Between north 4th ring road and north 5th ring road	13.4	3.11/44.4	iii
Zizhuyuan Park	Between west 2nd ring road and west 3rd ring road	7.5	3.24/46.3	iii
Wanfeng Country Park	Between west 3rd ring road and west 4th ring road	9.2	3.34/47.7	iii
Xinglong Park	Between east 4th ring road and east 5th ring road	11.4	3.44/49	iii
Dongba Country Park	Between east 5th ring road and east 6th ring road	14.6	4.09/58.4	iii
Dongxiaokou Forest Park	Between north 5th ring road and north 6th ring road	14.4	4.14/59	iii
Nanhaizi Country Park	Between south 5th ring road and south 6th ring road	15.5	5.3/75.7	ii
Jintian Country Park	Between east 5th ring road and east 6th ring road	14.4	5.42/77.4	ii
Lüdi Park	Between west 5th ring road and west 6th ring road	16.3	5.65/80.7	i

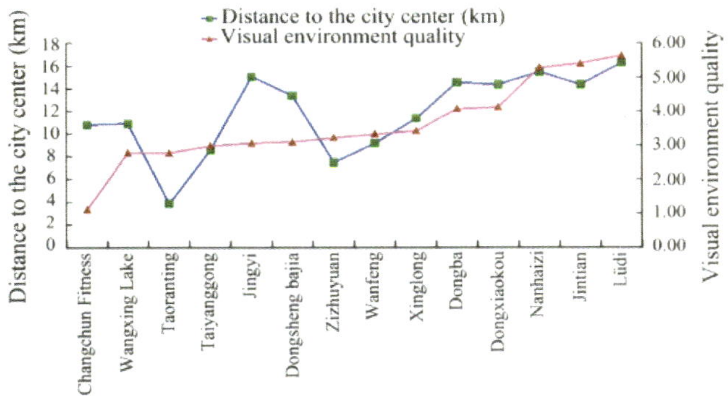

Fig. 2. The line graph of visual environment quality and distance to the city center

parks are mainly situated in the first green isolating areas, and the location of green isolating areas are mainly outside the 4th Ring Road in Beijing (the distance to the city center ≥8km). The park's attribute classification is too administrative, lacking in scientific basis, then we present two important bases to redefine the attribute of country park: (1) The visual environment quality of the country park reaches more than 4.2; (2) To have a certain distance to the city center, and the distance depends on different cities, being proportional to the city scale. For example, the country park to downtown area should be more than 8 km in Beijing. In accordance with the above two rules, among the 12 country parks selected, only the Lüdi, Jintian and Nanhaizi belong to the country park (Fig. 2).

Collecting and processing the field data of a park's visual environment to calculate the visual environment quality of each viewpoint. Through analysis and research, the factors which affect the visual environment quality of each viewpoint can be drawn. Then, the targeted improvement measures are proposed, so as to enhance the overall visual quality of the park. Taking Jingyi Country Park as an example, the viewpoints that impact on the overall visual quality of the park are: point 1, point 2, point 3 and point 8 (Fig. 3). The most obvious interference factors for the four points are outside of high-rise buildings. The high-rise buildings are not only close to the park, but their large body, as well as the impact of high-voltage tower. It can be covered by planting tall trees, and as much as possible to reduce the disturbances occurred in the vertical line of sight of visitors. Another way, micro-terrain design around the park boundary, to some degree, it may reduce the range of inter-visibility and visual disturbances, the interference of high-voltage tower and high-rise buildings, but it also can arise the green looking rate. So, the visual environment quality of park can be greatly improved, and the standard of country park can be reached.

The influence of building facilities on the visual environment, the sighting distance, visual disturbance frequency, inter-visibility and openness are especially evident. The denser of artificial buildings are, the higher of visual share and the degree of landscape feature alienation will be. Materials, volume, color, style and other aspects of the artificial buildings do not coordinate with the park natural rustic charm. At the same time, there are also other artificial facilities, such as billboards and pipelines, which bring negative influence on the visual environment quality of country park. During the field study, we found that most of the country parks have been built in the area of

Fig. 3. The four viewpoints of Jingyi Country Park
Point 1 (upper left), Point 2 (upper right), Point 3 (lower left), Point 8 (lower right).

high-voltage corridor. Because it locates in the park area, and the landscape striking degree is high, which brings serious visual pollution to the country park. The layout of vegetation within the park primarily affects inter-visibility and openness. The construction of country park was based on the first green isolating areas, and the main function of the original vegetation is the ecological protection, and lacks some considerations of landscape visual effect. The vegetation of some areas where adjacent to the main tour roads are too dense, blocking the line of sight, compressing the visual space, without concerning about the psychological perception of tourists, which gives a feeling of oppression to viewers.

Conclusions

1. To provide reference for planning control and land resource management in fringe area of central city. Firstly, "single-storey house be changed to high building" policy should be adjusted and need to be appropriate intensive layout. This policy has resulted in landscape fragmentation, and the continuity of country park ring and the integrity of landscape pattern are destroyed by the current implementation policy that the constructions are built in the original place and dispersed layout. Secondly, on the basis of country park ring that has been built between 4th ring road and 5th ring road, the thickness of the greenbelt should continue to widen between 5th ring road and 6th ring road, and make it a landmark in the urban fringe area. Thirdly, to some extent, the construction of country park limits the development direction of Beijing city fringe area, the distribution of country parks and the landscape pattern. There are mountainous areas in the west of Beijing city center, the historical and cultural protection area in the northwest, which owns a good substrate of greenbelt landscape, and there should be no large-scale construction in this area. The area that in the east and south of Beijing is in the intersection of the region of ecological protection and the urban-rural integration construction. The scale, speed and complexity of landscape change are unprecedented. Fourthly, it should maintain a certain width of green space among city center, fringe groups, and the peripheral of fringe groups. At least 200 meters visual buffer zones should be set up between construction area and landscape protection area.
2. To flexibly apply the following planning methods:
   (1) Protection. Avoiding the landscape which has a negative impact on the visual environment of park. Prohibiting super high-rise buildings so as to protect important skyline and landscape view corridor. Some country parks outside the 5th ring road should be taken protective measures firstly. For instance, Dongxiaokou Forest Park, Olympic Forest Park and Taiping Country Park are in the traditional north-south axis of Beijing city, which is an important visual channel for Beijing, and the inter-visibility of the visual corridor should be protected. Adjacent to the Yongding river, Lüdi Park is an important ecological corridor and a landscape corridor, but also an air duct, which plays a role in improving the air quality in the center city. Lüdi Park is the highest score in the grade of comprehensive evaluation of the visual environment quality among the 14 parks and is a key visual protection area. As the nearest natural wetland to city center, Nanhaizi Country Park keeps the country characteristic that country park should possess. To establish the control zone between the country park and other landscape areas, and avoid building too many artificial constructions around the periphery of the country parks. Then, the visual interference frequency will be reduced, and the continuity and inter-visibility of the green landscape will be protected.
   (2) Restriction. If the negative impact is inevitable, we must control its development and construction. The building that is near the park should strictly limit its number of floors and control its height. It has built some residential area in the north of the Dongba Country Park and the south of Jintian Country Park, like these country parks outside the 5th ring road, which should set up visual buffer zones between different landscape boundaries. By controlling the location, volume, shape, height of artificial construction and the distance to restrict the visual impact of artificial constructions on the country parks.
   (3) Avoidance. In order to enhance the aesthetic perception of tourist, the design for the park tour road should try to avoid the low quality of the visual environment of viewpoint, and makes tourism line through the high visual environment quality of the viewpoint as far as possible. In the important node of tour road of country park, like Jingyi Country Park and Dongshengbajia Country Park, ground cover, herbs, shrubs, trees should be cultivated to shape multi-level landscape to promote the aesthetic value of the country parks. In addition, making necessary adjustment of the direction of some tour roads to avoid interference of visual sensitive points of artificial buildings outside the country parks and improve the inter-visibility and openness of nodes.
   (4) Repair. For the existing visual pollution, we should take measures to repair the visual environment. Due to the continuous expansion of urban development to the periphery, the peripheral of country park has appeared the high-rise buildings and other negative landscape. Faced with such a situation, we take steps such as micro-topography, planting tall trees and other ways to weaken its visual impact. A few

country parks such as Xinglong Country Park and Jingyi Country Park are between 4th ring road and 5th ring road which need to adopt necessary measures for the boundary of parks to cover and conceal the visual contact of peripheral architecture. Taking micro-topography processing or planting some tall crown diameter trees, like poplar and ginkgo, to form the enclosure of internal landscape, at the same time, to improve the green looking ratio. The degree of visual influence of artificial high-rise buildings and high-voltage towers will be cut down.

(5) Reposition the park attribute: By protection, restriction, avoidance, repair and other measures, some country parks such as Wangxing Lake Country Park, Changchun Fitness Park, Taiyanggong Park and Wanfeng Country Park cannot change the visual impact of external artificial construction to the parks. These parks are relatively close to the city center and the surrounding areas have been completely urbanized, and the visual environment quality are under 4.2. Then, we suggest changing the attributes of these parks, and convert them into city parks.

Acknowledgments

In the writing process, we are very grateful to the help of Terry C. Daniel (Emeritus Professor of Psychology and Natural Resources,University of Arizona). He spent a lot of time and offered suggestions for revisions. We appreciate his help. This work was supported by the <Beijing Natural Science Foundation> under Grant [number 8142008].

References

Anderson, L. 1979. *Visual absorption capability in our national landscape*. USDA forest service general technical report, Nevada.

Ayad, Y. M. 2005. Remote sensing and GIS in modeling visual landscape change: a case study of the Northwestern arid coast of Egypt, *Landscape and Urban Planning* 73: 307–325. http://dx.doi.org/10.1016/j.landurbplan.2004.08.002

Beijing Municipal Commission of Urban Planning. 2009. *The green space system planning of Beijing* [online], [cited 20 April 2014]. Available from Internet: http://www.bjyl.gov.cn/zwgk/ghxx/gh/zcqgh/201010/t20101028_43423.html

Beijing Municipal People's Government. 2005. *Beijing City Master plan (2004–2020)* [online], [cited 20 April 2014]. Available from Internet: http://www.bjghw.gov.cn/web/static/catalogs/catalog_233/233.html

Binyi, L. 1990. The visual quality evaluation of landscape environment, *Urban Planning Forum* 68(4): 24.

Chuanfa, L.; Jianhua, Y. 2009. The synthetical evaluation of the visual environment of the beauty spot in Yuntaishan, *Areal Research and Development* 28(6): 126–130.

Cloquell-Ballester, V.; Torres-Sibille, A. C.; Cloquell-Ballester, V.; Santamarina-Siurana, M. C. 2012. Human alteration of the rural landscape: Variations in visual perception, *Environmental Impact Assessment Review* 32: 50–60. http://dx.doi.org/10.1016/j.eiar.2011.03.002

Daniel, T. C.; Vining, J. 1983. Methodological issues in the assessment of landscape quality, in I. Altman, J. F. Wohlwill (Eds.). *Behaviour and the natural environment*. New York: Plenum Press. http://dx.doi.org/10.1007/978-1-4613-3539-9_3

Dramstad, W. E.; Tveit, M. S.; Fjellstad, W. J.; Fry, G. L. A. 2006. Relationships between visual landscape preferences and map-based indicators of landscape structure, *Landscape and Urban Planning* 78: 465–474. http://dx.doi.org/10.1016/j.landurbplan.2005.12.006

Garré, S.; Meeus, S.; Gulinck, H. 2009. The dual role of roads in the visual landscape: a case-study in the area around Mechelen (Belgium), *Landscape and Urban Planning* 92: 125–135. http://dx.doi.org/10.1016/j.landurbplan.2009.04.001

Guxin, C.; Feng, W. 2009. Discussion on evaluation methods of landscape environment effects in construction of mining area, *Morden Mining* 9(7): 70–72.

Hernández, J.; García, L.; Ayuga, F. 2004a. Assessment of the visual impact made on the landscape by new buildings: a methodology for site selection, *Landscape and Urban Planning* 68: 15–28. http://dx.doi.org/10.1016/S0169-2046(03)00116-6

Hernández, J.; Garcia, L.; Ayuga, F. 2004b. Integration methodologies for visual impact assessment of rural buildings by geographic information systems, *Biosystems Engineering* 88(2): 255–263. http://dx.doi.org/10.1016/j.biosystemseng.2004.02.008

Hui, Z., Hua, Z.; Yansong, W.; Weitao, S. 2009. Approach on landscape visual impact assessment of project facilities in scenic areas, *Journal of Natural Resources* 24(1): 177–184.

Hurtado, J. P.; Fernandez, J.; Parrondo, J. L.; Blanco, E. 2004. Spanish method of visual impact evaluation in wind farms, *Renewable and Sustainable Energy Reviews* 8: 483–491. http://dx.doi.org/10.1016/j.rser.2003.12.009

Jianhua, Y. 2008. The impact of tourism on the scenic vision of Jigong mountains, *Journal of Fujian Forestry Science and Technology* 35(2): 215–218.

Jishuang, W.; Yong, X. 2001. Impact on the route flat, vertical curve radius of the line of sight requirements, *Journal of Highway and Transportation Research and Development* 18(6): 31–33. [online], [cited 24 April 2014]. Available from Internet: http://www.cqvip.com/QK/91479X/200106/12455321.html

Kaplan, A.; Taskin, T.; Onenc, A. 2006. Assessing the visual quality of rural and urban-fringed landscapes surrounding livestock farms, *Biosystems Engineering* 95(3): 437–448. http://dx.doi.org/10.1016/j.biosystemseng.2006.07.011

Kongjian, Y. 1991. Assessment of landscape sensitivity and impact resistability: with a case study of Wangxiangyan canyon in south MT.Taihang, *Geography Research* 10(2): 38–51. http://www.dlyj.ac.cn/CN/10.11821/yj1991020005

Krause, C. L. 2001. Our visual landscape. Managing the landscape under special consideration of visual aspects, *Landscape and Urban Planning* 54: 239–254. http://dx.doi.org/10.1016/S0169-2046(01)00139-6

Kyushik, O. 1998. Visual threshold carrying capacity (VTCC) in urban landscape management: a case study of Seoul, Korea, *Landscape and Urban Planning* 39: 283–294. http://dx.doi.org/10.1016/S0169-2046(97)00085-6

Meihong, Z.; Aihua, L. 2006. Entropy-weight-based integrative evaluation on the science and technology capability for the western provinces, *Mathematics in Practice and Theory*

36(12): 120–125. [online], [cited 3 May 2014]. Available from Internet: http://kns55.en.eastview.com/KCMS/detail/detailall.aspx?filename=ssjs200612019&dbcode=CJFQ&dbname=CJFD2006

Menegaki, M. E.; Kaliampakos, D. C. 2012. Evaluating mining landscape: a step forward, *Ecological Engineering* 43: 26–33. http://dx.doi.org/10.1016/j.ecoleng.2011.02.011

Ministry of Environmental Protection of the People's Republic of China. 2011. *Guideline for Technical Review of Environment Impact Assessment on Construction Projects* (HJ 616 – 2011). China Environmental Science Press.

Molina-Ruiz, J.; Martínez-Sánchez, M. J.; Pérez-Sirvent, C.; Tudela-Serrano, M. L.; Lorenzo, M. L. 2011. Developing and applying a GIS-assisted approach to evaluate visual impact in wind farms, *Renewable Energy* 36: 1125–1132. http://dx.doi.org/10.1016/j.renene.2010.08.041

Mortberg, U. M.; Balfors, B.; Knol, W. C. 2007. Landscape ecological assessment: a tool for integrating biodiversity issues in strategic environmental assessment and planning, *Journal of Environmental Management* 82: 457–470. http://dx.doi.org/10.1016/j.jenvman.2006.01.005

Nakamae, E.; Xueying, Q.; Tadamura, K. 2001. Rendering of landscapes for environmental assessment, *Landscape and Urban Planning* 54: 19–32. http://dx.doi.org/10.1016/S0169-2046(01)00123-2

Nohl, W. 2001. Sustainable landscape use and aesthetic perception preliminary reflections on future landscape aesthetics, *Landscape and Urban Planning* 54: 223–237. http://dx.doi.org/10.1016/S0169-2046(01)00138-4

Renhai, W. 1999. A preliminary study of landscape and visual impact assessment, *Research of Environmental Sciences* 12(3): 58–59 [online], [cited 7 May 2014]. Available from Internet: http://www.hjkxyj.org.cn/hjkxyj/ch/reader/view_abstract.aspx?file_no=19990315&flag=1

Rogge, E.; Nevens, F.; Gulinck, H. 2008. Reducing the visual impact of 'greenhouse parks' in rural landscapes, *Landscape and Urban Planning* 87: 76–83. http://dx.doi.org/10.1016/j.landurbplan.2008.04.008

Roth, M. 2006. Validating the use of Internet survey techniques in visual landscape assessment – An empirical study from Germany, *Landscape and Urban Planning* 78: 179–192. http://dx.doi.org/10.1016/j.landurbplan.2005.07.005

Tan, Z.; Yuhu, Y.; Chuanyan, S.; He, X.; Chunrong, J. 2000. Technology of landscape impact assessment, *China Environmental Science* 20(1): 86–90. http://118.145.16.227/Jweb_zghjkx/CN/abstract/abstract8791.shtml.2000.02.29

Tong, Q.; YaJuan, W.; Weihua, W. 2013. A review on visual landscape study in foreign countries, *Progress in Geography* 32(6): 975–983. http://dx.doi.org/10.11820/dlkxjz.2013.06.015

Wanhua, Q. 2010. *Management decision-making and application of entropy.* China Machine Press.

Xiaojun, D.; Honggang, W. 2002. Green ratia green space ratia green looking ratio, *New Architecture* (6): 75–76. [online], [cited 10 May 2014]. Available from Internet: http://www.abbs.com.cn/newarch/read.php?cate=6&recid=16409

Xiaojun, W. 1993. A review of visual resources management system and methodology in USA, *World Forestry Research* (5): 68–76. http://www.jnr.ac.cn/CN/abstract/abstract936.shtml.1993.10.28

Xüling, L.; Zhaoping, Y.; Xuegang, C. 2012. The impact of tourism on landscape vision in nature heritage area: case of Kanas nature reserve, *Industrial Survey* 2: 80–84.

Yang, Z.; Xiaolin, D. 1999. A study of highway landscape and visual impact assessment, *Journal of Xi'an Highway University* 19(4): 65–67. http://trid.trb.org/view.aspx?id=651508.1999.12.30

Ying, L.; Yong, H.; Xin, L.; Liqun, D. 2006. Planning of the controlled-construction area in Beijing: establishing urban expansion boundary, *City Planning Review* 30(12): 20–26.

Youbo, Z.; Yue, Y. 2007. The procedure and indicator system of the environmental impact assessment (EIA) of the general management plan (GMP) for Chinese scenic areas, *China Garden* (1): 49–52 [online], [cited 15 June 2014]. Available from Internet: http://kns.ccpd.cnki.net/kcms/detail/detail.aspx?filename=2008085134.nh&dbcode=CABS&dbname=cabs2008

Yuhu, Y. 2001. Study and application of the technology of visual impact assessment, *Environment and Exploitation* 16(2): 43–44 [online], [cited 17 August 2014]. Available from Internet: http://www.cqvip.com/QK/97521X/200102/5717567.html

Yu, Z.; Zongfang, Z. 2006. Selection method for credit risk assessing targets in commercial banks via entropy weights, *Journal of University of Electronic Science and Technology of China* 35(5): 23–30. http://dx.doi.org/10.3969/j.issn.1001-0548.2006.05.040

Chuanan LIU, M.S student, Department of Geography, College of Resource Environment and Tourism, Capital Normal University, Beijing, China. Main research areas: landscape design, landscape planning, landscape aesthetics. Publications: author of 2 journal articles.

Tong QI, Assistant Professor, Department of Geography, College of Resource Environment and Tourism, Capital Normal University, Beijing, China. Publications: author of 1 book, author of 12 journal articles, several scientific publications. Research interests: landscape, landscape geography, landscape aesthetics, landscape evaluation, geology.

Xiao MA, M.S student, Department of Geography, College of Resource Environment and Tourism, Capital Normal University, Beijing, China. Publications: author of 3 journal articles. Main research areas: landscape evaluation, landscape ecology, recreation preference.

SOIL QUALITY ATTRIBUTES RELATED TO URBANIZATION IN BRAZILIAN WATERSHED

Alexandre Marco da SILVA[a], Rodrigo Custodio URBAN[b], Luiz Augusto MANFRÉ[c], Michel BROSSARD[d], Marcelo Zacharias MOREIRA[e]

[a] *Department of Environmental Engineering, Campus Sorocaba, São Paulo State University, Av. Três de Março 511, Sorocaba, SP, Brazil*

[b] *Department of Sanitation and Environment, FEC, Campinas State University, Av. Albert Einstein, 951, Campinas, SP, Brazil*

[c] *Geoprocessing Laboratory, Polytechnic School, University of São Paulo, Avenue Professor Luciano Gualberto, Lane 3, 380 São Paulo, Brazil*

[d] *IRD, UMR 210 Eco&Sols, BP 64501, 34394 Montpellier cedex 5, France*

[e] *Centre for Nuclear Energy in Agriculture, University of São Paulo, Av. Centenário, 303 – Piracicaba, SP, Brazil*

Abstract. In this study we investigated the variation of soil attributes according to urban-related land cover categories. The study was carried out in an urbanized watershed located in the Brazilian subtropical region (Sorocaba Municipality, São Paulo). Soil samples were collected considering the land cover category for analysis of physical, chemical and isotopic attributes. The land cover influenced the soils attributes. Soils from wooded and grassed areas presented significant differences, especially for values of C isotopes. Soil bulk density was significantly altered. According to considered land cover mosaic in the study, we estimated 10,241.28 tons of C stored in the thickness 20 cm of the watershed (whole area), and this amount is almost a half of the total potential of C storing in the watershed. We stress that projects of planned land cover should effectively implemented in urbanized regions to effectively contribute in storing more C and improving the soil-related ecosystem services.

Keywords: environmental impact assessment, environmental sustainability, landscape management.

Introduction

Soils are a critical component of any terrestrial ecosystem and they have both inherent and dynamic properties that may vary according to regional physiographic features. However, human activities also alter the soil properties of many forms. Such alterations, usually degrading, have getting more and more importance over the time due the impoverishment of the soil quality (Alberti 2005). The establishment and the expansion of urban areas, when poorly designed, is an anthropogenic activity that expressively alters the original characteristics of the soils (Kahan 2008; Pickett, Cadenasso 2009; Hagan *et al.* 2012).

Urbanization is a human-induced process that has social, economic and environmental causes and implications (UN 2014). Urbanization frequently produces alterations in land cover (De Kimpe, Morel 2000; Scalenghe, Marsan 2009; Edmondson *et al.* 2011), usually resulting in a landscape characterized by mosaic of land cover patterns (Antrop 2004; Alphan, Güvensoy 2016).

One of the main factors that drive the quality of soil organic matter (SOM) is the kind of organic material that is deposited on the soil surface and that is slowly incorporated into the soil profile (Lorenz, Lal 2009). The modification of the land cover normally modifies the kind of organic material that is deposited on the soil surface and, consequently, changes the quality of SOM. Because of the process of land cover shifting, the native vegetation usually is removed and new species that might not otherwise co-exist, including exotic plant species, are inserted in the "new" environment, modifying many soil attributes and degrading the quality (Alberti 2005). Modifying the SOM signifies changing parts of the cycles of both carbon (C) and nitrogen (N) (Pouyat *et al.* 2006; Baltrenas *et al.* 2010) and such alterations are proved through analysis of the

Corresponding author: Alexandre Marco da Silva
E-mail: amsilva@sorocaba.unesp.br

stable isotopes of the C and N of the soil organic matter (Norra *et al.* 2005; Boeckx *et al.* 2006; Kaye *et al.* 2006).

Several urban-related activities voluntarily or involuntarily compact the soil. Compaction of the soil leads to increased bulk density (BD), and the BD usually becomes greater than natural areas, prejudicing, for example the soil porosity and water infiltration (Scharenbroch *et al.* 2005; Hagan *et al.* 2012). When the soil clods are getting more disaggregated, the porosity diminishes and the rates of soil loss caused by the rain-induced erosive process tend to augment. The erosion process plays an important role in the chemical impoverishment of the soil, which might signify diminution of concentration of some extractable elements in the soil (Morgan 2009), evaluated by means of analyses of the cation exchange capacity (CEC) or also the electrical conductivity (EC). EC is a measure relatively easy and cheap to be made and gives the idea of the soluble salts present in the soil (Seifi *et al.* 2010). Some forms of soil pollution trend to increasing the EC values (Saritha *et al.* 2014), while others, especially erosion-related problems, diminish the values of EC (Silva *et al.* 2015a).

Considering the effects of urbanization in the assessment of the budget of C in localities and regions that experience rapid urban expansions is of high importance, since there is the chance of identify hotspots of gains and losses of C (Jim 1998; Pouyat *et al.* 2002, 2006; Tao *et al.* 2014).

In the region of the Latin American continent the number of people living in urban areas has increased rapidly and the soils have been modified and converted into "Technosols". Currently 79% of the total population from Latin American continent and 85% from Brazilian territory live in urban areas (WB 2014). This urban swelling has happening without a suitable planning. Conversely, few studies have been developed in the urbanized areas of the Latin American continent in order to report aspects of the transformations in the soils of urban regions and related data are still meager.

Exemplifying, in a world context, there is a lot of papers regarding alterations of soil properties due the urbanization process, as Jim (1998) for Hong Kong; Norra *et al.* (2005) for Karlsruhe (Germany); Boeckx *et al.* (2006) for Gent (Belgium), Hagan *et al.* (2012) for Tampa (FL, USA) and/or by Wang *et al.* (2013) for Shanghai (China). However, for regions of Latin American continent, few papers are found. The papers published by Barrales *et al.* (2007) and Vela Correa *et al.* (2012) for Mexico City; Moura *et al.* (2006) for Teresina City (Capital City of Piauí State, Northeast Brazil); and López *et al.* (2006) for Buenos Aires (Argentina) are the main ones published. This characterizes a lack information and knowledge regarding biogeochemical mechanisms that happen in urban soils from Latin America and a high demand for new studies.

Hence, in this study we investigated the variation of the quality of soil and SOM attributes for land cover classes generated by the urbanization. We hypothesized that land cover change towards the urbanization provokes alteration in the soil properties. However, we do not know if the alterations in the region focused on this study the soils are similar to alterations reported in other regions (continents), in terms of typology and magnitude. Here we did not investigate the consequences of urbanization over time, but instead we focused on the effects of urbanization in a spatial context. Furthermore, because of the data regarding chemical and physical attributes in urban soils from Latin America are scarce; this paper constitutes a baseline to be compared with the later acquired information in urban soils from such region.

1. Location and environmental characteristics of study area

The watershed that attended the desired conditions for our investigation was one that lies on Sorocaba Municipality (Fig. 1). Sorocaba has 449 km^2 and about 629,000 inhabitants, being approximately 98% urban (IBGE 2013). The investigated watershed encompasses 277 hectares and has an average density of 3,000 inhabitants per km^2. Soil surface is predominantly impervious by asphalt or concrete. There is no highways or hard traffic avenues, but there is permanent traffic of cars, buses and small trucks. Land use is mainly for residential ends, with some small commercial establishments, as well as educational institutions. Grassy areas (or pasture) usually occur in places where the land use is less intense as in gardens and vacant lots. In such grassy areas there is no livestock. Brachiaria grass (*Brachiaria decumbens*, Stapf) is the main grass species that occurs along the grassy areas.

Summers are rainy and hot, whereas the winters are slight and dry. Annual average temperature is 21.4 °C. Annual rainfall depth is 1,309 mm. Regional relief is predominantly slight-waved. Bedrock is sedimentary and a significant portion is constituted by fine- to medium-grained sandstones (IGSP 2009). Soils are deep, mostly brown, with very low or null stoniness, and they are predominantly sandy loams and slightly acids. The original class is Ferralsols (Melfi *et al.* 2004). Cultivated, irrigated and/or fertilized lots do not take place along the watershed. In addition, neither lentic aquatic environments, nor landfills occur. Some earthworks are observed and they have the goal to adequacy of the topography of some places for residential constructions.

The original vegetation formation in the region was an ecotonal belt constituted by Atlantic Rain Forest (semideciduous forest) and Brazilian Savanna (Kronka *et al.* 2005). Most of the original vegetation has been removed for establishment of residential districts (houses in towns). Fire sometimes occurs in the region on dry seasons (July and August) and it usually occurs in grassy and/or wooded

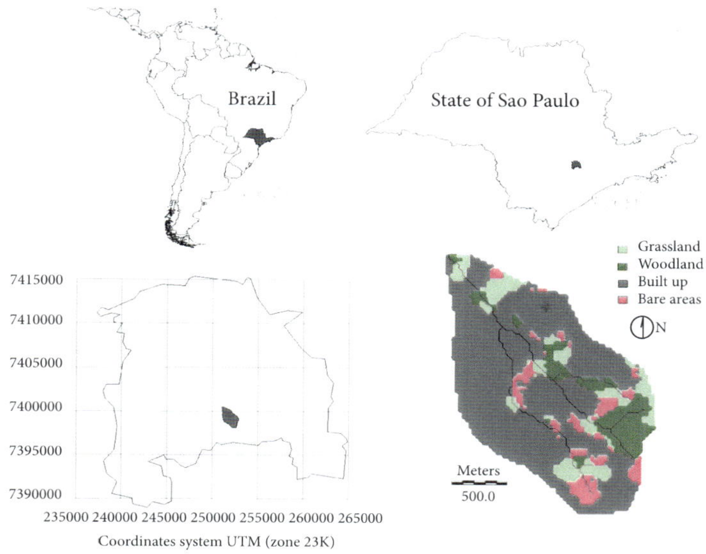

Fig. 1. Upper left: Brazilian states (highlighting State of Sao Paulo). Upper right: Boundaries of State of Sao Paulo (highlighting Sorocaba). Bottom left: Boundaries of Sorocaba (highlighting the studied watershed). Bottom right: Land cover map of studied map (modified from Silva 2012)

Table 1. Land cover categories considered for study, percentage of occurrence, brief characterization of each land cover category and number of soil samples taken in each land cover category (source: modified from Anderson 1976)

Land cover class and percentage of occurrence	Characterization	Number of soil samples collected according to each class of land cover (N = 40)
Woodland (10.7)	Small forest remnant patches in distinct degrees of conservation	10
Grassland (11.6)	Lawned areas covered with invasive grass and with distinct levels of conservation, usually occurring in vacant lots	9
Bare areas (8.2)	Areas with the soils poorly covered/protected, where soil exposure is clearly visible, typically used as domestic or public greenspaces	10
Built up parcels (69.5)	Urban or Built-up land is comprised of areas of intensive use with much of the land covered by structures. Also normally used as domestic or public greenspaces. They are private or public properties/places of various sizes, usually using decorative lawns in gardens of various sizes	11

sites. Currently, forested areas are reduced and scattered, with various levels of conservation. Some forest patches are riparian vegetation (Silva *et al.* 2013). The land cover categories were distinguished and identified according to USGS classification system (Anderson 1976). Table 1 provides specific information about each land cover class.

2. Procedures

2.1. Field sampling design and laboratorial analyses

Soil samples were taken following the land cover categories, aiming to capture the spatial heterogeneity of land cover (Vasenev *et al.* 2013). Forty undisturbed soil samples were collected throughout the watershed according to the occurrence of the four land cover categories.

All sampling points (henceforward SP) were georreferenced and they were always located at least 4 meters away from any fence lines to avoid possible edge effects (Silva 2012). They were collected in a thickness of 0–20 cm of soil, using a stainless steel cylinder (volume of 254 cm^3).

In the laboratory soil samples were oven-dried and weighted at 80 °C. The soil bulk density, i.e., the mass of a soil sample in a known volume (Jim 1998), was determined for each sample. Next, the samples were manually crushed and subdivided into three subsamples (Urban 2011). The first and second subsamples were sieved separately (both in mesh 2.00 mm) and the fraction <2.00 mm of the 1st subsample was delivered to Laboratory of Soil Analyses of Esalq – USP (Piracicaba, SP, Brazil) for analyzes of the sand fractions (coarse, medium and fine), silt and clay contents (Raij *et al.* 2001) and the Cation Exchange Capacity (CEC) (Ruggiero *et al.* 2002). The fraction of the 2nd subsample <2.00 mm was used for analyses of pH (H_2O) and electrical conductivity (EC). For pH and EC analyses, 30 ml of soil were mixed with 75 ml of distilled water in a glass beaker. After 30 minutes from the last agitation (Raij *et al.* 2001) and using a probe multi-parameter Oakton model PCS Test 35 previously calibrated, we measured the pH and the EC.

The 3rd subsample was also sieved (mash of 0.35 mm), and the portion <0.35 mm was sent to Laboratory of Isotopic Ecology of CENA-USP (Piracicaba, SP, Brazil) for quantification of the concentrations of C and N and their respective isotopic signatures. These determinations were done through gas chromatography after sample burning in an oxidant medium, by a Carlo Erba 1110 elemental analyzer conjugated to a Thermo Scientific Delta Plus isotopic ratio mass spectrometer.

The isotopic signatures of C and N were estimated using the Eq. (1) (Farquhar *et al.* 1989):

$$\delta^{13}C \text{ (or } \delta^{15}N) = [(R_{sa} - R_{st})\, R_{st}^{-1}]\, 1000, \quad (1)$$

where $\delta^{13}C$ (or $\delta^{15}N$) is the isotopic signal of the sample, R_{sa} is the $^{13}C/^{12}C$ (or $^{15}N/^{14}N$) ratio for the sample, R_{st} is the $^{13}C/^{12}C$ (or $^{15}N/^{14}N$) ratio for the standard.

For $^{13}C/^{12}C$, the reference material is the carbonate from a fossil belemnite from Pee Dee Formation (Farquhar *et al.* 1989). For $^{15}N/^{14}N$, the standard is atmospheric N_2 (Robinson 2001).

The soil samples were not treated with acid to remove the carbonates, because we wanted to analyze the total $\delta^{13}C$ of the top soil, determined by organic and inorganic C.

2.2. Estimative of the C stock

The amount of C stored in the soil was estimated using the Eq. (2), modified from Guo and Gifford (2002):

$$C_{st} = C\, BD\, h, \quad (2)$$

where C_{st} is the C stock (t ha^{-1}), C is the C concentration (g kg^{-1}), BD is the bulk density (g cm^{-3}), h is the thickness of the surface soil, in centimeters (0–20).

As the gravels did not occur in our samples, such fraction was not here quantified. The values of C stock were corrected to soil mass according to procedures and equations cited in Poeplau *et al.* (2011).

2.3. Statistical analyses

Statistical parameters were calculated using the database as a whole. In addition, the arithmetic average was determined for each attribute separately for each land cover category. The Kruskal–Wallis test was used for each variable to check the level of significance of the differences among the land cover classes. Spearman correlation test was performed in order to check the correlation level among the variables ($P = 0.05$).

The Fisher's linear discriminant analysis (F-LDA) was performed to check the similarity of the SP according to the land cover categories. The F-LDA is a statistical method that indicates which variables contribute most to group separation based on differences of the variables (Burns, R., Burns, R. 2008).

3. Results

3.1. Soil attributes, influence of land cover on soil attributes and Fisher's discriminant analysis

Sand, predominantly the coarse fraction, was major textural fraction for soils of the study area (Table 2). While the content of clay presented significant correlation only with other textural components of the soil, the silt fraction presented significant correlation with BD (direct correlation) and the sand fraction presented significant and inverse correlation with values of BD and $\delta^{13}C$.

Table 2. Summary statistics for studied attributes considering all samples (N = 40)

Attributes	Min	Max	Range	1st Q	Median	3rd Q	Mean	CV (%)
Coarse sand fraction (g kg^{-1})	135.0	567.7	432.8	259.7	366.2	418.7	342.9	30.2
Medium sand fraction (g kg^{-1})	90.5	698.0	607.5	168.7	249.8	292.1	249.6	42.4
Fine sand fraction (g kg^{-1})	28.8	432.2	403.5	92.5	134.5	183.0	157.4	57.1
Total sand fraction (g kg^{-1})	584.5	924.3	339.8	678.9	746.8	821.6	749.9	12.3
Clay fraction (g kg^{-1})	10.0	240.0	230.0	60.0	100.0	120.0	100.3	51.4
Silt fraction (g kg^{-1})	13.0	315.5	302.5	93.4	144.1	199.8	149.8	48.0
Soil bulk density (g cm^{-3})	0.64	1.90	1.26	1.26	1.49	1.64	1.42	22.1
N (g/kg)	0.1	3.3	3.2	0.5	1.0	1.8	1.2	73.4
C (g/kg)	0.3	39.4	39.1	6.4	13.8	21.1	15.1	71.6
C_{stock} (Mg ha^{-1})	0.8	75.4	74.7	14.4	32.3	52.3	33.3	68.1
$\delta^{13}C$ (‰)	–26.7	–15.9	10.8	–22.8	–20.1	–18.4	–20.8	15.8
$\delta^{15}N$ (‰)	2.6	11.8	9.3	5.1	6.1	7.5	6.5	27.7
C:N ratio	5.5	16.6	11.1	11.9	12.6	13.9	12.7	16.0
pH – H_2O	4.9	7.6	2.7	5.7	6.2	6.7	6.2	10.5
Cation Exch. Cap. (cmol kg^{-1})	22.5	233.3	210.8	48.9	76.9	120.0	89.8	62.8
Elect Conductivity (µS/cm^{-1})	33.9	577.0	543.1	109.4	177.5	270.0	203.9	61.8

For the BD the overall mean and median values were highly similar, and the highest value was 1.90 g.cm^{-3} (Table 2). Only one sample from woody site presented value higher than the overall mean. Most of the samples that presented values of BD higher than the value of the 3rd quartile are from built-up lands. The BD values were significantly and inversely (negatively) correlated with both C and N amounts.

The values of the amounts of C and N were highly correlated, as well as the values of the concentration of C and stock of C. Only one sample collected in woody site presented value of C concentration lower than the overall mean value. Overall mean value and range of values of pH stated the slight acidity of the soils, regardless the class of the category of land cover. The pH was the variable with lowest coefficient of variation (10.5%) and the mean and median values were identical.

Values of $\delta^{13}C$ ranged from −26.7 to −15.9‰. The values of the mean and median were similar, meaning that the data are equally divided around the mean. Based on the mean value, we can infer that the quality of the SOM in overall context of the study area is already modified. All samples presented positive values of $\delta^{15}N$ and more enriched than the standard (atmospheric air). The range of values of the $\delta^{15}N$ signature found in all samples, including that ones collected from wooded sites, suggested that the regional N cycle is not well conservative, because in a region where the cycle of the N is admittedly conservative, the values of $\delta^{15}N$ of the SOM are near of zero or even negative. An example is the soil from the non polluted forests located in Chile (Boeckx *et al.* 2005).

Most of the samples collected from wooded sites presented value of C:N lower than the overall mean, while all values of the SP collected from grassy sites were upper than the overall mean. The values of electrical conductivity (EC) ranged largely. The values of both cation exchange capacity and also EC presented high correlation with the amounts of C and N (Table 3). Usually in soils considered of good quality, the EC ranges between 200 and 1,200 µS/cm^{-1}, and they are highly correlated with the salt concentration in the soil, being the inverse of electrical resistivity (Brady, Weil 2004). The lowest limit value of this proposed classification is close of overall mean value found here. In our study area, 45% of the samples showed values upper than 200 µS/cm^{-1}.

From the ten variables here considered, eight presented some significant difference among the land cover categories (Table 4). For the values of BD, the mean values of the land cover classes built-up and bare area were approximately 38% higher than the mean value calculated for woody sites, being statistically different. For grassy areas the value was approximately 25% higher than the mean value of the reference area and insignificantly different from the others land cover classes.

Regarding the values of $\delta^{15}N$, we reported the tendency of the woody sites present lowest values, suggesting a SOM of better quality, while the samples from the class built-up presented the worse conditions (because of the highest values), although the values were statistically non different. For the pH values we reported, for sites from built-up category, the tendency of a slightly neutral when compared with the others land cover classes.

The soils from bare areas presented the highest mean value of BD, as well as smallest amounts of C and N, C stock and smallest value of EC. On the other hand, samples from grassy sites presented indicators slightly worse

Table 3. Spearman correlation matrix for soil attributes. Values in bold mean statistically significance at P = 5%

Variables	$Sand_m$	$Sand_c$	$Sand_t$	Clay	Silt	BD	N	C	C_{st}	^{13}C	^{15}N	C:N	pH	EC	CEC
$Sand_f$	**−0.45**	**−0.49**	−0.06	−0.10	0.13	−0.07	−0.27	−0.25	−0.30	0.06	−0.24	0.06	−0.09	−0.04	−0.08
$Sand_m$		0.17	**0.64**	−0.19	**−0.66**	−0.30	0.22	0.22	0.21	−0.21	−0.05	−0.01	0.19	0.12	0.13
$Sand_c$			**0.38**	**−0.36**	−0.21	0.02	0.19	0.17	0.22	−0.23	0.21	−0.11	0.29	0.07	0.01
$Sand_t$				**−0.64**	**−0.82**	**−0.35**	0.12	0.12	0.12	**−0.31**	−0.24	−0.06	0.26	0.12	0.13
Clay					0.16	−0.04	0.06	0.05	0.00	0.19	0.05	0.01	−0.28	0.06	0.04
Silt						**0.41**	−0.16	−0.15	−0.13	0.18	0.28	−0.01	−0.18	−0.20	−0.19
BD							**−0.56**	**−0.56**	**−0.51**	0.27	**0.48**	−0.09	0.08	**−0.62**	−0.60
N								**0.99**	**0.97**	**−0.35**	−0.15	0.29	0.07	**0.82**	**0.86**
C									**0.97**	**−0.33**	−0.17	**0.36**	0.06	**0.81**	**0.86**
C_{st}										**−0.33**	−0.14	**0.39**	0.06	**0.79**	**0.84**
^{13}C											0.02	0.19	−0.02	**−0.40**	**−0.39**
^{15}N												−0.06	**0.41**	−0.30	−0.26
C:N													0.06	0.23	0.27
pH														0.19	0.19
EC															**0.99**

Table 4. Soil attributes according to land cover classes (N = 40). For each variable, different letters mean difference significant at P = 5%. Identical letters mean no significant difference among the classes at P = 5%

Classes of land cover and number of sampling points	Wooded (10)		Grassy (9)		Built Up (11)		Bare Area (10)	
	Mean	CV (%)	Mean	CV (%)	Mean	CV (%)	Mean	CV (%)
BD (g cm^{-3})	1.13a	25.0	1.43ab	25.5	1.55b	14.6	1.56b	14.3
N (g kg^{-1})	2.2b	32.0	1.2ab	56.2	0.8a	67.0	0.5a	94.6
C (g kg^{-1})	27.4b	28.2	16.8ab	55.0	10.6a	70.7	6.2a	95.6
C_{stock} (Mg ha^{-1})	58.8b	19.1	37.9ab	54.0	24.0a	72.5	13.9a	96.7
$\delta^{13}C$ (‰)	−25.2a	6.2	−18.0b	7.3	−19.6b	11.0	−21.0ab	11.5
$\delta^{15}N$ (‰)	5.6a	17.6	6.2a	21.2	7.3a	34.8	6.6a	25.4
C:N	12.9ab	10.3	14.1b	6.3	11.9a	16.7	11.9a	23.5
pH – H_2O	6.1a	8.1	6.1a	9.6	6.6a	10.7	6.1a	12.4
Cation Exch. Cap. (cmol kg^{-1})	144.6b	32.0	85.8ab	70.5	83.4ab	56.2	45.5a	40.7
Elect. Conductivity (μS/cm^{-1})	359.7b	31.7	186.1ab	39.3	175.0a	47.9	95.7a	63.9

than samples taken from reference sites (i.e., higher BD, lower C and N amounts, and lower EC), but not significantly different for almost all variables, excepting for $\delta^{13}C$.

Also about the bare areas, for EC, although the values were statistically similar with the other land cover categories, we observe average value almost 50% smaller than the mean value of reference, woody sites. Samples from built-up areas also presented mean values of soil attributes significantly different from the mean values of reference sites for the attributes BD, C and N concentration, C stock and EC.

The diagram yielded through L-FDA (Fig. 2) confirms that soils under woody sites have characteristics different from the soils beneath anthropogenic categories of land cover. We note a region of SP overlapped in the regions that appear the SP collected in sites from human-induced land cover categories.

The confusion matrix (Table 5) supports the showed difference, because no one SP from woody site was misclassified. The samples from grassy sites were mostly successfully differentiated (illustrated by the higher values of percentage) and they occupied a specific region of the diagram. Similar situation is observed for SP from bare areas, although the SP of such category occupied another region of the diagram. On they turn, SP from built-up lands occupied intermediary region of the diagram and a high level of confusion and misclassification (63.6%) is observed. Hence, the overall mean percentage of success in the classification was 80.3%.

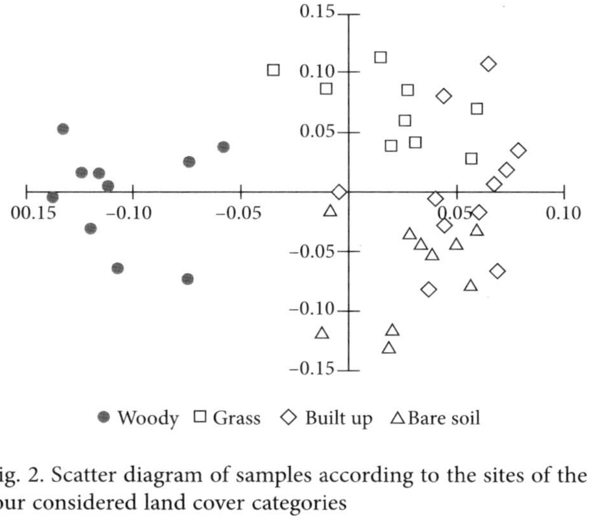

Fig. 2. Scatter diagram of samples according to the sites of the four considered land cover categories

Table 5. Confusion matrix for the estimation sample. The number in each cell corresponds to amount of samples that were correctly classified by the F-LDA

from \ to	Wood-land	Grass-land	Built up	Bare areas	Total	% correct
Woodland	10	0	0	0	10	100.0
Grassland	1	7	1	0	9	77.8
Built up	0	2	7	2	11	63.6
Bare areas	0	0	2	8	10	80.0
Total	11	9	10	10	40	(average) 80.3

4. Discussion

4.1. Soil attributes and influence of land cover

The BD is a measure of soil compaction and the soil compaction is commonly encountered in urban landscapes (De Kimpe, Morel 2000; Hagan *et al.* 2012). This trend is confirmed in our study. Land cover influenced critically the values of soil bulk density (BD). Increasing the BD means compress the soil and this alters some essential ecosystem services played by the soil, as water infiltration and storage capacity (Quraishi, Mouazen 2013). Maintaining the soil structure to mitigating the soil compaction is a critical ecological service required for both urban and agriculture soils. For a urbanized area, soil compaction for built-up and bare lands might be not especially critical to plant growth (root penetration), since usually we do not expect crop yield (like in agricultural areas), but alterations in infiltration capacity and disruption of soil aggregation are types of soil degradation that should be avoided because of the influence on infiltration of water into the soil profile and runoff generation (Gregory *et al.* 2006; Silva *et al.* 2013).

The fact of the significant inverse correlation among BD and both C and N strongly suggest that the alterations in the BD impacted the C and N concentrations, possibly because the reduction of the pore size, changing the dynamic of the C and N of the soil aggregates. This consequently reduces the water accessibility to microorganisms and reduces the capacity of storing C and N (Silva 2012; Silva *et al.* 2015a). The vegetation cover is a critical component that drives the BD and consequently the soil porosity (Jim 1998) and intensifying the arboreal cover in strategic locals is an important environmental service required as a part of the reengineering of the urban land cover design. For Latin America urban regions intensifying the arboreal cover is a mandatory activity, because the rate the migration of people from rural to urban regions is high (Grau, Aide 2008; WB 2014) and removing the vegetation due expansion of urban areas usually is an automatic consequence.

For soils under wooded sites, the average C stock estimated was 58.8 t ha^{-1} and for soils under grassy sites the average was 37.9 t ha^{-1}. Comparatively, Silva *et al.* (2009) in a rural area of Sorocaba (approximately 20 km far from the area of this study) found in soils under wooded sites an average value of 55.6 t ha^{-1} of stored C and 28.9 t ha^{-1} in soils under grassy sites in this same farm. The crucial difference among the grassy lands is that in the farm occurred a mismanaged livestock, but not in the urban-grassy sites. The trampling of animals and the excessive grass consumption by the cattle prejudices the formation of a mulch layer, and the ground surface becomes unprotected, stimulating the C and N losses from the soil by oxidation of SOM and soil erosion (Silva *et al.* 2015b). Further, we show that soils under forest even surrounded by urban-related land cover categories and subjected to several kinds of external pressures, have the potential to stock C in quantities similar to forest fragments located in non urbanized regions and this is a critical property of the urban forests that must be exalted in terms of planning of urban landscapes. This higher value of stored C in urban sites agrees with the trend reported by Edmondson *et al.* (2012) who stated that the urban soils store significantly more C than arable soils.

Taking into account the mosaic of land cover here reported the overall weighted mean for entire watershed of C stored was 28.5 t ha^{-1}. Considering the area of the watershed, the resulting value was 8,210.2 tons of C stored in the 20 cm topsoil for the entire watershed. This value was approximately 48.5% lower than the value considered, supposing that the area were entirely woody covered (16,934.4 tons) and interpreted as the maximum amount of C that the region is capable to store in the topsoil. Presuming that the watershed were entirely occupied exclusively by the built-up category (as the downtown of many medium- or big-sized cities), the value would be of 6,912.0 tons, meaning 15.8% less than the value estimated for current scenario and 59.2% less if all watershed were totally forest-covered. These predicted results evidence once more the necessity of a correct urban design, in order to preserve some essential ecosystem services.

Soils have a propensity to be chemically altered when unfamiliar materials are applied to them and this usually increases the value of soil electrical conductivity (EC) (Seifi *et al.* 2010). On the other hand, other degrading processes, as soil erosion, trend to provoke the loss of nutrients and by consequence decreases the EC values (Moebius-Clune *et al.* 2011). Values of EC exceptionally low or extremely high might cause difficulties for plants increasing and survivor. In our study area we note an expressive diminution of EC values in SP from human-induced land cover categories comparing with values from reference area. Hence, while in urban soils an enrichment of the soil could be supposed due to possibility of a usual contamination, for our study area we detected that the trend is an impoverishment due mismanagement of the ground cover and a forced exposition of the soil to weathering.

The C:N ratio reflects the kind of material that is deposited on the ground surface and also its vulnerability for decomposition, and indicates how the nutrient cycles are regulated in the soil (Edmondson *et al.* 2014). The range of values of C:N in urban soils is usually high, because natural and anthropogenic organic particles might occur together (Lorenz, Lal 2009). In our study area, in the built-up and of bare soil sites, which are areas where we supposed that none kind material is added voluntarily (as solid wastes), the mean value of C:N was lower than in the soils underneath woody areas. For soils from grassy vegetation, where we have a constant deposition of materials richer in fiber, the C:N is higher than in the woody sites. Comparatively, in a rural area approximately 60 km from Sorocaba, Silva *et al.* (2015b) found values of C:N of 14.4 in soils from organic farms and 14.3 in soils from conventional farms (in both there is plowing and addition of some kind of soil amendment), and both were discretely higher when compared with adjacent, forested soils.

Regarding the discrimination against the ^{13}C by some groups of plants, Vagen *et al.* (2006) suggest, for tropical soils whose value of $\delta^{13}C$ is $< -19.5‰$ the classification is C_3, and for soils with values of $\delta^{13}C \geq -19.5‰$ they fall in the C_4 group. Dividing our overall dataset according to this criterion, seventeen samples are classified as C_3. No one sample taken in woody sites was classified as C_4. One SP collected in a grass covered local was classified as C_3, as well eight from the ten samples collected in bare areas. For the samples classified as C_3, 43% were taken from woody sites and only one sample from grassy ground. The overall average for study area is –20.4‰, categorized as C_3, indicating that although an important fraction of the land cover of the study is already altered, an important fraction of the original SOM still remains in the soil, even in soils with human-made classes of land cover.

Constant soil disturbances caused by human activities stimulate large losses of ^{14}N, leaving behind a ^{15}N enriched residual (Boeckx *et al.* 2005). Although we noted the different categories of land cover did not produced significant difference in the $\delta^{15}N$ of the soils, the increasing in values of $\delta^{15}N$ and synchronized decreasing in the N concentration in the anthropogenic land cover classes states that after the disturbance, the regimen of loss of mineral N is intensified. Comparatively, while in our urbanized study site the enrichment of $\delta^{15}N$ was averaged in 20% for man-induced land covers in relation to reference sites, in rural areas (arable lands) in Ibiuna (SP, 60 km far from Sorocaba) the enrichment was averaged 38% (Silva *et al.* 2015a), suggesting that in the rural area the N cycle is more opened than in our urbanized study area, because the usual high positive correlation among ^{15}N enrichment of soil N and high humification of soil organic matter (Balieiro *et al.* 2012).

The pH is a determining factor for chemical reactions in the soil and the processes of soil acidification increment the mobility of some elements (Barrales *et al.* 2007). For our study area this attribute does not appear to be a concern, because of genetic reasons, the soil is naturally and slightly acidic and the differences in the land covers did not provoked significant alteration in the pH. For another factor that supposed influence the soil pH, the pH of the rainwater, Antunes *et al.* (2011) reported mean values of pH for rainwater of 6.6, classified as slightly acidic and typical for the region. Comparatively, low coefficient of variation of soil pH (12.8%) was also observed by Pouyat *et al.* (2007) for urban soils in the City of Baltimore, Maryland, USA.

4.2. Fisher's discriminant analyses and a complementary-integrative discussion

Human-driven landscapes constitute a mosaic with many edges and cores and consequently, distinct sites with patterns of disturbance and management that will affect soil features through time. The result is a mosaic of soil patches with particular features changing according to the land cover categories (Pickett, Cadenasso 2009). This trend is observed for our study area and supported by the complete and clear separation of the SP collected in woody sites from the other ones resulted by FDA (Fig. 2).

In the Figure 2 we see that the data from the three anthropogenic land cover formed one group of points with partial overlapping. Comparatively Pouyat *et al.* (2007), using soil data from different land cover categories and applying a statistical multivariate analysis, also show, in the scatter plot, a region where the soils under forested sites are concentrated in a separate region of the graphic, while the soils under human-induced land cover categories are not overlapped with forested sampling points, but highly overlapped among them. Hence, it seems that it is a common characteristic observed between urban regions from underdeveloped (Brazil, Latin America) and developed (USA) regions.

Changing land cover for urbanization creates hotspots for losses in carbon stocks. Conversely, such areas have potential to sequester and store C when impervious surfaces in existing urbanized areas such as pavements, parking lots, public squares are substituted by vegetated surfaces (e.g., bricks with gaps between them that allow the growing of herbaceous plants with edaphic biomass stored in root system of the plants) to minimize the impacts of soil sealing on storing C) (Tao *et al.* 2014). In other words, we have a set of subjects to be considered in a project aiming a territorial reorganization.

The electrical conductivity (EC) is a useful environmental descriptor, since it permits an integrate analysis of the role of the soil and land cover as potential retainer or exporter of ions, enabling to establish a budget among the rate of input of ions (by means of analysis of EC of the rain water), the stored ions in the soil and output/exportation by the superficial river network (Silva 2012). Comparatively, for urban watersheds in USA, values of EC were 30 times higher in the river water than in the rain water (Pellerin *et al.* 2008). For a rural area in Ibiuna (Brazilian Southeastern region) Silva *et al.* (2015a) reported increasing of two times. For the watershed here studied the augmentation was approximately 10 times (Fig. 3).

For the region of our studied watershed, major ion deposited through rain water is the potassium (Antunes *et al.* 2011), while major ions in the soil are (decreasing order) calcium > magnesium > potassium, regardless the land covers (Urban 2011). The same order was reported in the river water (Urban *et al.* 2010). Even taking into accounting the effect of dumping of domestic sewage into

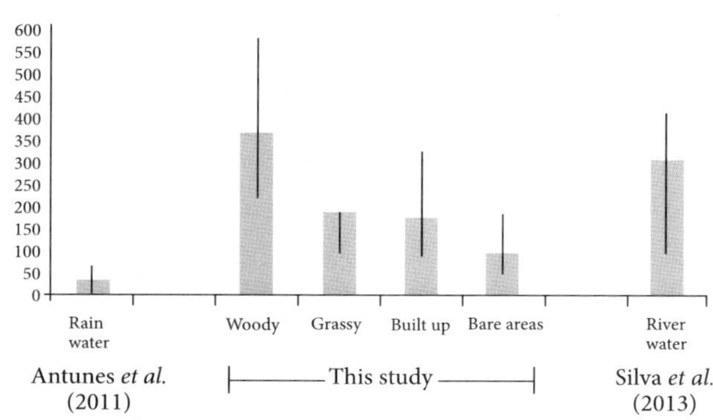

Fig. 3. Variation of values of electric conductivity (in μS/cm^{-1}) according to the compartments in study area

the streams of the area here studied (Silva *et al.* 2013), when we compare the data of EC of the soil and river water, we can infer that fewer ions are transported from terrestrial to aquatic systems when we compare soil under forested regions with the other anthropogenic land cover categories.

Although there are a considerable number of contaminated sites in urbanized regions of Latin America, especially in Sao Paulo State (Brazil) and such contaminated areas are mostly related with problems of gas stations (www.cetesb.sp.gov.br), in most urbanized places from Latin America, the main source of contamination in soils seems to be vehicular traffic (López *et al.* 2006; Moura *et al.* 2006). Hence, the necessity of a correct design of urban zoning is once again highlighted, in order to reach the equilibrium between development and conservation.

The effects of increasing of urbanized areas are far to be completely elucidated, meaning that urban soils constitute an open frontier for environmental research (Vrscaj *et al.* 2008). Hence, other studies like this one should be carried out and reliable data should be systematically surveyed. One of the reasons is that the effect is highly variable according to geographic region (Edmondson *et al.* 2011; Hagan *et al.* 2012), the environmental characteristics of the region (Baltrenas *et al.* 2010; Templer *et al.* 2012), as well as the modality and intensity of urbanization (Alberti 2005; Boeckx *et al.* 2006).

The urbanization cannot be treated as an indomitable phenomenon. Researchers have detected, and we confirmed in this study, which the main problem is not the urbanization, but the urbanization incorrectly designed and/or executed (Alphan, Güvensoy 2016). This is especially true for many megacities. Projects of urban expansion incorrectly formulated and/or implemented diminish the buffer capacity and resilience of the local ecosystem, and they do not guarantee the environmental sustainability. Regarding soils and water bodies, keep suitable forested areas in adequate places is an important mechanism of avoiding chemical unbalances and disruptions in the ecosystems, either by enrichment (increasing in concentration of products and materials, meaning contamination) or impoverishment (leaching, erosion).

Controlling the pollution is a critical issue that should be considered in projects of urban expansion, in order to avoid the emanation of contaminated areas and water bodies. The maintenance of a suitable design of the land cover also is an imperative factor in terms of making the soil a strategic compartment of storing C and others elements or even making the soil a truly source of pollutants, with implications for global biogeochemical cycles.

Indeed, appropriate generation and use of information is one of the pillars for a successful sustainable urbanization process. For example, analyzing many urban regions in Latin America through the Google Earth System (https://www.google.com/earth/), we clearly note a faint occurrence of forest patches in urban regions, as well as riparian vegetation along the river networks, and this appear to be a trend for urban regions of the continent, whatever is the size of the city.

Hence, projects considering re-vegetation of the riparian stretches (aiming to create riparian linear parks, for instance) should be prioritized in order to improve of the protection the water courses, incrementing the area forest-covered in the cities, improving the local climatic conditions, and contribute in the regulation of the N cycling in a watershed scale. In other words, the land use in urban regions needs to be more efficient, in order to let that the soil exerts, in fact, the ecosystem services and roles that the soils should exert. The data surveyed in study support this statement.

Conclusions

The urban-related land cover classes critically increased the soil bulk density and altered the quality of C of SOM. Urban-related land cover categories entail modifications that sometimes make the urban soils similar to soils from a rural region (for example soils under grassy sites). Other land cover categories produce specific features of urbanized regions (for example, soil beneath built-up regions).

The amount of C stored in the layer 0–20 cm of the soil for the entire watershed is almost a half of the total potential of the watershed and forested patches stored C in similar amount in relation to forest patches from non urbanized regions. This confirms that urban soils of the studied area are able to contribute in storing C if a project of land cover development is adequately implemented. The scenario of the studied watershed is not better because of the poorly designed land cover currently existing. To improve such design, natural soils should be preserved in strategic places where urban growth encroaches into pristine areas and new forested areas should be created following some principles suggested by landscape ecologists aiming to get the landscape ecologically more functional (Antrop 2004).

For our study area, the values of electrical conductivity of the soils are augmented in approximately 10 times when they are compared values among rainfall and river waters. Forested sites are retainers of ions, while other sites with human-induced land cover classes are exporters.

Soils from urban settlements should not be ignored in regional and global C evaluations, highlighting the urgent need to extend the measurement of C stocks to other urbanized regions of the Latin America. The maintenance or recovering of the soil quality is an issue that should be considered as priority in a process of (re)engineering of the territorial organization, in order to help mitigating the degrading effects of the (disorganized) urbanization.

Acknowledgements

This work was partially supported by the Brazilian agencies FAPESP (under grant number 2009/02182-3) and CAPES (grant number not provided).

References

Alberti, M. 2005. The effects of urban patterns on ecosystem function, *International Regional Science Review* 28: 168–192. https://doi.org/10.1177/0160017605275160

Alphan, H.; Güvensoy, L. 2016. Detecting coastal urbanization and land use change in Southern Turkey, *Journal of Environmental Engineering and Landscape Management* 24: 97–107. https://doi.org/10.3846/16486897.2015.1113976

Anderson, J. R. 1976. *A land use and land cover classification system for use with remote sensor data*, Vol. 964. US Government Printing Office.

Antrop, M. 2004. Landscape change and the urbanization process in Europe, *Landscape and Urban Planning* 67(1–4): 9–26. https://doi.org/10.1016/S0169-2046(03)00026-4

Antunes, M. L. P.; Antunes, R. H.; Fornaro, A.; Astolfo, R. 2011. Composição química da água de chuva e deposição atmosférica na cidade de Sorocaba (SP), in *26th Brazilian Congress of Environmental and Sanitation Engineering*, 2011, Porto Alegre Rio Grande do Sul, Brazil.

Balieiro, F. D. C.; Benites, V. D. M.; Caiafa, A. N.; Alves, B. J. R.; Fontana, A.; Canellas, L. P. 2012. Vegetation influence on organic matter source of black soils from high altitude rocky complexes traced by ^{13}C and ^{15}N isotopic techniques, *Catena* 99: 97–101. https://doi.org/10.1016/j.catena.2012.06.004

Baltrenas, P.; Pranskevičius, M.; Lietuvninkas, A. 2010. Investigation and assessment of dependences of the total carbon on pH in Neris regional park soil, *Journal of Environmental Engineering and Landscape Management* 18: 179–187. https://doi.org/10.3846/jeelm.2010.21

Barrales, C. C.; Bauer, L. I.; Santos, A. T.; González, R. C. 2007. Cadmio, plomo, níquel y zinc en suelos del Parque Desierto de los Leones, *Terra Latinoamericana* 25: 95–103.

Boeckx, P.; Meirvenne, M. van; Raulo, F.; Cleemput, O. 2006. Spatial patterns of $\delta^{13}C$ and $\delta^{15}N$ in the urban topsoil of Gent, Belgium, *Organic Geochemistry* 37: 1383–1393. https://doi.org/10.1016/j.orggeochem.2006.04.015

Boeckx, P.; Paulino, L.; Oyarzun, C.; Cleemput, O.; Godoy, R. 2005. Soil $\delta^{15}N$ patterns in old-growth forests of southern Chile as integrator for N-cycling, *Isotopes in Environmental and Health Studies* 41: 249–259. https://doi.org/10.1080/10256010500230171

Brady, N. C.; Weil, R. R. 2004. *Elements of the nature and properties of soils*. New Jersey, USA: Prentice Hall.

Burns, R.; Burns, R. 2008 *Business research methods and statistics using SPSS*. Extra chapter (number 25: Discriminant Analysis) [online], [cited 25 April 2013]. Sage Pub. Available from Internet: http://www.uk.sagepub.com/burns/website%20material/Chapter%2025%20-%20Discriminant%20Analysis.pdf

De Kimpe, C. R.; Morel, J. L. 2000. Urban soil management: a growing concern, *Soil Science* 165: 31–40. https://doi.org/10.1097/00010694-200001000-00005

Edmondson, J. L.; Davies, Z. G.; Gaston, K. J.; Leake, J. R. 2014. Urban cultivation in allotments maintains soil qualities adversely affected by conventional agriculture, *Journal of Applied Ecology* 51: 880–889. https://doi.org/10.1111/1365-2664.12254

Edmondson, J. L.; Davies, Z. G.; McCormack, S. A.; Gaston, K. J., Leake, J. R. 2011. Are soils in urban ecosystems compacted? A citywide analysis, *Biology Letters* 7: 771–774. https://doi.org/10.1098/rsbl.2011.0260

Edmondson, J. L.; Davies, Z. G.; McHugh, N.; Gaston, K. J.; Leake, J. R. 2012. Organic carbon hidden in urban ecosystems, *Scientific Reports* 2. https://doi.org/10.1038/srep00963

Farquhar, G. D.; Ehleringer, J. R.; Hubick, K. T. 1989. Carbon isotope discrimination and photosynthesis, *Annual Review of Plant Physiology* 40: 503–537. https://doi.org/10.1146/annurev.pp.40.060189.002443

Grau, H. R.; Aide, M. 2008. Globalization and land-use transitions in Latin America, *Ecology and Society* 13(2): 16 [online], [cited 23 September 2014] URL: http://www.ecologyandsociety.org/vol13/iss2/art16

Gregory, J. H.; Dukes, M. D.; Jones, P. H.; Miller, G. L. 2006. Effect of urban soil compaction on infiltration rate, *Journal of Soil and Water Conservation* 61: 117–124.

Guo, L. B.; Gifford, R. M. 2002. Soil carbon stocks and land use change: a meta-analysis, *Global Change Biology* 8: 345–360. https://doi.org/10.1046/j.1354-1013.2002.00486.x

Hagan, D.; Dobbs, C.; Timilsina, N.; Escobedo, F.; Toor, G. S.; Andreu, M. 2012. Anthropogenic effects on the physical and chemical properties of subtropical coastal urban soils, *Soil Use and Management* 28: 78–88. https://doi.org/10.1111/j.1475-2743.2011.00379.x

IBGE. 2013. *Cidades* [online], [cited 15 January 2015]. Federal Brazilian Institute for Geography and Statistics, Rio de Janeiro. Available from Internet: http://www.cidades.ibge.gov.br/xtras/home.php

IGSP. 2009. *Geological map of Sorocaba Municipality*: Report. Scale: 1:50,000. Geological Institute of São Paulo State. 59 p. (in Portuguese).

Jim, C. Y. 1998. Urban soil characteristics and limitations for landscape planting in Hong Kong, *Landscape and Urban Planning* 40: 235–249. https://doi.org/10.1016/S0169-2046(97)00117-5

Kahan, A. Y. 2008. *Nitrogen and Carbon biogeochemistry in soil and vegetation along an indirect urban-rural gradient in southeastern Michigan*: Dissertation. University of Michigan.

Kaye, J. P.; Groffman, P. M.; Grimm, N. B.; Baker, L. A.; Pouyat, R. V. 2006. A distinct urban biogeochemistry? *Trends in Ecology and Evolution* 21: 192–199. https://doi.org/10.1016/j.tree.2005.12.006

Kronka, F. J. N. *et al.* 2005. Monitoramento da vegetação natural e do reflorestamento no estado de São Paulo, in *Proceedings of XII Brazilian Symposia of Remote Sensing*, Goiânia, Goiás, Brazil, 1569–1576.

López, S. C.; Perelman, P. E.; Rivara, M.; Castro, M. A.; Faggi, A. 2006. Características del suelo y concentración de metales a lo largo de un gradiente de urbanización en Buenos Aires, Argentina, *Multequina* 15: 69–80.

Lorenz, K.; Lal, R. 2009. Biogeochemical C and N cycles in urban soils, *Environment International* 35: 1–8. https://doi.org/10.1016/j.envint.2008.05.006

Melfi, A. J.; Montes, C. R.; Carvalho, A.; Forti, M. C. 2004. Use of pedological maps in the identification of sensitivity of soils to acidic deposition: application to Brazilian soils, *Annals of Brazilian Academy of Sciences* 76: 139–145. https://doi.org/10.1590/S0001-37652004000100012

Moebius-Clune, B. N.; van Es, H. M.; Idowu, O. J.; Schindelbeck, R. R.; Kimetu, J. M.; Ngoze, S.; Lehmann, J.; Kinyangi, J. M. 2011. Long-term soil quality degradation along a cultivation chronosequence in western Kenya, *Agriculture, Ecosystems and Environment* 141: 86–99. https://doi.org/10.1016/j.agee.2011.02.018

Morgan, R. P. C. 2009. *Soil erosion and conservation*. John Wiley & Sons.

Moura, M. C. S.; Lopes, A. N. C.; Moita, G. C.; Neto, J. M. M. 2006. Estudo multivariado de solos urbanos da cidade de Teresina, *Química Nova* 29: 429–435. https://doi.org/10.1590/S0100-40422006000300005

Norra, S.; Handley, L. L.; Berner, Z.; Stüben, D. 2005. ^{13}C and ^{15}N natural abundances of urban soils and herbaceous vegetation in Karlsruhe, Germany, *European Journal of Soil Sciences* 56: 607–620. https://doi.org/10.1111/j.1365-2389.2005.00701.x

Pellerin, B. A. *et al.* 2008. The application of electrical conductivity as a tracer for hydrograph separation in urban catchments, *Hydrological Processes* 22: 1810–1818. https://doi.org/10.1002/hyp.6786

Pickett, S. T. A.; Cadenasso, M. L. 2009. Altered resources, disturbance, and heterogeneity: a framework for comparing urban and non-urban soils, *Urban Ecosystems* 12: 23–44. https://doi.org/10.1007/s11252-008-0047-x

Poeplau, C.; Don, A.; Vesterdal, L.; Leifeld, J.; Van Wesemael, B. A. S.; Schumacher, J.; Gensior, A. 2011. Temporal dynamics of soil organic carbon after land use change in the temperate zone-carbon response functions as a model approach, *Global Change Biology* 17: 2415–2427. https://doi.org/10.1111/j.1365-2486.2011.02408.x

Pouyat, R. V.; Groffman, P.; Yesilonis, I. D.; Hernandez, L. 2002. Soil carbon pools and fluxes in urban ecosystems, *Environment Pollution* 116: 107–118. https://doi.org/10.1016/S0269-7491(01)00263-9

Pouyat, R. V.; Yesilonis, I. D.; Nowak, D. J. 2006. Carbon storage by urban soils in the United States, *Journal of Environmental Quality* 35: 1566–1575.

Pouyat, R. V.; Yesilonis, I. D.; Russell-Anelli, J.; Neerchal, N. K. 2007. Soil chemical and physical properties that differentiate urban land-use and cover types, *Soil Science Society of America Journal* 71: 1010–1019. https://doi.org/10.2136/sssaj2006.0164

Quraishi, M. Z.; Mouazen, A. M. 2013. A prototype sensor for the assessment of soil bulk density, *Soil and Tillage Research* 134: 97–110. https://doi.org/10.1016/j.still.2013.07.011

Raij, B.; Andrade, J. C.; Cantarella, H.; Aquaggio, J. A. 2001. *Análise química para avaliação da fertilidade de solos tropicais*. Instituto Agronômico de Campinas, Campinas. 284 p.

Robinson, D. 2001. d^{15}N as an integrator of the nitrogen cycle, *Trends in Ecology and Evolution* 16: 153–162. https://doi.org/10.1016/S0169-5347(00)02098-X

Ruggiero, P. G. C.; Batalha, B. A.; Pivello, V. R.; Meirelles, S. T. 2002. Soil vegetation relationships (Brazilian savannah) and semi deciduous forest, Southeastern Brazil, *Plant Ecology* 160: 1–16. https://doi.org/10.1023/A:1015819219386

Saritha, V.; Vuppala, N. S.; Prashanthi, K.; Anjum, A. 2014. Soil properties governed by Municipal Solid Waste-Contemporary and enduring, *Agriculture and Soil Sciences* 1: 42–49.

Scalenghe, R.; Marsan, F. A. 2009. The anthropogenic sealing of soils in urban areas, *Landscape and Urban Planning* 90: 1–10. https://doi.org/10.1016/j.landurbplan.2008.10.011

Scharenbroch, B. C.; Lloyd, J. E.; Johnson-Maynard, J. L. 2005. Distinguishing urban soils with physical, chemical, and biological properties, *Pedobiologia* 49: 283–296. https://doi.org/10.1016/j.pedobi.2004.12.002

Seifi, M. R.; Alimardani, R.; Sharifi, A. 2010. How can soil electrical conductivity measurements control soil pollution?. *Research Journal of Environ Earth Science* 2: 235–238.

Silva, A. M. 2012. *Efeito da mudança da cobertura da terra para fins de urbanização na quantidade e qualidade do C e do N da faixa superficial dos solos de uma microbacia em Sorocaba – SP*: Habilitation for Adjunct Professor Thesis. São Paulo State University.

Silva, A. M.; Manfre, L. A.; Monteiro, T.; Moreira, M. Z.; Livingston, S.; Ducatti, C.; Brossard, M.; Sobral, A. C.; Urban, R. C. 2015a. Vulnerability of soil and its associated organic matter to degradation in a subtropical region: A study from São Paulo State, Brazil, *Environmental Quality Management* 24: 65–90. https://doi.org/10.1002/tqem.21402

Silva, A. M.; Manfre, L. A.; Urban, R. C.; Silva, V. H. O.; Manzatto, M. P.; Norton, L. D. 2015b. Organic farm does not improve neither soil, or water quality in rural watersheds from southeastern Brazil, *Ecological Indicators* 48: 132–146. https://doi.org/10.1016/j.ecolind.2014.07.044

Silva, A. M.; Nogueira, D. P.; Ikematsu, P.; Silveira, F. M.; Bomback, M.; Alves, S. H.; Paula, F. P.; Camargo, P. B. 2009. Carbon stocks and isotopic composition of the organic matter in soils covered by native vegetation and pasture in Sorocaba, SP, Brazil, *International Journal Environmental Research* 3: 435–440.

Silva, A. M.; Santos, A. R.; Fernandes, R. A.; Urban, R. C.; Carvalho, R. M.; Manfre, L. A.; Durrant, S. F. 2013. Hydrosedimentologic imbalance in an urbanized watershed, *Acta Limnologica Brasiliensia* 25: 140–149. https://doi.org/10.1590/S2179-975X2013000200005

Tao, Y.; Li, F.; Wang, R.; Zhao, D. 2014. Effects of land use and cover change on terrestrial carbon stocks in urbanized areas: a study from Changzhou, China, *Journal of Clean Production* 103: 651–657. https://doi.org/10.1016/j.jclepro.2014.07.055

Templer, T. H. *et al.* 2012. Sinks for nitrogen inputs in terrestrial ecosystems: a meta-analysis of ^{15}N tracer field studies, *Ecology* 93: 1816–1829. https://doi.org/10.1890/11-1146.1

United Nations (UN). 2014. *World Urbanization Prospects* (highlights) [online], [cited 3 July 2014]. Available from Internet: http://esa.un.org/unpd/wup/Highlights/WUP2014-Highlights.pdf

Urban, R. C. 2011. *Análise da relação entre a qualidade do solo, a qualidade da água e a fragilidade ambiental de uma microbacia de ocupação urbana crescente com relictos de vegetação natural localizada em Sorocaba – SP*: Dissertation. São Paulo State University.

Urban, R. C.; Silva, A. M.; Manfré, L. A. 2010. Ionic condition and input and output rates of some chemical solutes in Lavapés stream, Sorocaba, SP, Brazil, *Revista Ambiente & Água* 5: 265–277 (in Portuguese). https://doi.org/10.4136/ambi-agua.168

Vagen, T. G.; Walsh, M. G.; Sheperd, K. D. 2006. Stable isotopes for characterization of trends in soil carbon following deforestation and land use change in the highlands of Madagascar, *Geoderma* 135: 133–139. https://doi.org/10.1016/j.geoderma.2005.11.012

Vasenev, V. I.; Stoorvogel, J. J.; Vasenev, I. I. 2013. Urban soil organic carbon and its spatial heterogeneity in comparison with natural and agricultural areas in the Moscow region, *Catena* 107: 97–102. https://doi.org/10.1016/j.catena.2013.02.009

Vela Correa, G.; López Blanco, J.; Rodríguez Gamiño, M. D. L. 2012. Niveles de carbono orgánico total en el Suelo de Conservación del Distrito Federal, centro de México, *Investigaciones geográficas* 77: 18–30.

Vrscaj, B.; Poggio, L.; Marsan, F. A. 2008. A method for soil environmental quality evaluation for management and planning in urban areas, *Landscape and Urban Planning* 88: 81–94. https://doi.org/10.1016/j.landurbplan.2008.08.005

Wang, Z.; Cui, X.; Yin, S.; Shen, G.; Han, Y.; Liu, C. 2013. Characteristics of carbon storage in Shanghai's urban forest, *Chinese Science Bulletin* 58: 1130–1138. https://doi.org/10.1007/s11434-012-5443-1

World Bank (WB). 2014. *World databank indicators* [online], [cited 5 June 2014]. Available from Internet: http://data.worldbank.org/indicator

Alexandre Marco da SILVA. Department of Environmental Engineering – Campus Sorocaba – São Paulo State University, Av. Três de Março, 511. Sorocaba – SP – Brazil.
Doctor. Professor at Campus of Sorocaba – Unesp. 64 published papers. Distinguished scholar of the Brazilian Council for Research and Technology (CNPq).

Rodrigo Custodio URBAN. Department of Sanitation and Environment – FEC – Campinas State University, Av. Albert Einstein, 951, Campinas – SP – Brazil. Doctor in Civil Engineering. Professor at Pontific Catholic University (Campinas, SP Brazil). He has 10 published papers.

Luiz Augusto MANFRÉ. Geoprocessing Laboratory – Polytechnic School – University of São Paulo – Avenue Professor Luciano Gualberto, Lane 3, 380, São Paulo, Brazil. Doctor in Civil Engineering. He has 13 published papers.

Michel BROSSARD. IRD, UMR 210 Eco&Sols, BP 64501, 34394 Montpellier cedex 5, France.
Doctor. Researcher IRD Montpelier – France. 53 published papers.

Marcelo Zacharias MOREIRA. Centre for Nuclear Energy in Agriculture (CENA – USP) – University of São Paulo, Av. Centenário, 303 – Piracicaba, SP, Brazil. Professor at CENA – USP, Piracicaba, SP, Brazil. He has currently 54 published papers.

9

DETECTING COASTAL URBANIZATION AND LAND USE CHANGE IN SOUTHERN TURKEY

Hakan ALPHAN, Laden GÜVENSOY

Department of Landscape Architecture, Cukurova University, Balcali Campus, 01330, Adana, Turkey

Abstract. One of the most important needs in contemporary landscape planning is quantitative land use/land cover (LU/LC) change information. The reason a strong emphasis is placed on landscape change information is that it serves as an ecological and geographical basis for preparing and implementing development plans in a more sustainable manner. Multi-temporal analysis of LU/LC changes on the Eastern Mediterranean coast of Turkey revealed that there is a marked preference for these areas primarily for building development. This paper demonstrates a methodology that relies on quantitative analysis techniques for assessing spatiotemporal changes in LU/LC in the case of the eastern Mediterranean coast of Turkey. In this respect, satellite image datasets (SPOT panchromatic, Landsat TM) acquired in 1989, 1995, 2001 and 2007 were enhanced. Resulting images were classified and compared to detect coastal urbanization and development trends. Post-classification change analyses were employed to quantify land cover conversions in three periods from 1985 to 1995, from 1995 to 2001 and from 2001 to 2007. This paper demonstrated that urban, agriculture and shrublands changed rapidly in this part of the Mediterranean coast.

Keywords: urbanization, coastal development, Mediterranean, land use, change detection.

Introduction

Urbanization in coastal regions is complex since rapid building development in these areas is generally driven by various land uses. Most often, residential development, industries and transportation appear as major forces that create changes. Change detection of urban land use/land cover (LU/LC) is one of the most extensively studied topics in landscape monitoring (Ramachandra *et al.* 2014; Yu *et al.* 2013; Xian *et al.* 2012; Sancar *et al.* 2009; Ayazli *et al.* 2010). Utility and importance of urban LU/LC change detection based on multi-temporal remotely sensed imagery is widely accepted (Stefanov *et al.* 2001; Kamh *et al.* 2012; Schneider 2012). This importance arises from the fact that LU/LC information can be used as an indicator for understanding human interactions in the environment and for assessing implications of change trends on current and future states of ecosystem health and functioning (e.g., Desalegn *et al.* 2014). Urbanization has serious impacts on economic development (Walker 2001). It also affects biogeochemical processes (Zhang *et al.* 2014) and causes environmental pollution (Fang *et al.* 2014).

Since urbanization interferes with many ecological systems and processes, geospatial information on its implications is essential for land use planning and management.

Various techniques and datasets have been used for mapping urban land cover and its change. Techniques based on image radiometry provide satisfactory results in classifying urban land cover and detecting related change phenomena. In this respect, multi-resolution segmentation and object based classifiers have been proven to be useful (Smiraglia *et al.* 2014). Sub-pixel image analysis methods and the hybrid approaches that combine pixel- and object-based classifications help to increase the efficiency of urban land cover mapping procedures (Yang *et al.* 2003; Villa *et al.* 2012). Change vector analysis and data mining of dense time stacks may be given as examples of change detection approaches that have been reported to be useful for conceptualizing urban change (He *et al.* 2011; Schneider 2012).

Many of the urban LU/LC change studies use medium-resolution images such as Landsat (Alberti *et al.* 2004) or SPOT (Chang *et al.* 2010) multispectral images. Combinations of images acquired by different aerial and/or satellite sensors (e.g., Landsat TM/ETM+ and Terra/ASTER) are also frequently used (Kamh *et al.* 2012;

Corresponding author: Hakan Alphan
E-mail: alphan@cu.edu.tr

Merlotto *et al.* 2012). Spatial and spectral characteristics of these images are generally satisfactory for urban studies. However, depending on the extent, spatial configuration and history of the development, older images with higher spatial and spectral characteristics may sometimes be necessary. In such cases, merging these image datasets (Kamh *et al.* 2012; Deng *et al.* 2005) is useful as it leads to spatially enhanced images. The main advantage of this approach is to provide images with increased spatial and spectral capabilities that allow easier and more accurate classification and mapping. Several authors adopted this approach as an image enhancement procedure using multispectral and panchromatic images from the same remote sensing platform such as SPOT (Shaban, Dikshit 2002) or IRS (Ahmad, Singh 2002). Several other authors also merged SPOT with Landsat images (Nunez *et al.* 1998, 1999; Zhou *et al.* 1998; Hosomura, Katho 1998; Aiazzi *et al.* 1996; Chavez *et al.* 1991; Wu *et al.* 2004)

As summarized above, urban development attracts attention as it is an important environmental indicator, which can be taken into consideration in environmental planning and management processes (Alphan, Sonmez 2015; Hepcan 2013; Salvati, Sabbi 2011; Miralles i Garcia 2007; Prem 2007). Datasets and analysis protocols for analyses are dependent upon both planning needs and landscape characteristics. Data type and change analysis procedures for use in a fragmented and/or rapidly changing landscape may differ from those to be used in a homogenous and more stable landscape. Detecting changes in landscapes of spatial and temporal heterogeneity require higher spatial detail and frequent observations (Kowe *et al.* 2015; Liu, Yang 2015; Taubenbock *et al.* 2012; Yiran *et al.* 2012).

Urbanization on the eastern Mediterranean coast of Turkey has occurred rapidly and it created serious environmental implications such as water pollution and destruction of agricultural areas.

Residential building development is an important phenomenon along the coast. This is mostly driven by tourism- and recreation-related reasons that have created very dense vertical development of large multi-storey buildings since 1980s (Alphan, Sonmez 2015). This part of the Turkish Mediterranean coast is highly accessible by highways from populated Mediterranean and Southeast Anatolian cities such as Adana, Kahramanmaraş and Gaziantep. As being one of the closest and easily accessible sea coasts to above mentioned centres, study area has attracted domestic tourism, which resulted with development of large apartment blocks used as summer apartments.

The fact that building development takes place on fertile agricultural lands is one of the most important environmental concerns. Therefore, mapping and monitoring of these trends hold a potential to demonstrate spatial behaviour of development, which may be considered as a key input for suitability analyses (e.g., suitability of land areas for building development) and land use change modelling (e.g., CA Markov Model). This information is critically important for development plans that aim to achieve an optimum spatial organization for coastal land uses.

The aim of this study is to analyse LU/LC change trends in the west of Mersin, one of the most rapidly developing coastal zones in the Mediterranean region of Turkey. In this respect, digital image enhancement and interpretation techniques were employed in order to demonstrate change processes and provide qualitative and quantitative information for development planning.

1. Materials and methods

1.1. Study area

The study area is situated in the west of the city of Mersin, a populated urban centre in the Eastern Mediterranean region of Turkey (Fig. 1). Typical Mediterranean climate with dry summers and mild winters prevail in the region. Agriculture and residential development are the most prominent land uses. The area has experienced various environmental problems due to development of multi-storey buildings serving as summer apartments (Alphan, Derse 2013). Most of these buildings occupy a narrow coastal strip that was transformed from prime agricultural areas during the last three decades.

1.2. Data sets

The history of the urbanization and the biophysical characteristics of landscapes in study area make it necessary to work over a long time span with the images of high spatial resolution. For this reason, satellite image datasets (SPOT panchromatic and Landsat TM) acquired in a time span between 1989 and 2007 were used for image classification and change detection. As a commercial high-resolution optical imaging Earth observation satellite system that operates from space, SPOT is short for Satellite Pour l'Observation de la Terre (Satellite for observation of Earth). SPOT program was initiated by the French space agency in the 1970s (CNES-Centre National d'Études Spatiales). Panchromatic and multispectral instruments on board the SPOT satellites have been acquiring high-resolution images of Earth since February 1986, following the launch of the SPOT 1 satellite (10 m Panchromatic, 20 m multispectral). SPOT datasets are particularly important for change studies that date back to 1980's and that require high spatial resolution.

The Landsat Program is the longest running project that aims to acquire satellite images of the Earth's surface. Several instruments (RBV, MSS, TM) carried by Landsat satellites 1-5 have provided images of the Earth. Landsat

7 and 8 that carry ETM+ and OLI instruments are still operational. Millions of images acquired by these instruments are a unique resource for global change research and applications in agriculture, cartography, geology, forestry and regional planning. The fact that, on April 21, 2008, the USGS announced plans to provide all archived Landsat scenes at no charge to all users makes Landsat datasets even more important for change studies.

Combined use of these image datasets for analysis in landscapes of high spatial and temporal diversity has a potential to provide increased spatial and spectral capabilities. For this reason, four SPOT panchromatic and four near-anniversary Landsat TM datasets were selected for data fusion. Acquisition dates of these images are 1987 (December), 1995 (May/June), 2001 (May, June) and 2007 (September/November). Change analyses that employ spectral band comparisons (e.g., image ratioing) on bi-temporal basis require images of the same season in order to avoid false interpretations during comparisons. However, post-classification approaches are not dependent upon seasonal consistency since images of the earlier and later dates are classified separately. Considering different season of image acquisitions, post-classification classification was performed.

1.3. Image fusion

LU/LC mapping procedures based on panchromatic imagery sometimes require enhancement operations such as data fusion in order to create images with higher quality and more information. In this approach, the main idea arises from the fact that panchromatic images have much finer resolution than do the multispectral images. Therefore, panchromatic and multispectral images from the same or different platforms are merged to convey more information on the resulting image. Several methods have been developed for this purpose. Commonly used methods include red-green-blue (RGB) to intensity-hue-saturation (HIS) (e.g., Welch, Ehlers 1987), principal components analysis (PCA) (e.g., Chavez *et al.* 1991) and Wavelet (e.g., Wu *et al.* 2004) transformation techniques. Sarp (2014) reported that PCA provides the highest spectral quality among other techniques such as IHS, Gramm-Schmidt and Brovey transformations.

SPOT panchromatic images with 10-meter spatial resolution are greyscale images covering visible proportion of the electromagnetic spectrum. These images provide information on visual variables such as size, shape, hue, texture, orientation, etc. Higher spatial resolution of panchromatic images has its advantages for object identification in terms of several variables such as size, shape and texture. However, limited spectral information contained in these datasets generally requires pre-processing for image enhancement. Image fusion combines relevant information (e.g. spectral and spatial information) from geo-referenced input images into a single image.

Since image fusion requires datasets with high spatial consistency, image datasets in this study were geometrically corrected and registered to UTM projection. During

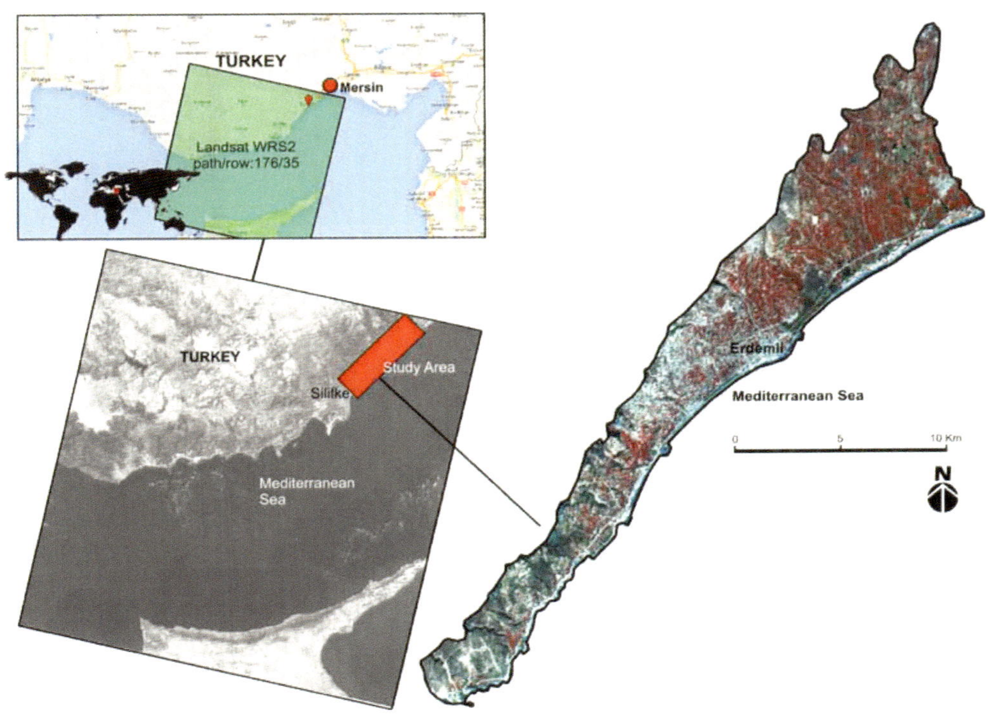

Fig. 1. Location of the study area

this process, resampling was done with RMS errors less than one pixel to obtain high spatial consistency.

Principal components merge were used for spatial enhancement. In this process it is assumed that the first principal component (PC) contains only overall scene luminance; all inter-band variation is contained in the other five PCs, and scene luminance in the short wave infrared (SWIR) bands is identical to visible scene luminance (Erdas 1999). Examples of high resolution panchromatic and multispectral input images and the results of this operation are given in Figure 2.

As shown in Figure 2, data merging operation substantially increased spatial quality of images. This is particularly important for discriminating small and scattered building patches from surrounding environment and for identifying agricultural patches, which can be confused with other land cover features such as vegetated (i.e. shrubland) or non-vegetated (i.e. bare areas) natural areas in input images.

1.4. Object-based image classifications

An object-based methodology was adopted for image classification. According to this methodology, image segmentation and classification were performed for each of the images of 1989, 1995, 2001 and 2007.

Segmentation is a process that groups adjacent pixels of a homogeneous spectral similarity. Specifically, during this process a watershed delineation approach is employed to partition input imagery based on their variance. A derived variance image is treated as a surface image allocating pixels to particular segments based on variance similarity. Across space and over all input bands, a moving window assesses this similarity and segments are defined according to a stated similarity threshold. Various similarity tolerances and other factors (e.g. weight mean and variance) were experimented to produce most representative segmentation results. During segmentation, a 3x3 window size was used. Weight mean and variance factors were 0.3. A similarity tolerance of 20 was set during this procedure

Segments can then be classified using a majority rule classifier based on a majority class within a segment (Eastman 2012). Thematic classifications were based on segmentation results. Sets of training segments were created for each of the dates in order to define spectral signatures prior to Maximum Likelihood classification. The classified maps were validated and compared on a bi-temporal basis. Post-classification change analyses were employed to quantify land cover conversions in three periods from 1985 to 1995, from 1995 to 2001 and from 2001 to 2007.

During segmentation and classification, "Segmentation" and "Segclass" modules of Idrisi 17.0 (The Selva Edition) was used. Flow of the classification procedure is given in Figure 3.

Fig. 2. Examples of merged images that show buildings and agriculture: (a) SPOT Panchromatic, (b) Landsat TM, (c) merged image

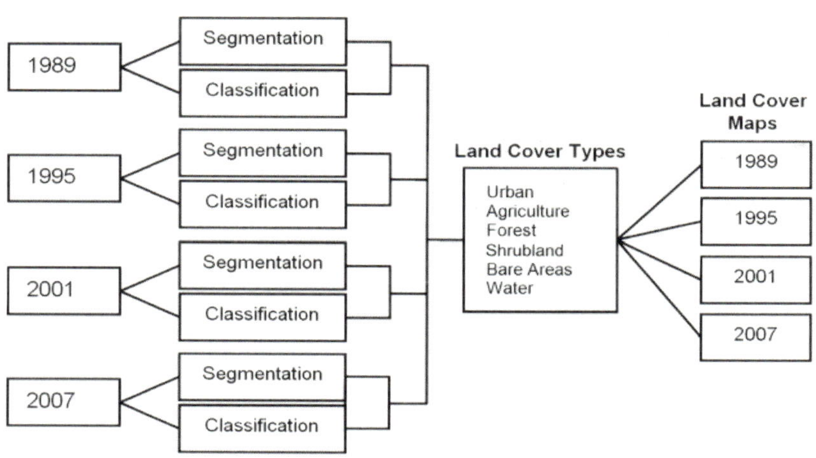

Fig. 3. Flow of the classification procedure

1.5. Accuracy assessment

Accuracies of individual classifications were evaluated in order to validate the results. For this purpose, overall and Kappa accuracies were calculated using reference information that included 256 pixels of reference for each of the individual dates. Stratified random sampling technique was used for selecting reference pixels. Minimum number of sampling points was 16.

2. Results and discussion

2.1. Land cover mapping

Images of 1989, 1995, 2001 and 2007 were pre-processed and classified to create land cover maps (Fig. 4).

Resulting maps in Figure 4 were investigated and compared. The most prominent change phenomena in the study area were (1) conversions from/to agriculture areas, and (2) urbanization. Characteristics of these changes can be summarized as follows:

(1) Agriculture-related changes such as agricultural encroachment over natural areas is predominantly observed in N/NW of the study area. As shown in Figure 4, expansion of agriculture over shrubland, resulted in a decrease of this cover type. But more importantly, shrubland cover became more fragmented after 1995. Conversions from agriculture to urban areas were also observed on the coast.

(2) Rapid urbanization is evident in the NE coast of the study area. Figure 4 suggests that these changes are more dramatic between 1989 and 2001.

2.2. Validating classification results

The results that indicate the level of consistency between observed (i.e., reference data) and estimated (i.e., land cover maps) land cover is given in Table 1.

Table 1. Accuracies for individual classifications (OA: Overall accuracy (%); UA: user's accuracy (%); PA: producer's accuracy (%); K: Overall Kappa)

	1989		1995		2001		2007	
LU/LC Type	PA	UA	PA	UA	PA	UA	PA	UA
Urban	100	94.7	74.1	100	95.0	100	76.9	87.0
Agriculture	99.2	97.5	99.1	94.6	99.2	91.7	93.8	95.3
Forest	100	100	100	93.8	100	93.8	100	100
Shrubland	98.4	100	95.4	95.4	93.0	100	93.8	95.7
Bare	90.5	100	82.6	82.6	68.0	89.5	79.0	65.2
Water	100	100	100	95.0	100	100	100	90.0
	OA: 94.53 K: 0.9183		OA: 94.14 K: 0.9194		OA: 94.53 K: 0.9183		OA: 91.80 K: 0.8812	

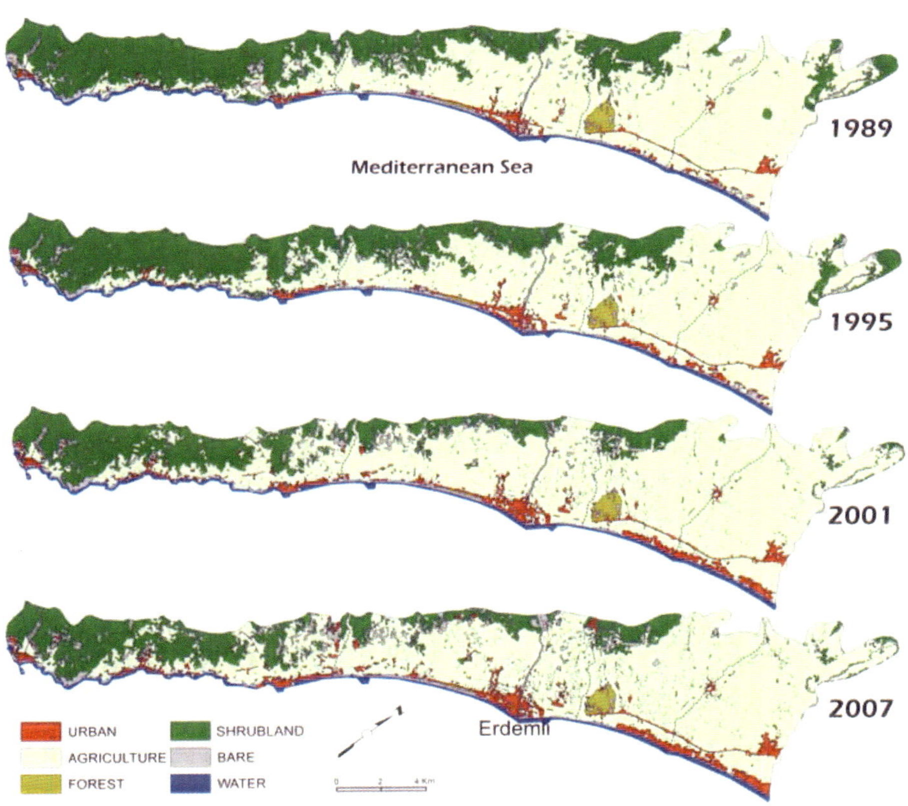

Fig. 4. Changes of LU/LC between 1989 and 2007

Inspection of the accuracy results showed that overall accuracies and Kappa values were over 90% in most cases (Table 1). Water surfaces had the highest accuracy values. Accuracy calculations suggested that urban areas were effectively mapped. Accuracies for urban areas in 2007 classification are slightly lower than those for the other dates but still around 90%. Buildings in the study area occur either as small and scattered or relatively larger aggregated patches. These pattern types are highly variable across space and time. High classification accuracies showed that, despite this variation, use of enhanced images provided opportunities for mapping this variability effectively. The fact that an object-based approach was adopted also helped to produce very accurate classifications.

2.3. Quantifying from-to changes

Cross tabulations of the earlier and later date LU/LC statistics for three periods (1989–1995, 1995–2001, 2001–2007) are given in Tables 2, 3 and 4.

Analysis results showed that area coverage of various LU/LC types is disproportionate. Agriculture and shrubland are the most extensive LU/LC types in all time periods. Agriculture covered 8842 ha of land, which made up 57% of the study area (15378 ha) in 1989. This increased 9994.65 ha in 2007, corresponding to 65% of total area. Shrubland cover that occupied 4600 and 2877 ha of land in 1989 and 2007, respectively, has the second largest extent. Bare areas covered 769 ha, almost twice as large as urban coverage (424 ha) in 1989. However, this changed in 2007, due to the facts that urban areas increased almost

Table 2. Quantitative changes in Period 1 (1989–1995)

		1995 (ha)						
		Urban	Agriculture	Forest	Shrubland	Bare	Water	Total 1989
1989	Urban	424.4	0.0	0.0	0.0	0.0	0.0	424.4
	Agriculture	113.2	8761.8	0.0	2.0	22.5	0.0	8899.4
	Forest	0.0	0.0	144.6	0.0	0.8	0.0	145.4
	Shrubland	12.0	24.2	0.0	4587.7	31.4	0.0	4655.4
	Bare	35.6	55.8	0.0	10.7	667.3	0.0	769.4
	Water	0.1	0.7	0.0	0.0	0.0	483.6	484.4
	Total 1995	585.3	8842.4	144.6	4600.5	722.0	483.6	15378.4

Table 3. Quantitative changes in Period 2 (1995–2001)

		2001 (ha)						
		Urban	Agriculture	Forest	Shrubland	Bare	Water	Total 1995
1995	Urban	585.3	0.0	0.0	0.0	0.0	0.0	585.3
	Agriculture	156.4	8612.3	0.0	0.0	73.8	0.0	8842.4
	Forest	0.3	2.7	133.3	0.0	8.3	0.0	144.6
	Shrubland	21.8	1234.9	0.0	3078.4	265.3	0.0	4600.5
	Bare	64.5	240.4	0.0	0.0	417.1	0.0	722.0
	Water	0.2	0.0	0.0	0.1	0.0	483.4	483.6
	Total 2001	828.5	10090.3	133.3	3078.5	764.5	483.4	15378.4

Table 4. Quantitative changes in Period 3 (2001–2007)

		2007 (ha)						
		Urban	Agriculture	Forest	Shrubland	Bare	Water	Total (2001)
2001	Urban	828.5	0.0	0.0	0.0	0.0	0.0	828.5
	Agriculture	146.5	9865.6	0.0	42.8	35.4	0.0	10090.3
	Forest	1.5	0.3	130.5	0.0	1.0	0.0	133.3
	Shrubland	22.4	117.1	0.0	2834.1	104.8	0.0	3078.5
	Bare	22.8	11.6	0.0	0.0	730.1	0.0	764.5
	Water	6.3	0.0	0.0	0.0	0.0	477.1	483.4
	Total (2007)	1028.0	9994.7	130.5	2876.9	871.3	477.1	15378.4

2.5 fold and reached 1028 ha and that bare areas increased at a much slower rate and reached 871 ha.

Cross-tabulations in Tables 2–4 provide information on "from-to" conversions, which showed that most of the changes took place in agricultural and built-up areas. Shrubland and a few patches of pine forest make up the natural vegetation in the study area. Some of these forest patches are included in a natural site. A smaller percentage of forest cover in the coastal zone corresponds to the areas used as camping sites. These areas are strictly protected from building development. Therefore no development occurred in these areas during the study period. Small amount of the change observed in forest areas is outside of the protected locations in upper areas. Conversions from water also exist in cross tabulations. These were mostly due to land reclamation (i.e. land fill) projects on the coast.

As previously mentioned, agriculture-related changes comprised conversions from/to agriculture. Urbanization mostly took place at the expense of agriculture, bare areas and shrubland. Figure 5 shows change trends in the forms of conversions "from agriculture" and "to urban" areas.

As shown in Figure 5, urbanization on agriculture areas (i.e. conversions from agriculture to urban areas) exists in all periods. Although highest amount of this type of change is observed in Period 2 (1995–2001), it persisted in Period 3 (2001–2007) at similar rates. The fact that higher rates of urbanization takes place on agriculture areas is an important finding worthy of consideration and it requires attention in development planning for the region.

Other changes depicted in Figure 5, such as those from agriculture to shrubland and bare areas also increased. High rates of conversions from agriculture to bare areas in Period 2 should be interpreted carefully owing to the fact that seasonality impacted map comparisons (e.g., 1989–1995, 1995–2001, etc.) differently. Change detection for the Periods 1 and 3 were based on the comparisons of winter and early summer images. However, Period 2 was assessed using early summer images for both the earlier and later dates. This did not cause any particular problem during classifications except for the case of open-field agriculture that show contrasting spectral responses between seasons due to presence (or absence) of crops. As a result, some agriculture fields without crops were classified as bare areas. These locations appeared as agriculture-to-bare (or vice versa) conversions in winter-summer comparisons. Therefore, conversion between agriculture and bare areas is affected by seasonality. In some locations, even the comparisons of the same seasons suffered from confusion between agriculture and bare areas due to different crop types (e.g. different harvest times) or changes between annual and permanent crops over time.

In the light of above given explanations one may expect that agriculture-to-bare conversions in Figure 5 (left) may be less than estimated, since a proportion of this type of conversion is actually agriculture-to-agriculture (i.e., crops-to-bare soil after harvest) conversions. On the other hand, bare-to-urban conversions in Figure 5 (right) may be interpreted that a certain amount of bare areas in this conversion type is in fact agriculture. These suggest that destruction of

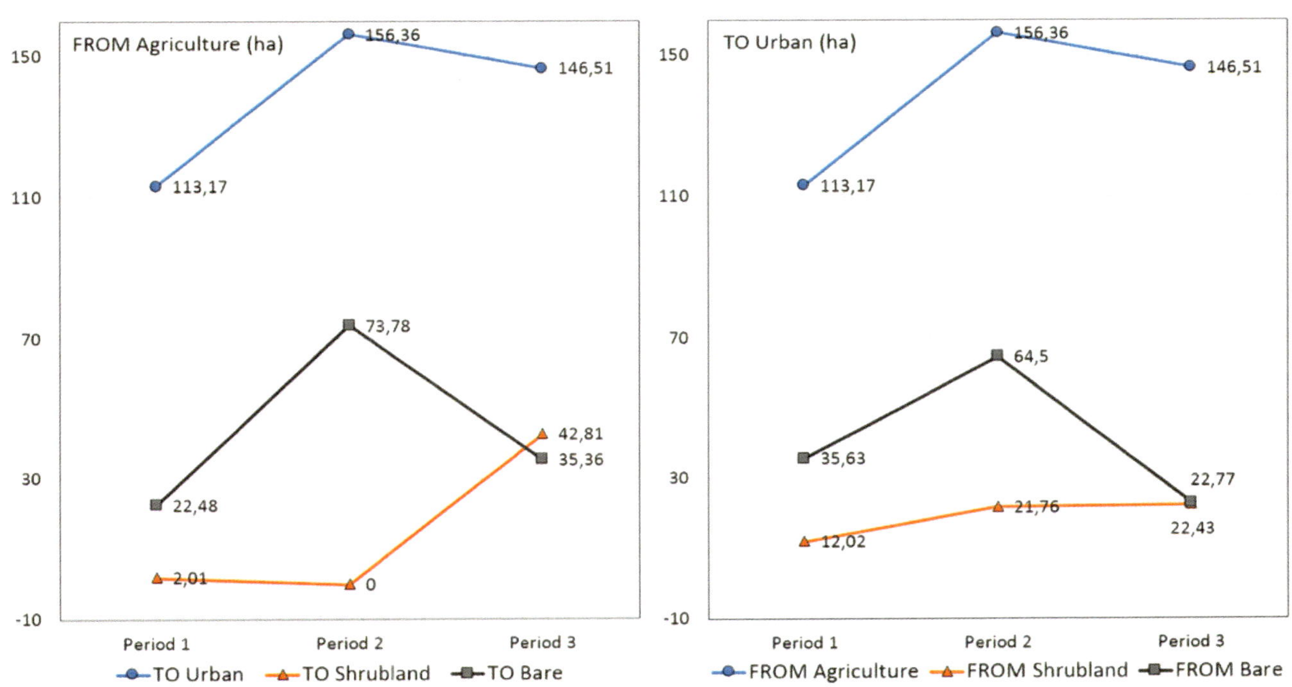

Fig. 5. Agriculture- and urban-related conversion trends across three time periods

agriculture areas by urbanization is even higher than calculated values due to overestimation of bare areas.

Conversion from agriculture to shrubland is due to land abandonment especially in upper areas, where limestone outcrops and patches of natural vegetation exist. Increasing amounts of conversions from shrubland to urban in Figure 5 (right) is the evidence that urbanization tends to grow in the west of the area, where shrubland dominates the landscape.

Alphan, Yilmaz (2005) reported that agriculture has increasingly grown over marginal areas and urban development occurred at the expense of prime croplands in the eastern Mediterranean coast of Turkey. Other parts of the Turkish Mediterranean coast such as Kusadasi and Edremit bay faced similar problems resulting from development (Kara *et al.* 2013; Irtem *et al.* 2005). Intense human uses also cause complex LU/LC changes around the Mediterranean basin (Salvati, Sabbi 2011). The analyses showed that study area is no exception in terms of diversity of the LU/LC and the magnitude of its change. Tourism is one of the main drivers for development on the coast. One of the easily recognizable impacts of building development on the coast is the change of visibility in coastal landscapes (Alphan, Sonmez 2015; Musaoglu *et al.* 2004). However, many other direct or indirect impacts such as those that are caused by land-based pollutants carried into the Mediterranean coastal zone also exist. Building development on the coast consume prime soils and it pushes agriculture towards marginal areas. This process is an important cause of agricultural intensification that may lead to biodiversity losses (Pekin 2013). Residential and touristic development is considered one of the key conservation challenges in Turkey. Significant conservation efforts have focused on protecting the species' nesting sites (Sekercioglu *et al.* 2011). Areas designated as archaeological and natural sites contribute to protecting important coastal areas from development. Analysis results in this study showed that such designations helped to protect coast from development. As it can clearly be seen in Figure 4, east coast of the town of Erdemli, which is a natural site, remained intact (i.e., no urban development between 1989 and 2007).

Despite the fact that several protection statuses control building development on the coast, development over unprotected coastal areas is still an important issue. Integrated management of coastal landscape requires a comprehensive inventory of biophysical features, landscape patterns and processes. Prem (2007) described main concepts and approaches of landscape management. Vulnerability analysis, landscape character assessment, modelling and simulation of change were described as some important topics for management. However, in absence of comprehensive information, priority areas may be determined using the areas of likely development suggested by future development models.

Several authors used remotely sensed images for landscape mapping and bi-temporal change detection purposes in coastal landscapes. For example, CORONA-ASTER or Landsat TM-ASTER image pairs were classified and compared (Hepcan 2013; Kesgin, Nurlu 2009) for change detection. However, this study used merged images with 10 m spatial resolution in order to maximize consistency between images and minimize possible classification and change detection errors due to resolution difference.

Conclusion

This paper dealt with tourism-oriented building development and LU/LC changes in a coastal zone, where land morphology and human preferences resulted with linear development with a scattered pattern of buildings. The need for higher resolution of spatial information that arises from spatial characteristics of this development made it necessary to employ image enhancement techniques. Therefore, SPOT panchromatic and Landsat images were integrated prior to image analyses (i.e., image classification and change detection).

Selection of seasonally consistent images is an important criterion for change detection in order to reduce potential negative impacts of different variables such as vegetation phenology, crop pattern, snow cover and soil wetness on image processing. However, it becomes increasingly difficult as a longer time span that includes more than two images is selected for change analyses.

Mapping the state of the LU/LC in this coastal zone in 1980's is of critical importance to effectively demonstrate magnitude and extent of change in 1980's due to the fact that integration of the country's economy into expanding world economy caused serious landscape-level environmental changes in this period. Therefore, an 18-year period between 1989 and 2007 was used in change detection to cover 1980's in change analyses. Time series included four images that are 6 years apart. In fact, the start of this period was the earliest available date for SPOT (panchromatic) imagery.

Images of the start and end dates of the 18-year study period are winter images, while other two images were early-summer images. For this reason, post-classification comparison approach was adopted for change detection. Unlike pre-classification, this approach relies on individual image classifications and pairwise comparisons. This approach eliminates difficulties arising from sensor differences or environmental variability, since every image is classified independently. However, some confusion between agriculture and bare areas still existed.

This paper demonstrated that urban, agriculture and shrublands changed rapidly in the Eastern Mediterranean coast of Turkey. Spatial configuration of LU/LC suggests that urban patches are likely to grow on agriculture areas.

It also shows that possible expansion of agriculture will take place over shrublands. As depicted in the LU/LC maps, shrublands became fragmented between 1989 and 2007. They may become even more fragmented due to agricultural expansion that creates important implications for biodiversity and wildlife. Development scenarios may be modelled and possible development areas can be taken as priority areas for further analyses in order to identify and protect valuable resources and contribute to sustainable management.

This paper used a methodology for analysing change trends resulting from development on the coast with spatially enhanced images. As demonstrated in this paper, SPOT panchromatic images can be incorporated with Landsat datasets to take the advantages of high spatial and spectral resolution in a single image dataset. Spatially enhanced images were successful for mapping the extent of developed areas and the other LU/LC types. The quantitative information derived from image classifications and change detection can be used as inputs for modelling future development and help to create development planning and resource management solutions.

Acknowledgments

This work was supported by Turkish Scientific and Technological Research Council (TUBITAK) under the grant number CAYDAG 111Y253.

References

Ahmad, R.; Singh, R. P. 2002. Comparison of various data fusion for surface features extraction using IRS PAN and LISS-III data, *Land Surface Characterization and Remote Sensing of Ocean Processes* 29(1): 73–78. http://dx.doi.org/10.1016/s0273-1177(01)00631-7

Aiazzi, B.; Alparone, L.; Baronti, S.; Carla, R. 1996. A pyramid approach to fusion of landsat TM and SPOT-PAN data to yield multispectral high-resolution images for environmental archaeology, *Remote Sensing for Geography, Geology, Land Planning, and Cultural Heritage* 2960: 153–162. http://dx.doi.org/10.1117/12.262459

Alberti, M.; Weeks, R.; Coe, S. 2004. Urban land-cover change analysis in Central Puget Sound, *Photogrammetric Engineering and Remote Sensing* 70(9): 1043–1052. http://dx.doi.org/10.14358/PERS.70.9.1043

Alphan, H. 2005. Perceptions of coastline changes in river deltas: southeast Mediterranean coast of Turkey, *International Journal of Environment and Pollution* 23(1): 92–102. http://dx.doi.org/10.1504/IJEP.2005.006398

Alphan, H.; Derse, M. A. 2013. Change detection in Southern Turkey using normalized difference vegetation index (NDVI), *Journal of Environmental Engineering and Landscape Management* 21(1): 12–18. http://dx.doi.org/10.3846/16486897.2012.663091

Alphan, H.; Sonmez, F. 2015. Mapping availability of sea view for potential building development areas, *Environmental Monitoring and Assessment* 187(7).

Alphan, H.; Yilmaz, K. T. 2005. Monitoring environmental changes in the Mediterranean coastal landscape: The case of Cukurova, Turkey, *Environmental Management* 35(5): 607–619. http://dx.doi.org/10.1007/s00267-004-0222-7

Ayazli, I. E.; Batuk, F.; Kleinschmit, B. 2010. Simulating landuse changes driven by a third bosphorus bridge, in *Special Joint Symposium of ISPRS Technical Commission IV & AutoCartoin conjunction with ASPRS/CaGIS 2010 Fall Specialty Conference*, 15–19 November 2010, Orlando, Florida.

Chang, N. B.; Han, M.; Yao, W.; Chen, L. C.; Xu, S. G. 2010. Change detection of land use and land cover in an urban region with SPOT-5 images and partial Lanczos extreme learning machine, *Journal of Applied Remote Sensing* 4.

Chavez, P. S.; Sides, S. C.; Anderson, J. A. 1991. Comparison of 3 different methods to merge multiresolution and multispectral data-landsat TM and SPOT panchromatic, *Photogrammetric Engineering and Remote Sensing* 57(3): 295–303.

Deng, J. S.; Wang, K.; Li, J.; Feng, X. L.; Huang, J. Q. 2005. Integration of SPOT-5 and ETM+ images to detect land cover change in urban environment, in *IGARSS 2005: IEEE International Geoscience and Remote Sensing Symposium*, 25–29 July 2005, Seoul, Korea. Vols 1–8, Proceedings, 2284–2286.

Desalegn, T.; Cruz, F.; Kindu, M.; Turrión, M. B.; Gonzalo, J. 2014. Land-use/land-cover (LULC) change and socioeconomic conditions of local community in the central highlands of Ethiopia, *International Journal of Sustainable Development & World Ecology* 21(5): 406–413.

Eastman, J. R. 2012. *IDRISI Selva Manual.* Manual Version 17.01. Clark Labs, Clark University. 324 p.

Erdas. 1999. *Field guide, revised and expanded.* 5th ed. Atlanta, Georgia: ERDAS, Inc. 672 p.

Fang, H.; Xu, Y.; Ye, Z.; Zhang, Z.; Pan, S.; Deng, L.; Luo, Z.; Chen, S. 2014. Impact of urbanization on nutrients and heavy metal pollution of Napahai Wetland, Shangri-La County, China, *International Journal of Sustainable Development & World Ecology* 22(2): 117–126. http://dx.doi.org/10.1080/13504509.2014.918908

He, C. Y.; Wei, A. N.; Shi, P. J.; Zhang, Q. F.; Zhao, Y. Y. 2011. Detecting land-use/land-cover change in rural-urban fringe areas using extended change-vector analysis, *International Journal of Applied Earth Observation and Geoinformation* 13(4): 572–585. http://dx.doi.org/10.1016/j.jag.2011.03.002

Hepcan, C. C.; 2013. Quantifying landscape pattern and connectivity in a Mediterranean coastal settlement: the case of the Urla district, Turkey, *Environmental Monitoring and Assessment* 185(1): 143–155. http://dx.doi.org/10.1007/s10661-012-2539-7

Hosomura, T.; Katho, K. 1998. Optical image data fusion by using intensity operation on HIS transformation, in *Igarss '98-1998 International Geoscience and Remote Sensing Symposium*, 6–10 July 1998, Seattle, WA, USA. Proceedings Vols 1–5, 1318–1319.

Irtem, E.; Kabdasli, S.; Azbar, N.; 2005. Coastal zone problems and environmental strategies to be implemented at Edremit Bay, Turkey, *Environmental Management* 36(1): 37–47. http://dx.doi.org/10.1007/s00267-004-0062-5

Kamh, S.; Ashmawy, M.; Kilias, A.; Christaras, B. 2012. Evaluating urban land cover change in the Hurghada area, Egypt, by using GIS and remote sensing, *International Journal of Remote Sensing* 33(1): 41–68. http://dx.doi.org/10.1080/01431161.2010.550331

Kara, B.; Esbah, H.; Deniz, B. 2013. Monitoring and analyzing land use/land cover changes in a developing coastal town: a case study of Kusadasi, Turkey, *Journal of Coastal Research* 29(6): 1361–1372. http://dx.doi.org/10.2112/JCOASTRES-D-11-00140.1

Kesgin, B.; Nurlu, E. 2009. Land cover changes on the coastal zone of Candarli Bay, Turkey using remotely sensed data, *Environmental Monitoring and Assessment* 157(1–4): 89–96. http://dx.doi.org/10.1007/s10661-008-0517-x

Kowe, P.; Pedzisai, E.; Gumindoga, W.; Rwasoka, D. T. 2015. An analysis of changes in the urban landscape composition and configuration in the Sancaktepe District of Istanbul Metropolitan City, Turkey using landscape metrics and satellite data, *Geocarto International* 30(5): 506–519. http://dx.doi.org/10.1080/10106049.2014.905638

Liu, T.; Yang, X. J. 2015. Monitoring land changes in an urban area using satellite imagery, GIS and landscape metrics, *Applied Geography* 56: 42–54. http://dx.doi.org/10.1016/j.apgeog.2014.10.002

Merlotto, A.; Piccolo, M. C.; Bertola, G. R. 2012. Urban growth and land use/cover change at Necochea and Quequen cities, Buenos Aires province, Argentina, *Revista De Geografia Norte Grande* 53: 159–176. http://dx.doi.org/10.4067/S0718-34022012000300010

Miralles i Garcia, J. L. 2007. The introduction of the sustainability in the procedures regulated for urban planning: the case of the last act on urbanism, territory and landscape in the Valencian Community (Spain), *Sustainable Development and Planning III* 102(1–2): 505–513. http://dx.doi.org/10.2495/sdp070482

Musaoglu, N.; Seker, D. Z.; Kabdasli, S.; Kaya, S.; & Duran, Z. 2004. Using remote sensing and GIS for the assessment of visual attributes: a case study of the south coastal zone of Turkey, *Fresenius Environmental Bulletin* 13(9): 854–859.

Nunez, J.; Otazu, X.; Fors, O.; Prades, A.; Pala, V.; Arbiol, R. 1999. Image fusion with additive multiresolution wavelet decomposition. Applications to SPOT plus Landsat images, *Journal of the Optical Society of America a-Optics Image Science and Vision* 16(3): 467–474. http://dx.doi.org/10.1364/JOSAA.16.000467

Nunez, J.; Otazu, X.; Fors, O.; Prades, A.; Pala, V.; Arbiol, R. 1998. Data fusion of SPOT and LANDSAT images using additive multiresolution wavelet decomposition, *Image and Signal Processing for Remote Sensing* 4(3500): 202–213. http://dx.doi.org/10.1117/12.331865

Pekin, B. K. 2013. Effect of widespread agricultural chemical use on butterfly diversity across Turkish provinces, *Conservation Biology* 27(6): 1439–1448. http://dx.doi.org/10.1111/cobi.12103

Prem, M. 2007. Landscape management in the Mediterranean, in *Medcoast 07: Eighth International Conference on the Mediterranean Coastal Environment,* 13–17 November 2007, Alexandria, Egypt 211–222.

Ramachandra, T. V. Bharath, S.; Bharath, A. 2014. Spatio-temporal dynamics along the terrain gradient of diverse landscape, *Journal of Environmental Engineering and Landscape Management* 22(1): 50–63. http://dx.doi.org/10.3846/16486897.2013.808639

Salvati, L.; & Sabbi, A. 2011. Exploring long-term land cover changes in an urban region of southern Europe, *International Journal of Sustainable Development and World Ecology* 18(4): 273–282. http://dx.doi.org/10.1080/13504509.2011.560453

Sancar, C.; Turan, S.O.; Kadiogullari, A. L. 2009. Land use-cover change processes in Urban fringe areas: Trabzon case study, Turkey, *Scientific Research and Essays* 4(12): 1454–1462.

Sarp, G. 2014. Spectral and spatial quality analysis of pan-sharpening algorithms: A case study in Istanbul, *European Journal of Remote Sensing* 47: 19–28. http://dx.doi.org/10.5721/EuJRS20144702

Schneider, A. 2012. Monitoring land cover change in urban and pen-urban areas using dense time stacks of Landsat satellite data and a data mining approach, *Remote Sensing of Environment* 124: 689–704. http://dx.doi.org/10.1016/j.rse.2012.06.006

Sekercioglu, C. H.; Anderson, S.; Akcay, E.; Bilgin, R.; Can, O. E.; Semiz, G.; *et al.* 2011. Turkey's globally important biodiversity in crisis, *Biological Conservation* 144(12): 2752–2769. http://dx.doi.org/10.1016/j.biocon.2011.06.025

Shaban, M. A.; Dikshit, O. 2002. Evaluation of the merging of SPOT multispectral and panchromatic data for classification of an urban environment, *International Journal of Remote Sensing* 23(2): 249–262. http://dx.doi.org/10.1080/01431160010007088

Smiraglia, D.; Rinaldo, S.; Ceccarelli, T.; Bajocco, S.; Salvati, L.; Ricotta, C.; Perin, L. 2014. A cost-effective approach for improving the quality of soil sealing change detection from Landsat imagery, *European Journal of Remote Sensing* 47: 805–819. http://dx.doi.org/10.5721/EuJRS20144746

Stefanov, W. L.; Ramsey, M. S.; Christensen, P. R. 2001. Monitoring urban land cover change: An expert system approach to land cover classification of semiarid to arid urban centers, *Remote Sensing of Environment* 77(2): 173–185. http://dx.doi.org/10.1016/S0034-4257(01)00204-8

Taubenbock, H.; Esch, T.; Felbier, A.; Wiesner, M.; Roth, A.; Dech, S.; 2012. Monitoring urbanization in mega cities from space, *Remote Sensing of Environment* 117: 162–176. http://dx.doi.org/10.1016/j.rse.2011.09.015

Villa, P.; Boschetti, M.; Bianchini, F.; Cella, F. 2012. A hybrid multi-step approach for urban area mapping in the Province of Milan, Italy, *European Journal of Remote Sensing* 45(2): 333–347. http://dx.doi.org/10.5721/EuJRS20124529

Walker, R. 2001. Urban sprawl and natural areas encroachment: linking land cover change and economic development in the Florida Everglades, *Ecological Economics* 37(3): 357–369. http://dx.doi.org/10.1016/S0921-8009(01)00178-1

Welch, R.; Ehlers, W. 1987. Merging multiresolution SPOT HRV and Landsat TM data, *Photogrammetric Engineering & Remote Sensing* 53(3): 301–303.

Wu, Y.; Li, M.; Liao, G. S. 2004. Multiresolution wavelet decomposition to merge Landsat TM and SPOT panchromatic data, in *3rd International Conference on Computational Electromagnetics and Its Applications*, 1–4 November 2004, Beijing, China.

Xian, G.; Homer, C.; Bunde, B.; Danielson, P.; Dewitz, J.; Fry, J.; Pu, R. L. 2012. Quantifying urban land cover change between 2001 and 2006 in the Gulf of Mexico region, *Geocarto International* 27(6): 479–497. http://dx.doi.org/10.1080/10106049.2011.652675

Yang, L. M.; Xian, G.; Klaver, J. M.; Deal, B. 2003. Urban land-cover change detection through sub-pixel imperviousness mapping using remotely sensed data, *Photogrammetric Engineering and Remote Sensing* 69(9): 1003–1010. http://dx.doi.org/10.14358/PERS.69.9.1003

Yiran, G. A. B.; Kusimi, J. M.; Kufogbe, S. K. 2012. A synthesis of remote sensing and local knowledge approaches in land degradation assessment in the Bawku East District, Ghana, *International Journal of Applied Earth Observation and Geoinformation* 14(1): 204–213. http://dx.doi.org/10.1016/j.jag.2011.09.016

Yu, X. Y.; Zhang, A. D.; Hou, X. Y.; Li, M. J.; Xia, Y. X. 2013. Multi-temporal remote sensing of land cover change and urban sprawl in the coastal city of Yantai, China, *International Journal of Digital Earth* 6: 137–154.

Zhang, F. T.; Zhang, J. Y.; Shao, J. X. 2014. Karst urban land use/cover change research - in Guiyang city as an example, *Environmental Engineering* 1–4(864–867): 2764–2767.

Zhang, P.; Tang, C.; MA, S.; Yuan, H.; Gao, H.; Fan, W. 2011. Using Markov chains to analyze changes in wetland trends in arid Yinchuan Plain, China, *Mathematical and Computer Modelling* 54(3–4): 924–930. http://dx.doi.org/10.1016/j.mcm.2010.11.017

Zhou, J.; Civco, D. L.; Silander, J. A. 1998. A wavelet transform method to merge landsat TM and SPOT panchromatic data, *International Journal of Remote Sensing* 19(4): 743–757. http://dx.doi.org/10.1080/014311698215973

Hakan ALPHAN. Dr, Professor at the Department of Landscape Architecture, Faculty of Agriculture, Cukurova University, Turkey. He gained his Landscape Architecture Ph.D. in 2004 from the Institute of Science, Cukurova University, Turkey. Publications: Author/co-author of over 60 scientific publications, including peer-reviewed journal papers, book chapters and papers in conference proceedings. Research interests: remote sensing and GIS, landscape-level environmental monitoring, change detection, landscape pattern and environmental processes.

Laden GÜVENSOY. She gained her M.Sc. from Institute of Science, Department of Landscape Architecture, Cukurova University, Turkey in 2014. Research interests: remote sensing and GIS.

10

ASSESSING URBAN GREEN SPACE DISTRIBUTION IN A COMPACT MEGACITY BY LANDSCAPE METRICS

Huilin LIANG[a], Di CHEN[b], Qingping ZHANG[a]

[a]*College of Landscape Architecture, Nanjing Forestry University, No. 159, Longpan Road, Nanjing, Jiangsu, China*

[b]*College of Architecture and Urban Planning, Tongji University, No. 1239, Siping Road, Shanghai, China*

Abstract. The pattern and structure of urban green space (UGS) plays a significant role in the landscape and ecological quality (LEQ) of UGS, especially in a compact city with limited space. Based on landscape metrics, this study proposes an innovative method to quantify the effects of UGS pattern and structure on LEQ. Taking Shanghai, China as the study area, we calculated all landscape-level spatial metrics in FRAGSTATS, used correlation analysis in SPSS for data reduction, and adopted factor analysis and cluster analysis to statistically analyze the metrics and assesse the LEQ of UGS. These methods bridge the research gap of UGS distribution assessment for LEQ value by landscape metrics. Results showed that new districts usually have higher LEQ of UGS than old towns. Of the 17 districts in Shanghai, Chongming has the highest LEQ of UGS and Hongkou has the lowest. For the UGS pattern and structure, the eight old towns are similar, in contrast to the new districts of Chongming and Pudong, which are more dissimilar than the other districts for LEQ of UGS. The findings could help compact cities having limited UGS to develop and achieve better LEQ.

Keywords: urban green space (UGS), landscape and ecological quality (LEQ), landscape pattern, compact city, landscape metric, environmental sustainability.

Introduction

Urban green space (UGS) is important for urban sustainability (Haq 2011), and provides cities a wide range of ecosystem services (Wolch *et al.* 2014). These spaces may support urban ecological integrity (Andersson 2006), provide food (Groenewegen *et al.* 2006), improve microclimate regulation (Neuenschwander *et al.* 2014), control pollution (Escobedo *et al.* 2011), filter air (Gill *et al.* 2007), clean water, attenuate noise, and replenish groundwater (Thompson 2002; Sherer 2003; James *et al.* 2009). Considering cities are becoming increasingly hotter, congested, crowded, and polluted (Blanco *et al.* 2009), which could be migrated by UGS, the environment benefits of UGS need more concern and attention for urban sustainability.

Most landscapes provide a multitude of functions (i.e., regulation, habitat, production, information and carrier functions), which have been divided into three types of values: ecological, social-cultural and economic value (de Groot 2006). Urban green space is a spatial structure that is related to various ecological processes (McGarigal, McComb 1995), and has spatial ecological effects of its landscape patterns and functions within its boundaries and beyond (Chang *et al.* 2011). As remnants of a cultural landscape with rich biodiverse habitats (Barthel *et al.* 2005), UGS provides significant contribution to ecosystem services (Goddard *et al.* 2010). UGS structure is crucial for biodiversity maintenance (Antrop 2005). Moreover, the spatial configuration of UGS significantly affects the magnitude of land surface temperature (Kong *et al.* 2014) and has been used frequently to assess the urban heat island effect (Chen *et al.* 2014). Due to accelerating urbanization in compact cities, urban planners have an imperative to make limited UGS address the deteriorating environmental quality and to provide greater ecological benefits by optimizing UGS distribution. To encourage sensible choices that promote sustainable development, information on the spatial distributions of landscape functions and services is needed (Hermann *et al.* 2014). The ecological value of UGS is determined by the integrity of habitat and regulation functions of UGS according to ecological metrics, such as diversity, complexity and rarity (de Groot *et al.*

Corresponding author: Qingping Zhang
E-mail: Zhangqingping2015@163.com

2003). Thus assessing UGS structure for landscape and ecological value using relative landscape metrics could promote better urban planning and improve development decision making for landscape protection and improvement (Buyantuyev *et al.* 2010; Wu *et al.* 2011).

A large body of research has tried to analyze and assess UGS structure and pattern based on spatial metrics. Some studies focused on the effects of UGS spatial pattern on land surface temperature by using landscape metrics (Li *et al.* 2012; Hamada *et al.* 2013; Li *et al.* 2013; Kong *et al.* 2014). Some studies focused on the correlation between biodiversity and UGS spatial character based on landscape metrics (Schindler *et al.* 2008; Kong *et al.* 2010; La Rosa *et al.* 2013; Walz 2015). Some studies focused on analyses of specific landscape patterns by using landscape metrics, such as UGS fragmentation (Tian *et al.* 2011; Fan, Myint 2014) and heterogeneity (Plexida *et al.* 2014). Some studies focused on the dynamics of UGS in urbanization using spatial metrics (Zhou, Wang 2011; Byomkesh *et al.* 2012; Qian *et al.* 2015). Compared with other softwares such as Patch Analyst (Rempel *et al.* 1999), APACK (Saura, Torne 2009) and QRULE (Gardner 1999), FRAGSTATS (McGarigal *et al.* 2002) is comprehensive, powerful, easy to use, and is the most widely used package for landscape pattern analysis (McGarigal, McComb 1995; Turner *et al.* 2001). Most of the relevant research used numerous landscape indicators to quantify UGS pattern and structure. However, few studies quantify the effects of UGS pattern and structure on landscape and ecological quality (LEQ); an exception is the research of Tian *et al.* (Tian *et al.* 2014) who selected some landscape indices to analyze the ecological quality of UGS landscape patterns in the compact city of Hong Kong.

To rectify the shortcomings of currently available assessment techniques of UGS landscape and ecological quality, we used multiple software to study the contribution that UGS pattern and structure makes to the LEQ of UGS. The study was conducted on Shanghai, a compact megacity. Our study had three objectives: (1) introduce a new approach for evaluating the contribution of spatial pattern and structure to LEQ of UGS; (2) use the new approach to analyze the structure and distribution of UGS in the compact city of Shanghai; (3) account for the correlation between UGS pattern and distribution and its influences, thereby assisting policy makers and planners to develop intervention programs in UGS planning and development.

1. Materials and methods

1.1. Study area

Shanghai (121°50′E, 31°40′N) is located at the mouth of the Yangtze River (Chang Jiang) (Fig. 1) and is one of the largest and most densely populated cities in China. The city covers an area of 634,050 ha, has a registered population of 18.6 million and consists of 17 districts, including eight old towns and nine new districts (Fig. 1). Belonging to the subtropical moist marine climate zone, the city experiences four distinct seasons and receives sufficient rainfall and large amounts of sunshine. Compared with the cities with more than 300 m² per capita green cover (Fuller, Gaston 2009), such as Lie'ge (Belgium), Oulu (Finland) and Valenciennes (France), the quantity of UGS in Shanghai is really scarce, as its per capita green cover was only 13.38 m² in 2014. Shanghai is a congested and compact megacity with meager ground UGS, and is in great need of enhancing the efficiency with which UGS provides environmental benefits.

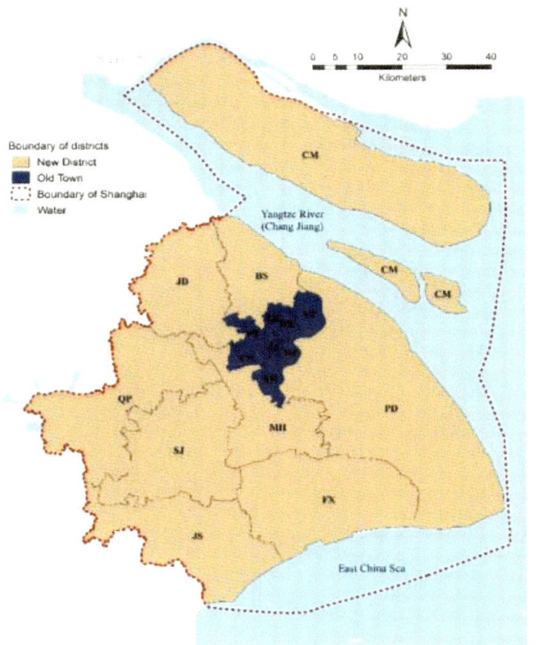

Abbreviation: Huangpu (HP); Xuhui (XH); Changning (CN); Jiang'an (JA); Putuo (PT); Zhabei (ZB); Hongkou (HK); Yangpu (YP); Pudong (PD); Minhang (MH); Baosha (BS); Jiading (JD); Jinshan (JS); Songjiang (SJ); Qingpu (QP); Fengxian (FX); Chongming (CM)

Fig. 1. Study area

1.2. Data preparation

The detailed methodology of this study is shown in Figure 2. The data used in this study were purchased from the Chinese government (http://ngcc.sbsm.gov.cn/). These data included satellite images acquired in 2014 (resolution of 0.5×0.5 m), high quality land-use maps (resolution of 0.5×0.5 m, compiled in 2014), land survey data, and district boundary maps. The UGS in this study included four categories: public green land, forest, garden and agriculture. The boundaries of different land uses (the four UGS categories, as well as water and built-up land) were manually digitized from the high quality land-use

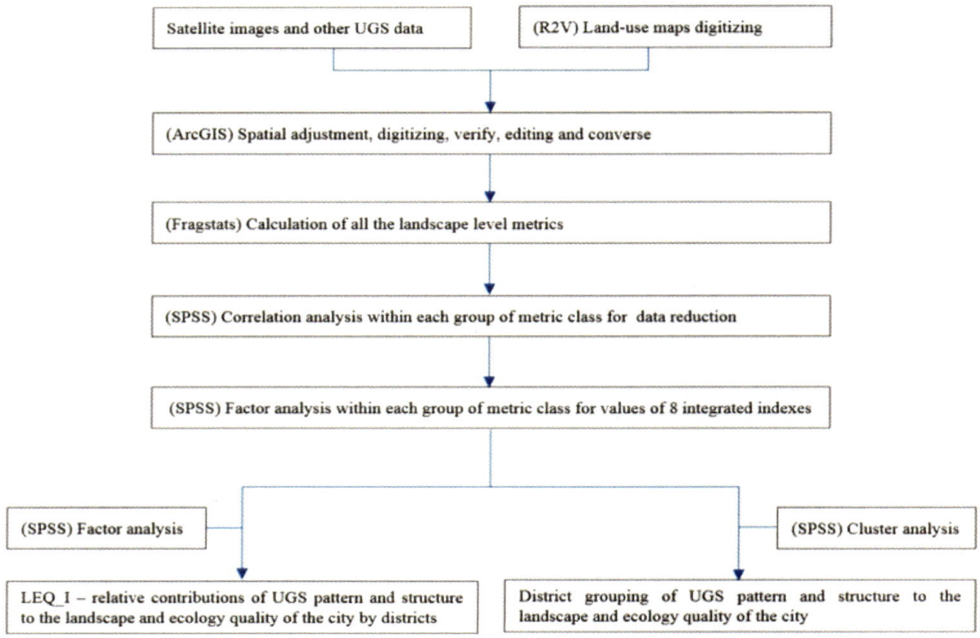

Fig. 2. Methodology chart

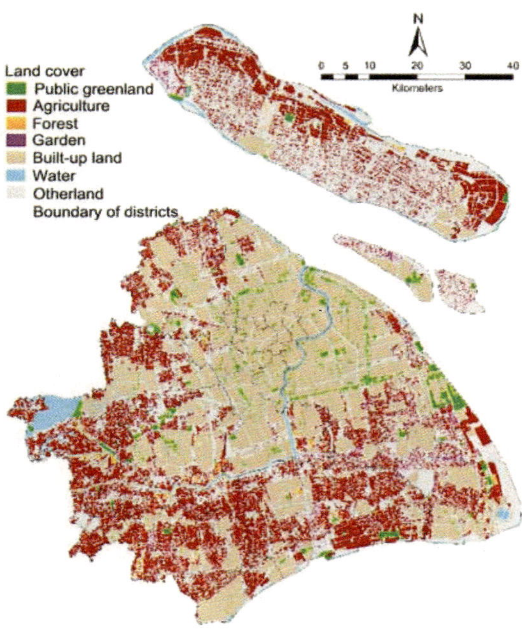

Fig. 3. Land use distribution in Shanghai

maps and labeled piece by piece using the software R2V for Microsoft Windows (version 5.5, Able Software Corp., Lexington, MA, USA) to output area feature files (shapefile). These area feature files were input to ArcGIS (version 10.2, ESRI, Redlands, CA, USA) software for spatial adjustment with reference to the satellite images. The satellite images were also visually interpreted and field trips were performed to check, modify, refine and verify the land use data digitized using R2V. All land use data digitized in ArcGIS were assigned a feature property of land use type (public green land, forest, garden, agriculture, water, built-up land or other land) and a corresponding type ID (Fig. 3). The combined "layer" with all seven land use data and different type IDs was converted from a feature to a raster format in ArcGIS and output as TIFF files of 17 districts with a cell size of 5 m, which was suitable for processing using the FRAGSTATS software.

1.3. Landscape metric calculation

For greater number of indices, the software of FRAGSTATS (version 4.2) was used in this research for landscape metric calculation. To quantify the landscape pattern and green structure of the city, all the 116 landscape metrics at the landscape level in FRAGSTATS were chosen to analyze the land use pattern of UGS by district. The metrics calculated include eight categories, e.g. area, shape, edge and contrast, core area, proximity, subdivision, contagion, and diversity (Table 1). Detailed information for the calculated metrics can be found elsewhere (McGarigal *et al.* 2012).

Table 1. Landscape metrics calculated in Fragststs (McGarigal *et al.* 2012)

Class	Acronym	Metric name
Area	TA	Total area
	LPI	Large patch index
	AREA	Mean patch area
	GYRATE	Radius of gyration
Shape	PAFRAC	Perimeter-area fractal dimension

End of Table 1

Class	Acronym	Metric name
	PARA	Perimeter-area ratio
	SHAPE	Shape index
	FRAC	Fractal dimension index
	CIRCLE	Related circumscribing circle
	CONTIG	Contiguity index
Edge & Contrast	TE	Total edge
	ED	Edge density
	CWED	Contrast-weighted edge density
	TECI	Total edge contrast index
	ECON	Edge contrast index
Core area	TCA	Total core area
	NDCA	Number of disjunct core area
	DCAD	Disjunct core area density
	CORE	Core area
	DCORE	Disjunct core area
	CAI	Core area index
Proximity	ENN	Euclidean nearest neighbor distance
	PROX	Proximity index
	SIMI	Similarity index
	CONNECT	Connectance index
Subdivision	NP	Number of patches
	PD	Patch density
	DIVISION	Landscape division index
	SPLIT	Splitting index
	MESH	Effective mesh size
Contagion/ Interspersion	CONTAG	Contagion
	IJI	Interspersion juxtaposition index
	PLADJ	Proportion of like adjacencies
	AI	Aggregation index
	LSI	Landscape shape index
	COHESION	Patch cohesion index
Diversity	PR	Patch richness
	PRD	Patch richness density
	RPR	Relative patch richness
	SHDI	Shannon's diversity index
	SIDI	Simpson's diversity index
	MSIDI	Modified Simpson's diversity index
	SHEI	Shannon's evenness Index
	SIEI	Simpson's evenness index
	MSIEI	Modified Simpson's evenness index

1.4. Statistical analysis

Using the FRAGSTATS calculations of all 116 landscape-level UGS metrics for each of 17 districts as a basis, statistical analyses were conducted using the software SPSS (version 22, IBM Corp., Armonk, NY, USA) for the assessment of UGS pattern and structure. Correlation analysis was used within each class of landscape metrics for data reduction. Spearman's rank correlation coefficient (r_s) was calculated for all pairwise correlations among the variables (i.e., the metrics in this study) in each class of the pairs of metrics having a rank correlation coefficient equal to or higher than 0.9 ($|r_s| \geq 0.9$), only one metric was retained (Griffith *et al.* 2000; Torras *et al.* 2008). For the pairs of highly correlated metrics, the metrics commonly used in greenspace pattern analysis literature were selected.

After data reduction, factor analysis (Riitters *et al.* 1995) was performed for the remaining metrics within each group for different metric classes. By using the extraction method of principal component analysis, the varimax rotation method with Kaiser normalization, and the factor-score calculation method of regression, additional variances, non-correlated factors, component score coefficients, and regression factor scores were obtained. Then the integrated index value of each metric class by district was formulated as Eqs (1)–(8).

$$ARE\_I_d = \sum_{i=1}^{n} \left(ARE\_E_{di} \times ARE\_FS_{di} \right). \quad (1)$$

Equation (1) yields the index value for area (*ARE*) in a given district *d*; I_d is the integrated index value of each metric class; E_{di} is the eigenvalue of each component; and FS_{di} is the regression factor score of each factor. Eqs (2)–(8) yield index values for edge and contrast (*EDG*), shape (*SHA*), core area (*COR*), proximity (*PRO*), subdivision (*SUB*), contagion (*CON*) and division (*DIV*).

$$EDG\_I_d = \sum_{i=1}^{n} \left(EDG\_E_{di} \times EDG\_FS_{di} \right); \quad (2)$$

$$SHA\_I_d = \sum_{i=1}^{n} \left(SHA\_E_{di} \times SHA\_FS_{di} \right); \quad (3)$$

$$COR\_I_d = \sum_{i=1}^{n} \left(COR\_E_{di} \times COR\_FS_{di} \right); \quad (4)$$

$$PRO\_I_d = \sum_{i=1}^{n} \left(PRO\_E_{di} \times PRO\_FS_{di} \right); \quad (5)$$

$$SUB\_I_d = \sum_{i=1}^{n} \left(SUB\_E_{di} \times SUB\_FS_{di} \right); \quad (6)$$

$$CON\_I_d = \sum_{i=1}^{n} \left(CON\_E_{di} \times CON\_FS_{di} \right); \quad (7)$$

$$DIV\_I_d = \sum_{i=1}^{n} \left(DIV\_E_{di} \times DIV\_FS_{di} \right). \quad (8)$$

Table 2. Landscape metrics selected after data reduction

Class (number of metrics)	Landscape metric (acronym)
Area (8)	Total area (TA); Large patch index (LPI); Mean patch area (AREA_MN); Area-weighted mean patch area (AREA_ AM); Median patch area (AREA_MD); Mean radius of gyration (GYRATE_ MN); Standard deviation in radius of gyration (GYRATE_SD); Coefficient of variation in radius of gyration (GYRATE_CV)
Shape (19)	Perimeter-area fractal dimension (PAFRAC); Area-weighted mean perimeter area ratio (PARA_ AM); Median perimeter area ratio (PARA_MD); Range in perimeter area ratio (PARA_RA); Mean shape index (SHAPE_MN); Area-weighted mean shape index (SHAPE_AM); Median shape index (SHAPE_MD); Range in shape index (SHAPE_ RA); Standard deviation shape index (SHAPE_ SD); Mean fractal dimension index (FRAC_ MN); Area-weighted mean fractal dimension index (FRAC_AM); Median fractal dimension index (FRAC_MD); Range in fractal dimension index (FRAC_RA); Standard deviation in fractal dimension index (FRAC_SD); Area-weighted mean related circumscribing circle (CIRCLE_AM); Median related circumscribing circle (CIRCLE_ MD); Standard deviation in related circumscribing circle (CIRCLE_SD); Coefficient of variation in related circumscribing circle (CIRCLE_CV); Area-weighted mean contiguity index (CONTIG_AM)
Edge & Contrast (3)	Edge density (ED); Range in edge contrast index (ECON_RA); Coefficient of variation in edge contrast index (ECON_CV)
Core area (9)	Total core area (TCA); Disjunct core area density (DCAD); Mean core area (CORE_MN); Median core area (CORE_MD); Standard deviation in core area (CORE_SD); Coefficient of variation in core area (CORE_CV); Median disjunct core area (DCORE_MD); Range in core area index (CAI_RA); Standard deviation in core area index (CAI_SD)
Proximity (10)	Mean Euclidean nearest neighbor distance (ENN_MN); Median Euclidean nearest neighbor distance (ENN_MD); Range in Euclidean nearest neighbor distance (ENN_RA); Standard deviation in Euclidean nearest neighbor distance (ENN_ SD); Coefficient of variation in Euclidean nearest neighbor distance (ENN_CV); Mean proximity index (PROX_MN); Area-weighted mean proximity index (PROX_AM); Median proximity index (PROX_ MD); Coefficient of variation in proximity index (PROX_ CV); Connectance index (CONNECT)
Subdivision (4)	Number of patches (NP); Patch density (PD); Splitting index (SPLIT); Effective mesh size (MESH)
Contagion/ Intersper-sion (4)	Contagion (CONTAG); Aggregation index (AI); Landscape shape index (LSI); Patch cohesion index (COHESION)
Diversity (2)	Patch richness (PR); Patch richness density (PRD)

With the eight integrated index values of each metric class by district obtained above, factor analysis was performed again to get the integrated index value of all metric classes by district for *UGS* pattern and structure analysis among the 17 districts. The integrated index value of *UGS* was formulated as Eq. (9):

$$LEQ\_I_d = \sum_{i=1}^{n}\left(UGS\_E_{di} \times UGS\_FS_{di}\right), \qquad (9)$$

where in a given district d, LEQ_I_d is the integrated index value of all metric classes.

To further analyze and compare the UGS pattern and structure of different districts, hierarchical cluster analysis of all the integrated index values of different metric classes by district was performed. In the analysis, centroid clustering and squared Euclidean distance were used to group the districts for which integrated index values were related and similar.

2. Results

2.1. UGS distribution by district

The area of different types of UGS by districts was shown in Figure 3, which presents the general characteristics of UGS distribution for different types in different districts in Shanghai. Areas of the four types of UGS in new districts of the city are all significantly larger than that in old towns. Pudong has a larger area of public green land, forest and garden than most of the other districts. Chongming has a larger area of forest, garden and agriculture than most of the other districts. Baoshan and Minhang have less forest, garden and agriculture area than the other new districts.

2.2. UGS pattern and structure characteristics by metric class

Using the correlation analysis procedure within each metric class, 1029 metric correlations were examined for all pairs of metrics within all groups, of which 143 were found to be significant ($|r_s| \geq 0.9$). Through data reduction, the original set of landscape-level metrics was reduced from116 to 59 (Table 2). The integrated scores of each landscape metric class by district, obtained through factor analyses of the 59 selected metrics within eight groups, are given in Figure 4, which shows the relative contributions of different aspects of UGS pattern and structure to the LEQ of UGS in different districts.

The new districts usually have higher LEQ of UGS subdivision, core area, and patch edge and contrast than the old towns (Fig. 4). Of the 17 districts, Fengxian and Jinshan have the highest LEQ of UGS patch edge and contrast and the eight old towns all have the lowest quality. Pudong has the highest LEQ of UGS subdivision and Changning has the lowest. Chongming has the highest LEQ of UGS core area and Jing'an has the lowest. Thus, the relative contributions of UGS patch subdivision, core area and edge to LEQ of UGS in new districts has been improved through district development. Furthermore, the characteristics of UGS patch edge in old towns are exactly the same.

In contrast, the new districts usually have lower LEQ of UGS diversity than the old towns (Fig. 4). Of the 17 districts, Jing'an has the highest LEQ of UGS diversity and the nine new districts all have much lower values, which are similar to each other. Although new districts generally have higher patch richness (PR) of UGS than the old towns, the patch richness density (PRD) of UGS in districts is much lower than in the old towns. Thus, the relative contributions of UGS diversity to LEQ of UGS have been lowered due to district development; the mono-distribution of UGS type might be the primary cause.

There were no significant difference of LEQ of UGS area, shape, proximity and contagion metrics between new districts and old towns. Of the 17 districts, Chongming has the highest LEQ of UGS area and Minhang has the lowest. Changning has the highest LEQ of UGS shape and Jing'an has the lowest. Jing'an has the highest LEQ of UGS proximity and Fengxian has the lowest. Chongming has the highest LEQ of UGS contagion and Huangpu has the lowest. Thus, the relative contributions of UGS area, shape, proximity and contagion metrics to LEQ of UGS are not directly related to district development, although some indexes of these four groups are related to district development.

2.3. LEQ of UGS by district

Tables 3 and 4 show results of factor analysis of the eight integrated indexes for the 17 districts. Three notable components were extracted, with the first contributing 45.38% variance of initial eigenvalues, the second contributing 26.03%, and the third contributing 18.66%. The first component was characterized by a high negative loading of DIV_I and high positive loading of EDG_I and SUB_I. Thus, the first component represents the LEQ of UGS in new districts. The second component was characterized

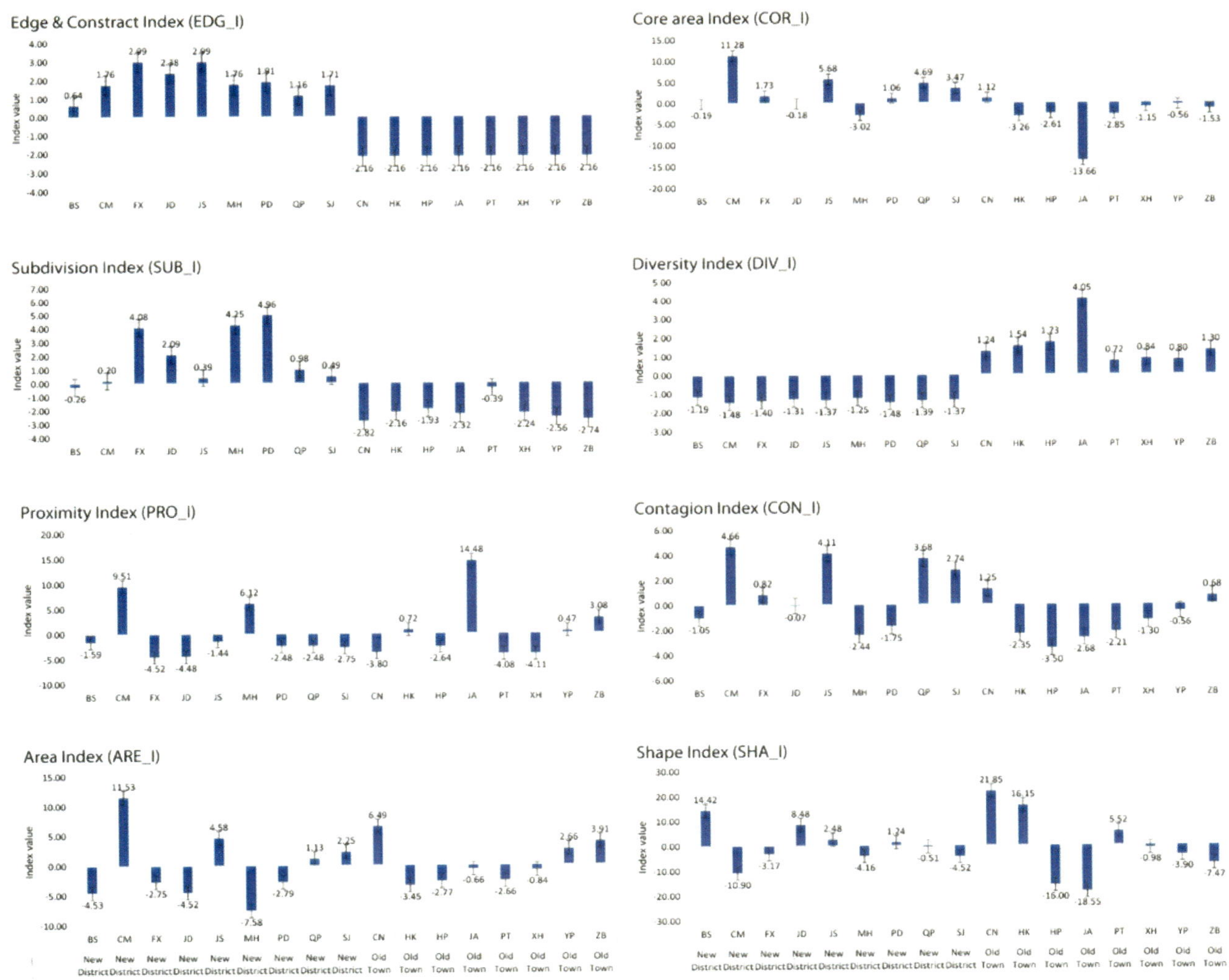

Note. The abbreviations of districts are explained in Figure 1.

Fig. 4. Index value of each metric class by district

by a high positive loading of ARE_I, CON_I and COR_I. Thus, the second component represents the UGS having large area in old towns. The third component was characterized by a high positive loading of PRO_I and high negative loading of SHA_I, and was correlated to UGS with small area in old towns.

Figure 5 shows the relative contributions of UGS pattern and structure to the LEQ of the city by district.

Table 3. Factor loadings in factor analysis with varimax method

Component	Initial Eigenvalues			Extraction sums of squared loadings			Rotation sums of squared loadings		
	Total	% of Variance	Cumulative %	Total	% of Variance	Cumulative %	Total	% of Variance	Cumulative %
1	3.630	45.376	45.376	3.630	45.376	45.376	2.951	36.889	36.889
2	2.082	26.028	71.405	2.082	26.028	71.405	2.551	31.892	68.781
3	1.493	18.664	90.069	1.493	18.664	90.069	1.703	21.288	90.069
4	0.457	5.712	95.780	–	–	–	–	–	–
5	0.171	2.141	97.921	–	–	–	–	–	–
6	0.110	1.374	99.295	–	–	–	–	–	–
7	0.039	0.481	99.776	–	–	–	–	–	–
8	0.018	0.224	100.000	–	–	–	–	–	–

Table 4. Rotated component matrix and variance explained for UGS pattern and structure characteristics for 17 districts in Shanghai

Integrated index for each metric category	Component		
	1	2	3
ARE_I	−0.282	0.919	0.186
CON_I	0.285	0.921	−0.017
COR_I	0.454	0.808	−0.252
DIV_I	−0.873	−0.338	0.286
EDG_I	0.939	0.243	0.007
PRO_I	−0.173	0.017	0.869
SHA_I	−0.042	0.014	−0.876
SUB_I	0.953	−0.181	0.001

Note. *Italic* numbers indicate component loadings >|0.8|. Components are characterized by the corresponding integrated indexes with bold number. The metrics class abbreviations are explained in methods part.

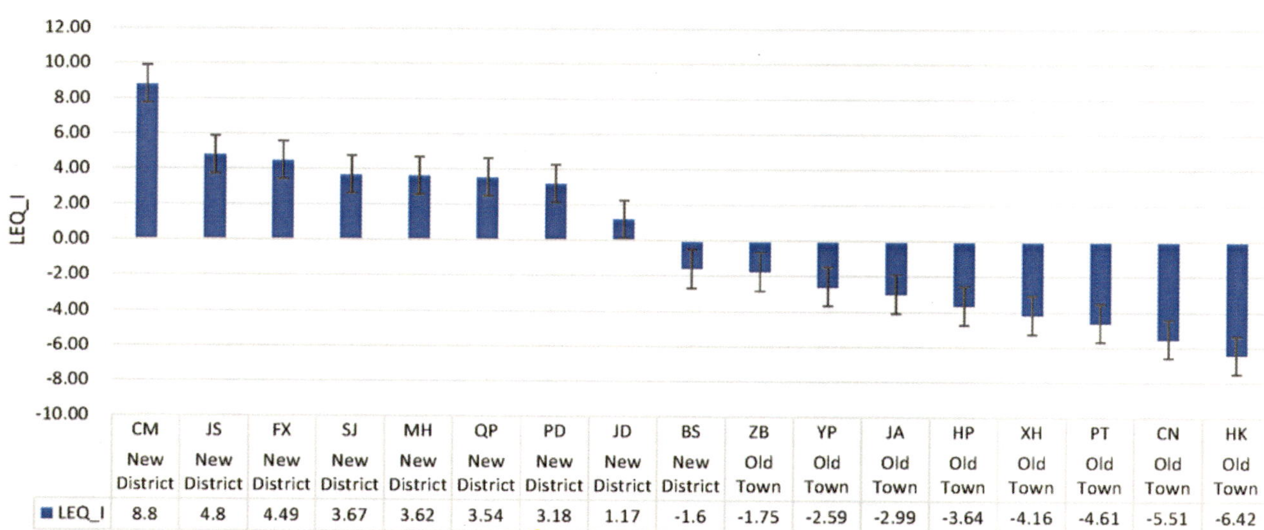

Note. The abbreviations of districts are explained in Figure 1. "LEQ_I" is explained in methods part.

Fig. 5. Landscape ecological quality index values for districts in Shanghai

The new districts usually have higher LEQ of UGS than the old towns. Of the 17 districts, Chongming has the highest LEQ of UGS and Hongkou has the lowest. The contribution of UGS pattern and structure to the LEQ of the city has been improved due to district development.

The district group results from cluster analysis of selected metrics are shown in Figure 6. Rescaled distance equal to "5" was considered suitable for grouping the districts into clusters. Three clusters were identified that represented different characteristic UGS patterns and structures of districts. The first cluster is related to the eight old towns in Shanghai. The third cluster is related to the new districts of Chongming and Pudong. The second cluster is related to the remaining seven new districts. Thus, the eight old towns are similar in their UGS pattern and structure, as are most of the new districts (except Chongming and Pudong). The new districts of Chongming and Pudong are dissimilar from all other districts (new and old) for LEQ of UGS.

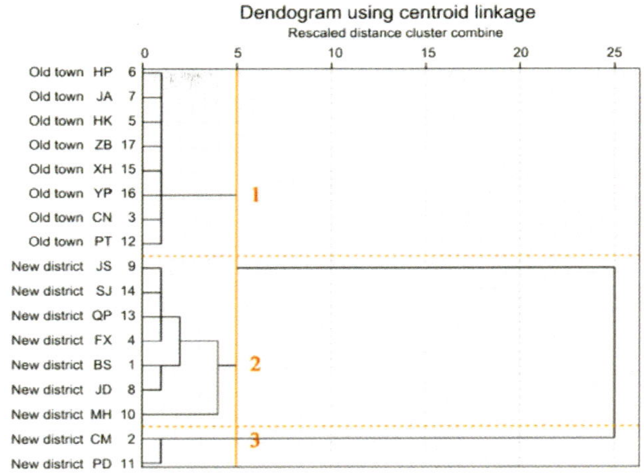

Note. The abbreviations of districts are explained in Figure 1.

Fig. 6. Cluster analysis of selected metrics

2.4. LEQ of UGS and per capita green cover

Figure 7 shows the paradoxical relationships between LEQ_I and per capita green cover. Green cover in the figure refers to public green land and forest. It seems a positive relationship between LEQ_I and per capita green cover, as new districts have both higher green cover per capita and better UGS LEQ than old towns. This phenomenon seems conceivable and reasonable. However, the relationship between LEQ_I and per capita green cover within the group of new districts or old towns pretended to be negative. For new districts, as green cover per capita increases, LEQ_I decreases. For old towns except Hongkou, as green cover per capita increases, LEQ_I decreases. This phenomenon might be counterintuitive. It showed that districts of the same development with lower green cover per capita usually have better UGS LEQ.

3. Discussion

In the calculation of integrated landscape metrics, because of agriculture, forest and garden UGS types, the new districts usually have more types of UGS, especially for agriculture. Agricultural and forested lands usually occur as discrete, large areas distributed within districts. Thus, although new districts have more UGS types than old towns, the new districts also have lower patch richness density (PRD) and lower LEQ of UGS diversity.

In the pattern characteristic grouping of the districts, the fact that the eight old towns have similar character in integrated landscape metrics might be because they underwent a similar complex development process that resulted in a similar UGS quantity and distribution pattern.

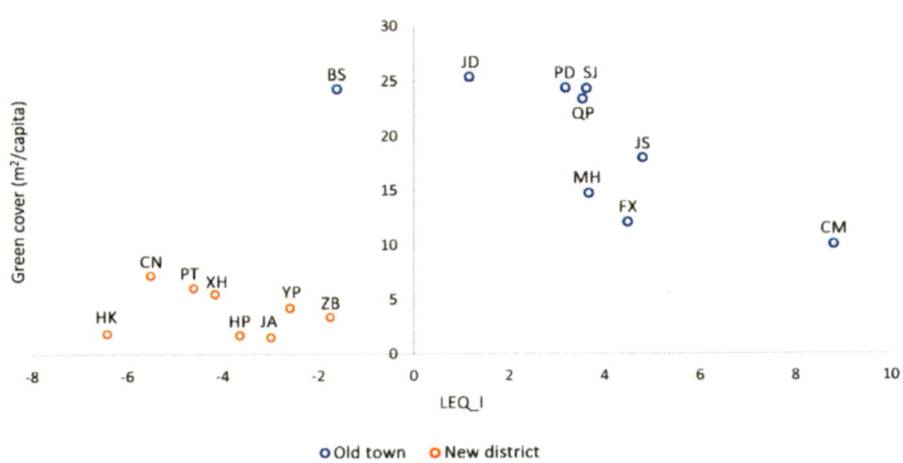

Note. The abbreviations of districts are explained in Figure 1. "LEQ_I" is explained in methods part.

Fig. 7. UGS LEQ and per capita green cover

That the seven new districts (except Chongming and Pudong) have similar character in integrated landscape metrics might be because they are similar in area (which is larger than other districts), with fewer restrictions posed by constructed buildings and with more agricultural and forested land. Chongming and Pudong are distinctive in UGS character, but for different reasons. Chongming is the least developed district in Shanghai and has a large area and the most agriculture. Conversely, Pudong is the most developed district, encompassing a large area with the most public green land in Shanghai.

The fact that the new districts have higher LEQ_I and per capita green cover than the old towns indicates that the causes of UGS in new districts with higher LEQ include both quantity and distribution of UGS. The negative relationship between LEQ_I and per capita green cover among both new districts and old towns indicates that the distribution of UGS largely determines its LEQ in these districts without a significant difference in quantity of UGS. Of the new districts, Chongming has the highest LEQ_I for UGS but lowest per capita green cover. This relationship might arise because per capita green cover is defined by per capita public green land cover and excludes agricultural land; Chongming has much more agriculture than the other districts. Of the old towns, Hongkou has lower LEQ_I and per capita green cover than the other districts indicating that both quantity and distribution of UGS are the causes of its lower LEQ. Thus, Hongkou not only needs to increase UGS (especially public green land) but also needs to optimize the distribution and structure of UGS.

This study provides some important contributions to the assessment of UGS. It fills a gap in the scientific literature by assessing the LEQ of UGS by qualification of UGS pattern on the basis of spatial metrics. Because a single landscape metric is inadequate, indices are related; thus, redundant information in multiple indices must be abridged for effective UGS pattern measurement, and selection of appropriate metrics is crucial. Previous studies identified some important landscape metrics for calculation by subjective factors (Zhou, Wang 2011; Tian *et al.* 2014). In contrast, this study calculated all indices using the FRAGSTATS software and selected appropriate metrics objectively through correlation analysis. This procedure not only ensures data integrity, but also avoids data redundancy using objective criteria. By employing factor analyses twice, this improved procedure reduced the data dimension of landscape indices to quantify UGS pattern and structure for each district. Therefore, this method provides a direct comparison of UGS LEQ among 17 districts. Moreover, by cluster analysis of all the integrated indexes of metric classes, the grouping analysis of UGS pattern and structure of the 17 districts also avoided the use of subjective factors.

However, this study has some limitations. It focused on one point in time instead of several times. Thus, the results can be used to reflect and analyze only the present situation, not the evolution of UGS pattern over time. For a time-evolution analysis, the statistical changes of UGS landscape metrics over a number of years are needed, requiring UGS measurements at multiple points in time. Thus, future studies should include data from multiple time points to obtain a more comprehensive understanding of UGS evolution over time in a compact city.

Conclusions

In summary, this research developed an improved method for UGS LEQ assessment and used the method to analyze UGS distribution, pattern and structure for the megacity of Shanghai. In addition, the study provides valuable insights for policy makers and planners striving to develop an ecological and sustainable city with limited UGS. Results from the research justify the following conclusions. The relative contributions of UGS area, shape, proximity and contagion metrics to LEQ of UGS are not directly related to district development. New districts usually have higher LEQ of UGS than the old towns. However, for new districts and old towns respectively, districts with lower green cover per capita have better UGS LEQ. These results of the study in the compact city (e.g., Shanghai) not only assist in identifying the strengths and weaknesses of UGS distribution, pattern and structure, but also in optimizing UGS pattern and distribution to improve LEQ of UGS.

Author contributions

Conceived and designed the experiments: HL. Performed the experiments: HL DC. Analyzed the data: HL. Contributed materials/analysis tools: QZ. Contributed to the writing of the manuscript: HL. Revised and approved the final version of the paper: HL QZ.

References

Andersson, E. 2006. Urban landscapes and sustainable cities, *Ecology and Society* 11(1): 34. https://doi.org/10.5751/ES-01639-110134

Antrop, M. 2005. Why landscapes of the past are important for the future, *Landscape and Urban Planning* 70(1): 21–34. https://doi.org/10.1016/j.landurbplan.2003.10.002

Barthel, S.; Colding, J.; Elmqvist, T.; Folke, C. 2005. History and local management of a biodiversity-rich, urban cultural landscape, *Ecology and Society* 10(2): 10. https://doi.org/10.5751/ES-01568-100210

Blanco, H.; Alberti, M.; Forsyth, A.; Krizek, K. J.; Rodriguez, D. A.; Talen, E.; Ellis, C. 2009. Hot, congested, crowded and diverse: emerging research agendas in planning, *Progress in Planning* 71(4): 153–205. https://doi.org/10.1016/j.progress.2009.03.001

Buyantuyev, A.; Wu, J.; Gries, C. 2010. Multiscale analysis of the urbanization pattern of the Phoenix metropolitan landscape of USA: time, space and thematic resolution, *Landscape and Urban Planning* 94(3): 206–217. https://doi.org/10.1016/j.landurbplan.2009.10.005

Byomkesh, T.; Nakagoshi, N.; Dewan, A. M. 2012. Urbanization and green space dynamics in Greater Dhaka, Bangladesh, *Landscape and Ecological Engineering* 8(1): 45–58. https://doi.org/10.1007/s11355-010-0147-7

Chang, H.; Li, F.; Li, Z.; Wang, R.; Wang, Y. 2011. Urban landscape pattern design from the viewpoint of networks: a case study of Changzhou city in Southeast China, *Ecological Complexity* 8(1): 51–59. https://doi.org/10.1016/j.ecocom.2010.12.003

Chen, A.; Yao, L.; Sun, R.; Chen, L. 2014. How many metrics are required to identify the effects of the landscape pattern on land surface temperature?, *Ecological Indicators* 45: 424–433. https://doi.org/10.1016/j.ecolind.2014.05.002

De Groot, R. D. 2006. Function-analysis and valuation as a tool to assess land use conflicts in planning for sustainable, multi-functional landscapes, *Landscape and Urban Planning* 75(3): 175–186. https://doi.org/10.1016/j.landurbplan.2005.02.016

De Groot, R. D.; Perk, J. V.; Chiesura, A.; Vliet, A. 2003. Importance and threat as determining factors for criticality of natural capital, *Ecological Economics* 44(2): 187–204. https://doi.org/10.1016/S0921-8009(02)00273-2

Escobedo, F. J.; Kroeger, T.; Wagner, J. E. 2011. Urban forests and pollution mitigation: analyzing ecosystem services and disservices, *Environmental Pollution* 159(8): 2078–2087. https://doi.org/10.1016/j.envpol.2011.01.010

Fan, C.; Myint, S. 2014. A comparison of spatial autocorrelation indices and landscape metrics in measuring urban landscape fragmentation, *Landscape and Urban Planning* 121: 117–128. https://doi.org/10.1016/j.landurbplan.2013.10.002

Fuller, R. A.; Gaston, K. J. 2009. The scaling of green space coverage in European cities, *Biology Letters* 5(3): 352–355. https://doi.org/10.1098/rsbl.2009.0010

Gardner, R. H. 1999. RULE: a program for the generation of random maps and the analysis of spatial patterns, in J. M. Klopatek, R. H. Gardner (Eds.). *Landscape ecological analysis*: issues and applications. New York: Springer-Verlag. https://doi.org/10.1007/978-1-4612-0529-6_13

Gill, S.; Handley, J.; Ennos, A.; Pauleit, S. 2007. Adapting cities for climate change: the role of the green infrastructure, *Built Environment* 33(1): 115–133.

Goddard, M. A.; Dougill, A. J.; Benton, T. G. 2010. Scaling up from gardens: biodiversity conservation in urban environments, *Trends in Ecology and Evolution* 25(2): 90–98. https://doi.org/10.1016/j.tree.2009.07.016

Griffith, J. A.; Martinko, E. A.; Price, K. P. 2000. Landscape structure analysis of Kansas at three scales, *Landscape and Urban Planning* 52(1): 45–61. https://doi.org/10.1016/S0169-2046(00)00112-2

Groenewegen, P. P.; Van den Berg, A. E.; De Vries, S.; Verheij, R. A. 2006. Vitamin G: effects of green space on health, well-being, and social safety, *BMC Public Health* 6(1): 149. https://doi.org/10.1186/1471-2458-6-149

Hamada, S.; Tanaka, T.; Ohta, T. 2013. Impacts of land use and topography on the cooling effect of green areas on surrounding urban areas, *Urban Forestry and Urban Greening* 12(4): 426–434. https://doi.org/10.1016/j.ufug.2013.06.008

Haq, S. M. A. 2011. Urban green spaces and an integrative approach to sustainable environment, *Journal of Environmental Protection* 2: 601–608. https://doi.org/10.4236/jep.2011.25069

Hermann, A.; Kuttner, M.; Hainz-Renetzeder, C.; Konkoly-Gyuró, É.; Tirászi, Á.; Brandenburg, C.; Allex, B.; Ziener, K.; Wrbka, T. 2014. Assessment framework for landscape services in European cultural landscapes: an Austrian Hungarian case study, *Ecological Indicators* 37: 229–240. https://doi.org/10.1016/j.ecolind.2013.01.019

James, P.; Tzoulas, K.; Adams, M.; Barber, A.; Box, J.; Breuste, J.; Elmqvist, T.; Frith, M.; Gordon, C.; Greening, K. 2009. Towards an integrated understanding of green space in the European built environment, *Urban Forestry and Urban Greening* 8(2): 65–75. https://doi.org/10.1016/j.ufug.2009.02.001

Kong, F.; Yin, H.; James, P.; Hutyra, L. R.; He, H. S. 2014. Effects of spatial pattern of greenspace on urban cooling in a large metropolitan area of eastern China, *Landscape and Urban Planning* 128: 35–47. https://doi.org/10.1016/j.landurbplan.2014.04.018

Kong, F.; Yin, H.; Nakagoshi, N.; Zong, Y. 2010. Urban green space network development for biodiversity conservation: Identification based on graph theory and gravity modeling, *Landscape and Urban Planning* 95(1): 16–27. https://doi.org/10.1016/j.landurbplan.2009.11.001

La Rosa, D.; Privitera, R.; Martinico, F.; La Greca, P. 2013. Measures of Safeguard and Rehabilitation for landscape protection planning: a qualitative approach based on diversity indicators, *Journal of Environmental Management* 127: S73–S83. https://doi.org/10.1016/j.jenvman.2012.12.033

Li, X.; Zhou, W.; Ouyang, Z. 2013. Relationship between land surface temperature and spatial pattern of greenspace: what are the effects of spatial resolution?, *Landscape and Urban Planning* 114: 1–8. https://doi.org/10.1016/j.landurbplan.2013.02.005

Li, X.; Zhou, W.; Ouyang, Z.; Xu, W.; Zheng, H. 2012. Spatial pattern of greenspace affects land surface temperature: evidence from the heavily urbanized Beijing metropolitan area, China, *Landscape Ecology* 27(6): 887–898. https://doi.org/10.1007/s10980-012-9731-6

McGarigal, K.; Cushman, S. A.; Neel, M. C.; Ene, E. 2002. *FRAGSTATS: spatial pattern analysis program for categorical maps* [online], [cited 14 March 2016]. Available from Internet: http://www.citeulike.org/group/342/article/287784

McGarigal, K.; Cushman, S.; Ene, E. 2012. *FRAGSTATS v4: spatial pattern analysis program for categorical and continuous maps* [online], [cited 14 March 2016]. Available from Internet: http://www.umass.edu/landeco/research/fragstats/fragstats.html

McGarigal, K.; McComb, W. C. 1995. Relationships between landscape structure and breeding birds in the Oregon Coast Range, *Ecological Monographs* 65(3): 235–260. https://doi.org/10.2307/2937059

Neuenschwander, N.; Hayek, U. W.; Grêt-Regamey, A. 2014. Integrating an urban green space typology into procedural 3D visualization for collaborative planning, *Computers, Environment and Urban Systems* 48: 99–110. https://doi.org/10.1016/j.compenvurbsys.2014.07.010

Plexida, S. G.; Sfougaris, A. I.; Ispikoudis, I. P.; Papanastasis, V. P. 2014. Selecting landscape metrics as indicators of spatial heterogeneity – a comparison among Greek landscapes, *International Journal of Applied Earth Observation and Geoinformation* 26: 26–35. https://doi.org/10.1016/j.jag.2013.05.001

Qian, Y.; Zhou, W.; Li, W.; Han, L. 2015. Understanding the dynamic of greenspace in the urbanized area of Beijing based on high resolution satellite images, *Urban Forestry and Urban Greening* 14(1): 39–47. https://doi.org/10.1016/j.ufug.2014.11.006

Rempel, R. S.; Elkie, P. C.; Carr, A. 1999. *Patch analyst user's manual: a tool for quantifying landscape structure*. Thunder Bay: Ontario Ministry of Natural Resources, Boreal Science, Northwest Science and Technology.

Riitters, K. H.; O'neill, R.; Hunsaker, C.; Wickham, J. D.; Yankee, D.; Timmins, S.; Jones, K.; Jackson, B. 1995. A factor analysis of landscape pattern and structure metrics, *Landscape Ecology* 10(1): 23–39. https://doi.org/10.1007/BF00158551

Saura, S.; Torne, J. 2009. Conefor Sensinode 2.2: a software package for quantifying the importance of habitat patches for landscape connectivity, *Environmental Modelling and Software* 24(1): 135–139. https://doi.org/10.1016/j.envsoft.2008.05.005

Schindler, S.; Poirazidis, K.; Wrbka, T. 2008. Towards a core set of landscape metrics for biodiversity assessments: a case study from Dadia National Park, Greece, *Ecological Indicators* 8(5): 502–514. https://doi.org/10.1016/j.ecolind.2007.06.001

Sherer, P. M. 2003. *Why America needs more city parks and open space*. Washington, DC: The Trust for Public Land.

Thompson, C. W. 2002. Urban open space in the 21st century, *Landscape and Urban Planning* 60(2): 59–72. https://doi.org/10.1016/S0169-2046(02)00059-2

Tian, Y.; Jim, C. Y.; Wang, H. 2014. Assessing the landscape and ecological quality of urban green spaces in a compact city, *Landscape and Urban Planning* 121: 97–108. https://doi.org/10.1016/j.landurbplan.2013.10.001

Tian, Y.; Jim, C.; Tao, Y.; Shi, T. 2011. Landscape ecological assessment of green space fragmentation in Hong Kong, *Urban Forestry and Urban Greening* 10(2): 79–86. https://doi.org/10.1016/j.ufug.2010.11.002

Torras, O.; Gil-Tena, A.; Saura, S. 2008. How does forest landscape structure explain tree species richness in a Mediterranean context?, *Biodiversity and Conservation* 17(5): 1227–1240. https://doi.org/10.1007/s10531-007-9277-0

Turner, M. G.; Gardner, R. H.; O'neill, R. V. 2001. *Landscape ecology in theory and practice*. New York: Springer.

Walz, U. 2015. Indicators to monitor the structural diversity of landscapes, *Ecological Modelling* 295: 88–106. https://doi.org/10.1016/j.ecolmodel.2014.07.011

Wolch, J. R.; Byrne, J.; Newell, J. P. 2014. Urban green space, public health, and environmental justice: the challenge of making cities 'just green enough', *Landscape and Urban Planning* 125: 234–244. https://doi.org/10.1016/j.landurbplan.2014.01.017

Wu, J.; Jenerette, G. D.; Buyantuyev, A.; Redman, C. L. 2011. Quantifying spatiotemporal patterns of urbanization: the case of the two fastest growing metropolitan regions in the United States, *Ecological Complexity* 8(1): 1–8. https://doi.org/10.1016/j.ecocom.2010.03.002

Zhou, X.; Wang, Y. 2011. Spatial-temporal dynamics of urban green space in response to rapid urbanization and greening policies, *Landscape and Urban Planning* 100(3): 268–277. https://doi.org/10.1016/j.landurbplan.2010.12.013

Huilin LIANG. Affiliation: College of Landscape Architecture, Nanjing Forestry University, Nanjing, Jiangsu, China. Scientific degree: Master's degree. Research interests: Environmental sustainability; Landscape management; Landscape planning. Number of publications: 3. Number of attended conferences: 2.

Di CHEN. Affiliation: College of Architecture and Urban Planning, Tongji University, Shanghai, China. Scientific degree: Master's degree. Research interests: Landscape management; Landscape planning. Number of publications: 5. Number of attended conferences: 2.

Qingping ZHANG. Affiliation: College of Landscape Architecture, Nanjing Forestry University, Nanjing, Jiangsu, China. Scientific degree: PhD. Research interests: Environmental sustainability; Landscape management; Landscape planning. Number of publications: 23. Number of attended conferences: 16.

11

A STUDY ON ECOTECT APPLICATION OF LOCAL CLIMATE AT A RESIDENTIAL AREA IN CHUNCHEON, KOREA

Qian WU, Hyun-Kil JO

Kangwon National University, Chuncheon, Republic of Korea

Abstract. The local climate data were investigated at a residential area in Chuncheon from May to October in 2013. In addition, the simulations were carried out by the Autodesk Ecotect Analysis tool. Site A had the lowest temperature in summer. The best building orientation was south by east 157.5°. Based on the photosynthetically active radiation (PAR) values day-neutral species should be planted around sites C and E, shade-tolerant species around site D and helioplous species around sites A and B. In addition, site A is the most comfortable place in the residential area with a predicted mean vote (PMV) value around 0. It was found that there were four main aspects to be considered for the residential designs which were building height, land cover, planting design and building orientation. Moreover, the results showed that local climate and shading conditions should be of great concern in design or improvement of residential areas.

Keywords: local climate, air temperature, Ecotect, thermal comfort, landscape management.

Introduction

Urban microclimate is significantly affected by the solar radiation and shading conditions, since solar radiation affects temperature (Andreou 2014). There are not enough green spaces and cool sinks in downtown area, and hampered airflow in the canyons which decrease the escape ability of the emitted infrared radiation in the atmosphere (Andreou 2014). In addition, geometry parameters such as building height and building orientation determine solar access and shading conditions. Hence, they play an important role in the thermal comfort of outdoor spaces and also the energy performance of buildings (Jouri, Miljana 2012). It has also been reported that urban geometry is much more crucial at the microscope than thermal behavior, materials and the albedo (Andreou 2014). Moreover, it is of high importance whether the research of microclimate can provide the design guidelines to the residential area at the beginning of the design (Andreou 2014). Relationship between solar access issues with orientation and latitude have been considered in the past (Paramita, Koerniawan 2013). The effects of geometry and orientation on solar access and shading conditions for different latitudes have been examined by some researchers (Bourbia, Awbi 2004). In addition, some studies reported the urban density, orientation and solar access, in an effort to investigate urban design options (Van Esch *et al.* 2012).

In this study, simulations were carried out using the Autodesk Ecotect Analysis tool (2011, US) which has high capability to be successfully used for the analysis of the building orientation, shading conditions, solar access, PAR value and thermal comfort. Ecotect is an environmental analysis tool which can simulate building performance right in the design stage. It has a wide array of detailed analysis functions with a visual and interactive display that can make the analysis results be found directly in the analytical picture (Yang *et al.* 2014). Therefore, Ecotect was selected as the analysis software of this study. The geometrical parameters of physical model and study area location were the variables imported into the software. The effect of geometry and shading conditions, building orientation and meteorological condition were previously considered in many studies (Andreou 2014), while there were four new points in the present study. First, combination of the measurement data and Ecotect simulations to analyze the thermal comfort in a residential area. Second, comprehensive analysis of the local climate weather data in Chuncheon. Third, building orientation, shading conditions, solar access, PAR values and thermal comfort were taken into consideration in residential area design.

Corresponding author: Hyun-Kil Jo
E-mail: jhk@kangwon.ac.kr

Fourth, a new connection between ecological theory and landscape design was developed. All the measurement data and Ecotect simulations were combined to analyze the microclimatic changes and assess the problems in residential area design.

The study area was a residential area in Chuncheon which is the capital of Kangwon province, South Korea. The lowest measured temperature was –27.9 °C (February 6, 1969), and the highest was 36.8 °C (Aug 5, 2012) (Korea Meteorological Administration 2011). Figure 1 shows the exact location of the residential area and Figure 2 shows the layouts of five measurement sites, which include 25 buildings that cover 8.12 ha. The area was located in the east part of downtown area in Chuncheon, consisting of four apartments, Chonggu, Booyoung, Jinheung, and Doosan. There were five measurement sites in the study area, each with 30 m diameters representing the same size area (706.9 m^2). The five measurement sites in the central part of each apartment and the central area in the whole study area were selected according to the different green cover percentages.

1. Methods

1.1. Data sources

The data used in this study were air temperature (Ta), relative humidity (RH), solar radiation (SR) and wind speed (V) in a residential area during the growing period, spanning from May to October in 2013. The field surveys involved microclimatic monitoring with the use of the portable weather measurement equipments. During the summer, the surface temperature (Ts) was measured for the asphalt, concrete, grass, soil and cream color wall in the study area.

At each measurement site, the portable weather data logging equipments were used for recording Ta and RH (Thermo-hygrometer TH-100, Japan), SR (LI-250 Light Meter, Japan), V (Anemometer TA 45, UK) and Ts (K-type Thermometer SK-1110, Japan). All the microclimatic factors such as Ta, RH, SR and V were measured on the height of 1.5 m away from the ground. The surveys took place every month during the growing period, but the surface temperature was just measured from June to August which is summer in Chuncheon. Each point was monitored 3 times (5th, 15th and 25th) every month in summer and once (15th) every month in spring and fall, for 24 hours a day at 3 h intervals.

1.2. Simulations

After creating 3D model of the study area confirmed by the geometrical parameters from field measurement, the location information was analyzed by Ecotect using the tool bar for the necessary functions (Fig. 3). The longitude and latitude were 127.8° E and 37.9° N, respectively. Finally, the shading and thermal comfort were calculated and visualized. Also, solar radiation was simulated over any period as long as the information of the area was input.

2. Results

2.1. Building information and land cover characteristics

There were 25 buildings in the study area including four management offices with just one floor. The building floors and heights were listed in Table 1. The biggest height

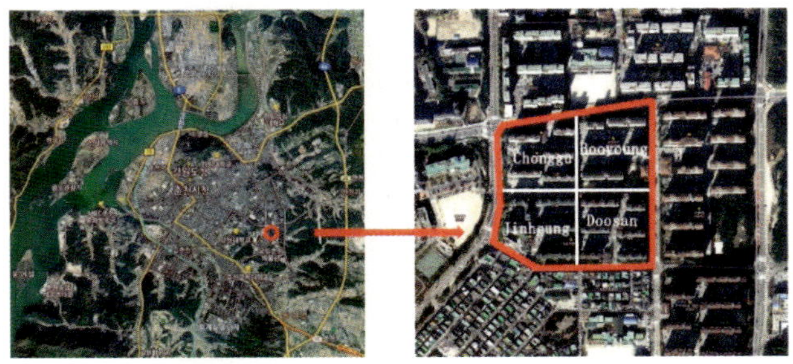

Fig. 1. Local study area in Chuncheon. The red line indicates the boundary of the study area including four apartments; Chonggu, Booyoung, Jinheung and Doosan

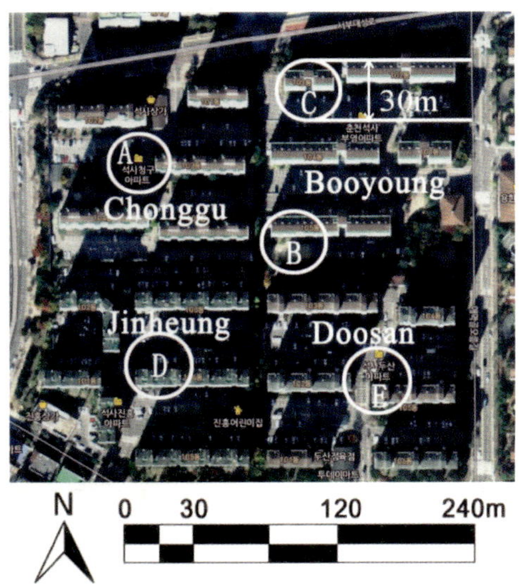

Fig. 2. Layouts of five measurement sites. All the measurement sites are equal-size circles with 30 m in diameter

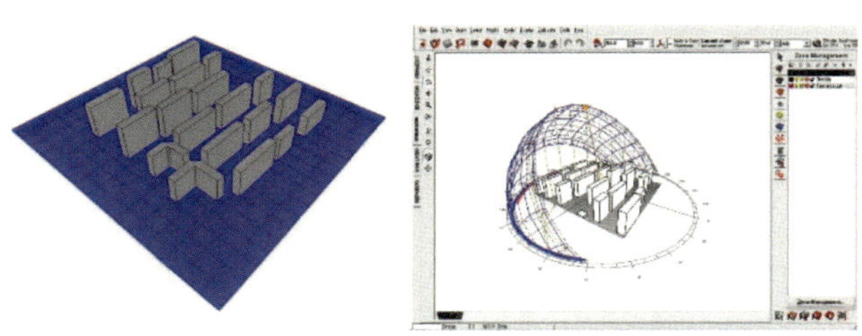

Fig. 3. The 3D model of the study area and the interface of Autodesk Ecotect Analysis tool (2011, US)

difference of the buildings was 21 m except the management offices. Table 2 summarizes the land cover and surface characteristics of each measurement site. It can be found that site A had the highest green ratio (43%) and site C had the lowest.

Table 1. Buildings information in the study area

Building floor	19	18	16	15	14	13	12	1
Number of buildings	1	4	2	7	3	3	1	4
Height (m)	57	54	48	45	42	39	36	4

2.2. Local climatic measure results

Figure 4a shows the Ta measure results from 6 am to 6 pm during May to October. Site C had the highest average Ta reaching 27.9 °C which was 1 °C higher than site A which was the lowest Ta (26.9 °C) at 3 pm. Comparing the Ta in site A and C with the same time that is the Ta difference ranged from 0.7–1 °C from 6 am to 6 pm and the biggest difference was at 3 pm. The Ta in site B was slightly lower than site C and higher than site D, with the differences was 0.2–0.4 °C and 0.2–0.6 °C, respectively. No significant differences were observed between sites D and E. Generally, site E was slightly lower than site D. Apparently, site C had the highest Ta, followed by site B, D, E and A.

The RH change trend was presented in Figure 4b. Not surprisingly, five sites showed similar general trend. Comparing the RH in the five sites at the same time, site A had the highest RH (62.3%) and site C was the lowest (59.3%). Site E was found to exhibit lower RH (0.3–1.7%) than site A followed by sites D and B. In addition, there was a slight difference between sites B and D (0.2–0.6%).

Figure 4c shows the wind speed differences in five sites. Comparing V in day time, the value of site A was higher than other areas and the average value was 0.57 m/s in site A. There is no significant difference among other four sites. Moreover, the obvious change trend was not found in V in the study area.

The T-test had been used to compare the weather measurement data and simulation. All the parameters' P values were greater than 0.05, which means there was no significant difference between measurement data and simulation. Therefore, Ecotect simulation was feasible in this study.

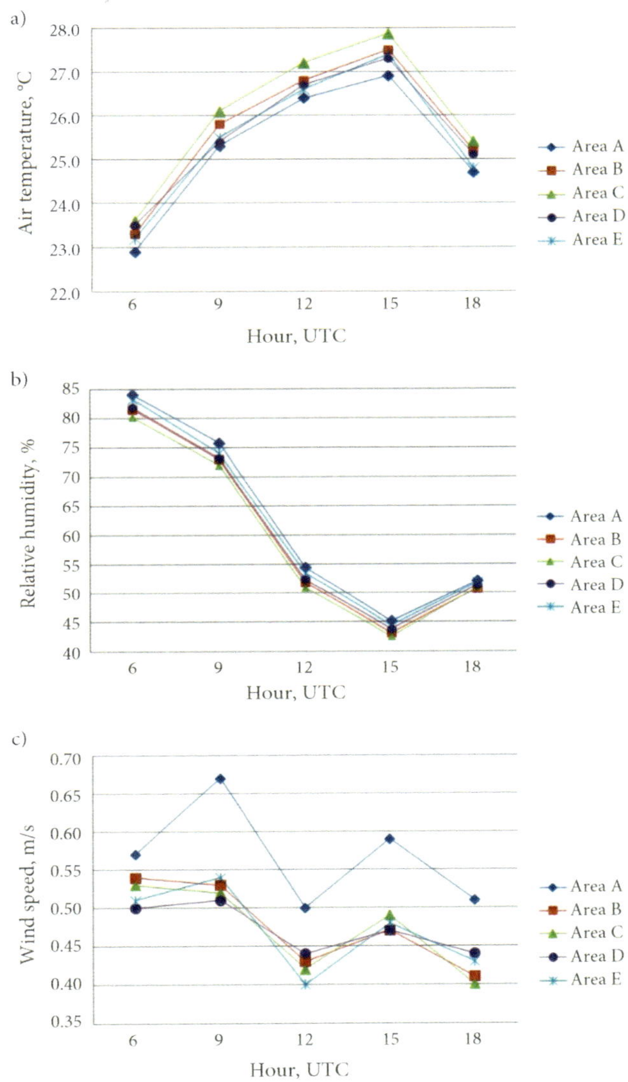

Fig. 4. Comparison of weather measurements:
a) Air temperature; b) Relative humidity; c) Wind speed

Table 2. Summary of land cover and surface characteristics of each measurement site

Site	Area (m^2)	Horizontal surface								Vertical surface
		Building (%)	Greenspace (%)					Concrete (%)	Asphalt (%)	Wall (m^2)
			Trees	Shrub	Grass	Soil	Total			
A	706.9	38	32	15	24	3	43	12	7	1267.4
B	706.9	49	8	5	6	10	22	15	14	1335.6
C	706.9	51	5	4	7	9	18	19	12	1604.4
D	706.9	49	10	7	7	8	25	15	11	1375.2
E	706.9	48	29	6	25	3	34	12	6	1623.6

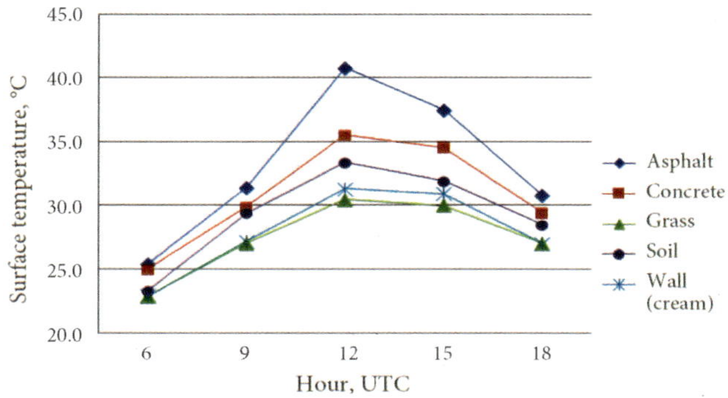

Fig. 5. Surface temperatures of five different surface types in summer time

2.3. Surface types and temperature

Figure 5 shows the five main types of surfaces in the study area including the horizontal and vertical surfaces. The day-time surface temperatures were presented. There were five types of surface in the study area which were asphalt, concrete, grass, soil and cream color wall. The highest surface temperatures were recorded at the mid-day temperature of the pavement in the exposed ground (reaching a maximum 40.9 °C absorbed in asphalt ground surface) and grass surface temperature was 30.4 °C, which its surface temperature started to rise from 9 am to 3 pm. It can be found that asphalt had the highest values and grass had the lowest values. Not surprisingly, site A had the highest green ratio and lowest hard surface including concrete and asphalt showed the lowest Ta in summer and site C was the opposite.

2.4. Shading conditions and PAR value

The shading analysis was carried out using the shading range tool. Figure 6 presents the shadow range of the whole study area from May to October. The violet red color rectangles were the building plans which rounded by the simulation of diurnal sun orbit. All the semitransparent grey shades were arranged by the Ecotect tool. Apparently, October had the most shade during the six months. Comparing the shade amount in five sites, site D had the most shade and site C had the least. It can be found that site A had more shade than sites E and B.

The average daily PAR was shown in Figure 7 and Figure 8 (summer time). The Insolation Analysis tool can simulate the PAR value from 0.40–5.40 MJ/m^{2}d. Figure 7 used eleven colors and 5.148 visible nodes showed the PAR values in the whole study area. It can be found that there was no 5.40 and 4.90 MJ/m^{2}d value in study area from May to October and also in summer time (June, July and August). The highest value was 4.4 MJ/m^{2}d in the west, southwest and east side and lowest value was 0.40 MJ/m^{2}d which is building area. The value was about 0.90–1.90 MJ/m^{2}d in site D which was the lowest in the five sites. Sites A's and B's value ranged from 2.40–3.40 MJ/m^{2}d. Another two sites C and E had the values between 2.40–2.90 MJ/m^{2}d.

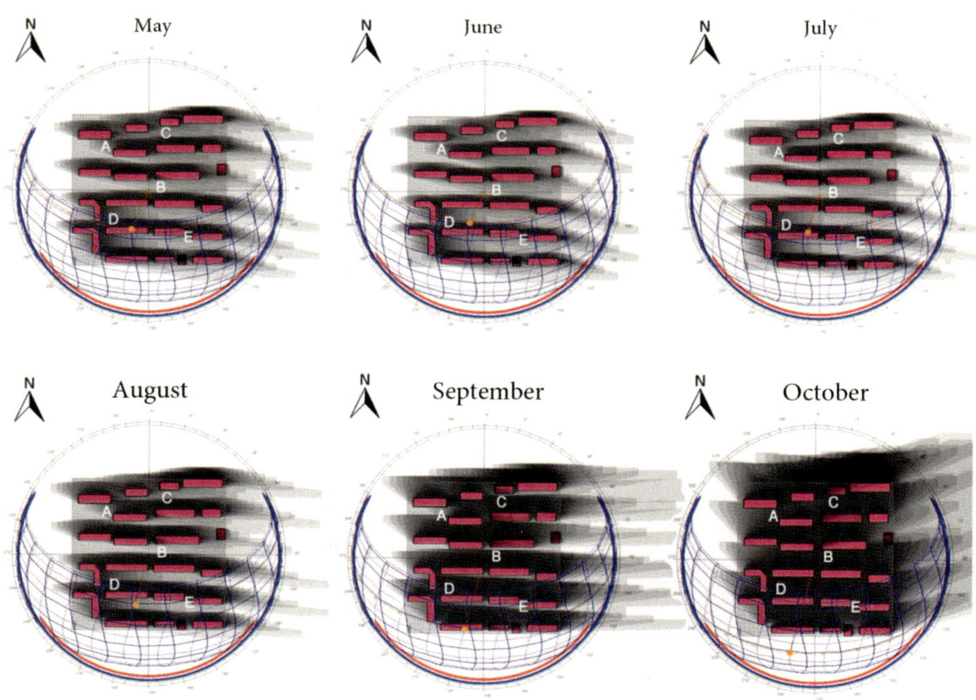

Fig. 6. Shading conditions simulations from May to October in the study area. The rectangles are the buildings' master plan and there are some semitransparent shadows around them

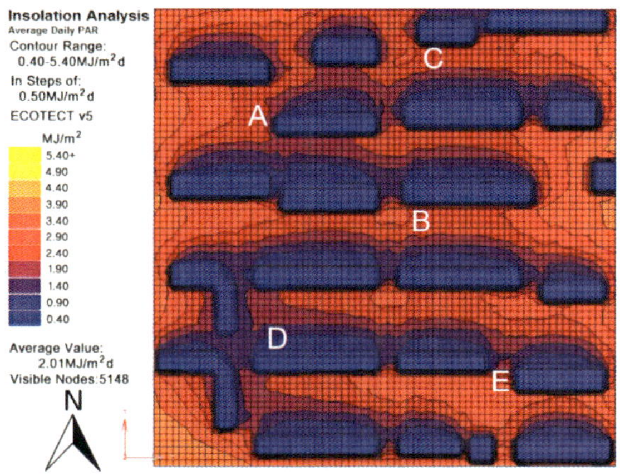

Fig. 7. PAR analysis from May to October

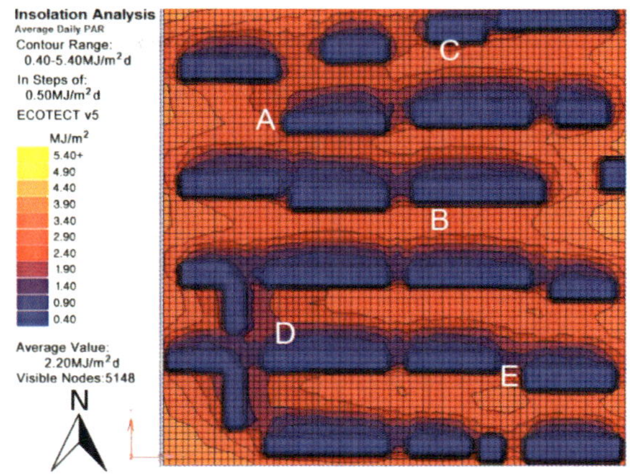

Fig. 8. PAR analysis from June to August

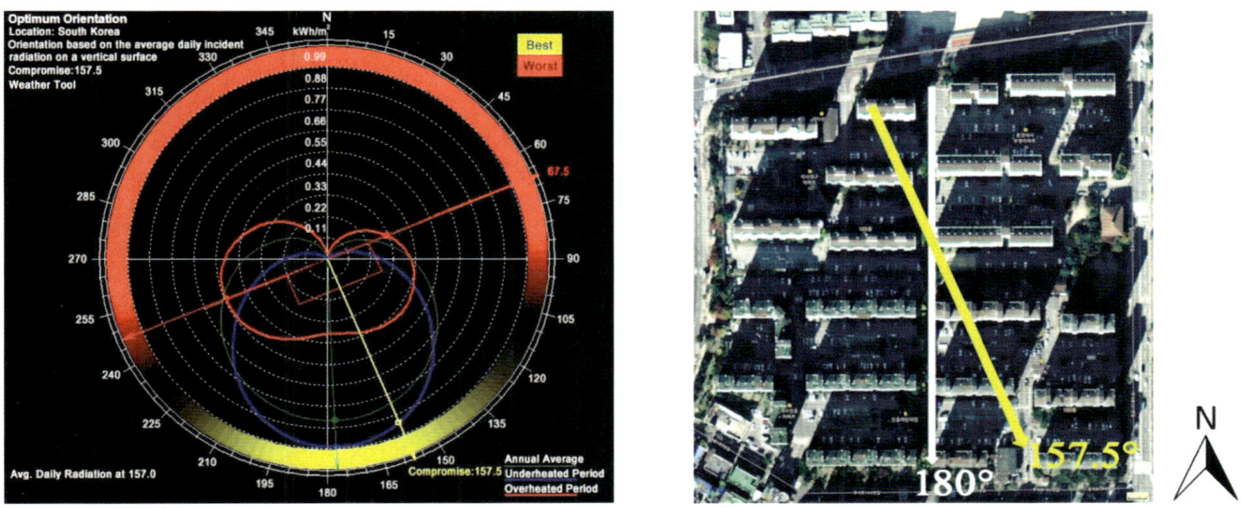

Fig. 9. Analysis of best building orientation. The yellow arrow shows the best building orientation (157.5°) and the white arrow was the present building orientation (south) in the study area

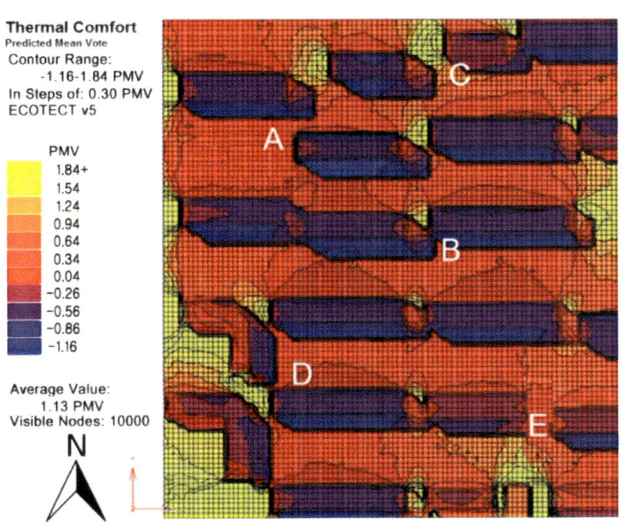

Fig. 10. Analysis of predicted mean vote in the study area in June

2.5. Best building orientation

According to the best building orientation analysis from the Autodesk Ecotect Analysis tool (Fig. 9), the green line was annual average range, the blue line was under heated range and the red line was overheated range. The yellow line was the best building orientation range. The white arrows in right side picture showed the primary building orientation which needs to be modified and the yellow arrows showed the best building orientation. In order to avoid the under heated and overheated situation, 157.5° (south by east 22.5°) was the best building orientation in study area and north by east 67.5° was the worst.

2.6. Thermal comfort

Figures 10, 11 and 12 show the PMV in summer (June, July and August) carried out by thermal comfort

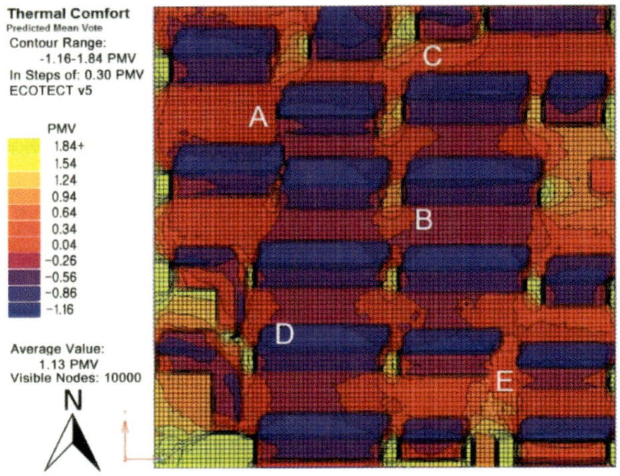

Fig. 11. Analysis of predicted mean vote in the study area in July

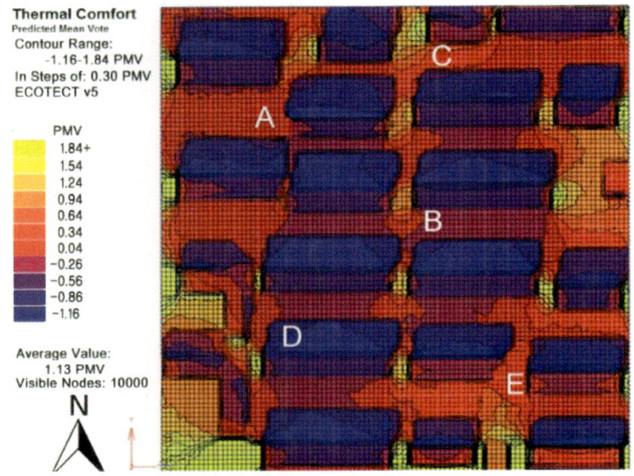

Fig. 12. Analysis of predicted mean vote in the study area in August

analysis tool. There were eleven colors and 10,000 visible nodes showing the PMV values. The PMV value in the study area ranged from −1.16 to 1.84 and the average value was 1.13. It can be found that most of the study area had the PMV value between 0.04 and 0.34. Moreover, the value was over 1.84 just in the southwest side and a little in east side in the study area. Site A's and E's average PMV was closest to 0 (0.34) and site C was the highest (0.64–1.24). When refer to the PMV around site D was between −0.86 and 0.34. Site B was ranged from −0.86 to 0.04.

3. Discussion

Site C was located at the parking lot with the low percentages of green cover (Table 2) and used more pavements compared with other sites, which is one of the reasons that it showed the highest temperature. Not surprisingly, site A proved to be the coldest site and had highest RH since it had the highest green ratio.

Although all the measurement data was recorded in the sunny days, there were many rainy days in July, which may had influenced the results of average RH data at five measurement sites. It is well known that evapotranspiration emissions of water vapour conduce to the humidity in urban area (Gup-yu *et al.* 2013). Therefore, site A, which had the highest green ratio showed more evapotranspiration and had the highest RH, and the lowest Ta value, while site C had the opposite trend. In addition, the RH diurnal variation curves show the same changing trend which was because of Ta, V and evapotranspiration.

The dominant wind was from northwest in the study area in summer (Korea Meteorological Administration 2011). But there were 21 high-rise buildings in the study area which could reduce the wind speed and changed the wind direction. Apparently, there was no high building in the west of site A to stop the airflow which resulted in it being the site with the highest wind speed which also one of the reasons that site A is the coolest space among all. However, the buildings around sites B, C, D and E almost had the similar height and same direction which could stop the wind flow in this area.

Another element influencing both air temperature and human thermal comfort is the radiant temperature which comes from surface (Limor *et al.* 2009). Materials used in the vertical and horizontal surfaces play a very important role in the urban thermal balance (Santamouris *et al.* 2011). Many studies have been carried out to analyze the optical and thermal characteristics of materials as well as their impact on the surface temperature (Hassan *et al.* 2014). It can be found that sites A and E had less asphalt surface than sites B and C that is one of the reasons they were cooler in summer.

Both vegetation and shading have the potential to reduce air temperature and cool the atmosphere (Bonan 2000). High shading percentage has a significant impact on the pedestrian thermal comfort as well as on the temperature (Capeluto *et al.* 2005). In other words, if the building orientation is 157.5° in the study area, all the building will face the wind blowing direction, which can be one of the cooling effects in hot summer. In the study area, all the buildings orientations were south which is not good for the wind flow in summer. If the building orientations could be changed to 157.5°, it would be easier to absorb the heat in winter and have appropriate ventilation in summer which can save the fuel and energy more efficiently.

Sites A and E had more shading time during summer, which is one of the reasons they were cooler than other sites. It can be found that sites B and D had the similar green ratio, but their temperatures were different because site D had more shade than site B, resulting in about 0.2–0.6 °C temperature difference between the two sites.

PAR designates the spectral range of solar radiation from 400 to 700 nanometers that belong to visible light, which can be absorbed by photosynthetic organisms to use in the process of photosynthesis (Andrew *et al.* 2006). Therefore, PAR can be one of the parameter to consider in planting design. The PAR range was from 0.40–5.40 MJ/m^{2}d and the area which the value was lower than 1.9 MJ/m^{2}d need to plant shade-tolerant species, 1.90–2.90 MJ/m^{2}d is better to plant the day-neutral species and over 2.90 MJ/m^{2}d is good for the heliophilous species. Therefore, it properly plants the shade-tolerant species around site D, the day-neutral species around sites C and E and heliophilous species around sites A and B.

Best building orientation is the most fundamental and generally, most easily addressed aspect of passive solar design which should use the ecology and energy saving technology to design (Morrissey *et al.* 2011). Appropriate design should consider key building parameters such as building orientation, plan proportion and building shape (Aksoy, Inalli 2006). Therefore, all the buildings orientation could be changed to south by east 22.5°; it would be more effective to make the cooling effect in the study area in summer.

One of the most widely used indexes is PMV, which predicts the mean thermal response of a large population of people (Ruey-Lung, Shiu-Ya 2011). There are seven scales of PMV (+3 = hot, +2 = warm, +1 = slightly warm, 0 = neutral, −1 = slightly cool, −2 = cool, −3 = cold) (Ruey-Lung, Shiu-Ya 2011). PMV has been included in the International Organization for Standardization ISO standard (PVM ISO 7730: 2005). Originally developed as an indoor thermal comfort index, PMV has also been commonly adopted in outdoor thermal comfort studies in which large groups of people were being surveyed (Cheng *et al.* 2012). Sites A's and E's PMV was most close to 0 (0.34), which means that sites A and E were most comfortable. In addition, site C was so hot in summer and the PMV value is largest (0.64–1.24). Even according to Ta, site E was cooler than sites B and D. But when refer to the PMV around site B is between −0.86 to 0.04, so site B has the higher comfort level.

Conclusions

1. This study was based on time series analysis of Ta, RH, SR, V and Ts that were carried out at five measurement sites in a residential area in Chuncheon. The monthly variations of the four factors were determined, showing that five measurement sites had different microclimate changes because of the land cover differences. Site C showed the highest mean diurnal air temperature (27.9 °C) at 3 pm because of the surroundings and site A was the coldest in summer because of the lowest Ta. In addition, hereby results address the relevance of the three factors in the microclimate.

2. The results of the study show that: A) Sites A and E had more shading time during summer and site D had more shade than site B which resulted in Ta to be 0.2–0.6 °C higher in site B than site D in summer; B) Shade-tolerant species may be useful for site D because the PAR value was lower than 1.90 MJ/m^{2}d which is good for shade-tolerant species to grow. In addition, day-neutral species should be planted in sites C and E because the PAR value was between 1.9–2.9 MJ/m^{2}d which is good for the day-neutral plants to grow. Heliophilous species can be planted around sites A and B according to the PAR value which is higher than 2.9 MJ/m^{2}d; C) The best building orientation was south by east 22.5° in the study area.

3. Based on the weather measurements data and simulations from the Autodesk Ecotect Analysis tool, the study introduced the suggestions for the modification design of this area. Firstly, after analyzing the difference between the five sites, it is not difficult to find that the sites which had green cover percentage between 30–40% were cooler that the other ones. It can be boldly conceived that if the green ratio is 30%–40% in the whole area uniformly, the air temperature will be lower and the comfort level will be higher. Secondly, there should be more shade-tolerant species around site D, while more day-neutral species should be added at sites C and E and heliophilous species can be planted around sites A and B. Meanwhile, changing the planting design by the PAR distribution diagram will be an applicable way to change the microclimate in the study area. Finally, the biggest difference height between adjacent buildings was 6 m around site A, resulting in faster airflow than other sites. On the contrary, all the buildings were almost the same height around site C, so airflow was so slow which had bad influences such as decrease the Ta because of the airflow obstruction. If the buildings have different heights and all the building orientation is perpendicular to the dominant wind direction, the best airflow condition may be possible.

4. Rapid economical development all over the world and the contemporary problems of energy insufficiency and waste. Therefore, ecological landscape management and architect design and sustainable development have become the most important factors of designers. Hence, the ecological design assistive tools such as Autodesk Ecotect Analysis tool will become the new research direction and will be widely used for various regional. Using such tools can improve the building design but also from the reducing energy consumption of buildings. The solar radiation analysis, solar orbit simulation and thermal comfort analysis can be displayed visually and specifically by Ecotect. Meanwhile, urban designer could make more considerations to various ecological methods in the begin stage of design the whole process of landscape management by analyzing them. In addition, it is significant to create an environmental-friendly, comfortable and energy saving

living environment by analyzing the initial design of the residential area, educational area, industrial area, etc.

Acknowledgements

The authors would like to express their gratitude to the Laboratory of Urban Ecosystem Planning in Kangwon National University for the financial support.

References

Aksoy, U. T.; Inalli, M. 2006. Impacts of some building passive design parameters on heating demand for a cold region, *Building and Environment* 41: 1742–1754. http://dx.doi.org/10.1016/j.buildenv.2005.07.011

Andreou, E. 2014. The effect of urban layout, street geometry and orientation on shading conditions in urban canyons in the Mediterranean, *Renewable Energy* 63: 587–596. http://dx.doi.org/10.1016/j.renene.2013.09.051

Andrew, O.; Susan, C.; Grimmond, B.; Hans-Peter, S.; Craig, A. W. 2006. Local-scale heterogeneity of photosynthetically active radiation (PAR), absorbed PAR and net radiation as a function of topography, sky conditions and leaf area index, *Remote Sensing of Environment* 103: 324–337. http://dx.doi.org/10.1016/j.rse.2005.09.021

Bonan, G. B. 2000. The microclimates of a suburban Colorado (USA) landscape and implications for planning and design, *Landscape and Urban Planning* 49: 97–114. http://dx.doi.org/10.1016/S0169-2046(00)00071-2

Bourbia, F.; Awbi, H. B. 2004. Building cluster and shading in urban canyon for hot dry climate part 2: shading simulations, *Renewable Energy* 29: 291–301. http://dx.doi.org/10.1016/S0960-1481(03)00171-X

Capeluto, I. G.; Yezioro, A.; Bleiberg, T.; Shaviv, E. 2005. From computer models to simple design tools: solar rights in the design of urban streets, in *the 9th International Building Performance Simulation Association (IBPSA),* 15–18 August 2005, Montreal, Canada, 131–138.

Cheng, V.; Ng, E.; Chan, C.; Givoni, B. 2012. Outdoor thermal comfort in sub-tropical climate: a longitudinal study based in Hong Kong, *International Journal of Biometeorology* 56: 43–56. http://dx.doi.org/10.1007/s00484-010-0396-z

Gup-yu, Q.; Hong-yong, L.; Qing-tao, Z.; Wan, C.; Xiao-jian, L.; Xiang-ze, L. 2013. Effects of evapotranspiration on mitigation of urban temperature by vegetation and urban agriculture, *Journal of Integrative Agriculture* 12: 1307–1315. http://dx.doi.org/10.1016/S2095-3119(13)60543-2

Hassan, R.; Essam, A.; Stephen, S. 2014. On the colours and properties of building surface materials to mitigate urban heat islands in highly productive solar regions, *Building and Environment* 72: 162–172. http://dx.doi.org/10.1016/j.buildenv.2013.11.005

Jouri, K.; Miljana, H. 2012. Solar energy as a design parameter in urban planning, *Energy Procedia* 30: 1143–1152. http://dx.doi.org/10.1016/j.egypro.2012.11.127

Korea Meteorological Administration (KMA) [online]. 2011 [cited 2 May 2011]. Available from Internet: http://www.kma.go.kr/weather/climate/extreme_daily.jsp?type=max_temp&mm=13&x=30&y=8

Limor, S. B.; David, P.; Evyatar, E. 2009. The cooling efficiency of urban landscape strategies in a hot dry climate, *Landscape and Urban Planning* 92: 179–186. http://dx.doi.org/10.1016/j.landurbplan.2009.04.005

Morrissey, J.; Moore, T.; Horne, R. E. 2011. Affordable passive solar design in a temperate climate: an experiment in residential building orientation, *Renewable Energy* 36: 568–577. http://dx.doi.org/10.1016/j.renene.2010.08.013

Paramita, B.; Koerniawan, M. D. 2013. Solar envelope assessment in tropical region building case study: vertical settlement in bandung, Indonesia, *Procedia Environmental Science* 17: 757–766. http://dx.doi.org/10.1016/j.proenv.2013.02.093

PMV ISO 7730: 2005. Ergonomics of the thermal environment-Analytical determination and interpretation of thermal comfort using calculation of the PMV and PPD indices and local thermal comfort criteria. 52 p.

Ruey-Lung, H.; Shiu-Ya, S. 2011. Building envelope regulations on thermal comfort in glass facade buildings and energy-saving potential for PMV-based comfort control, *Building and Environment* 46: 824–834. http://dx.doi.org/10.1016/j.buildenv.2010.10.009

Santamouris, M.; Synnefa, A.; Karlessi, T. 2011. Using advanced cool materials in the urban built environment to mitigate heat islands and improve thermal comfort conditions, *Solar Energy* 85: 3085–3102. http://dx.doi.org/10.1016/j.solener.2010.12.023

Van Esch, M. M. E.; Looman, R. H. J.; De Bruin-Hordijk, G. J. 2012. The effects of urban and building design parameters on solar access to the urban canyon and the potential for direct passive solar heating strategies, *Energy and Buildings* 47: 189–200. http://dx.doi.org/10.1016/j.enbuild.2011.11.042

Yang, L.; He, B.-J.; Ye, M. 2014. Application research of ECOTECT in residential estate planning, *Energy and Buildings* 72: 195–202. http://dx.doi.org/10.1016/j.enbuild.2013.12.040

Qian WU is currently carrying out her PhD at the Laboratory of Urban Ecosystem Planning in Kangwon National University. She graduated from college of Landscape Architecture of Northeast Forestry University (NEFU) in China and received her MSc in NEFU in 2012. Research interests: landscape architecture, urban ecosystem planning and greenspace planning.

Hyun-Kil JO received his PhD in University of Arizona in 1993. He was the vice-president of Korean Society of Environment and Ecology (2003–2005) and president of Ecological Landscape Association in the Korean Institute of Landscape Architecture (2003–2005). He is an editorial board of Landscape and Ecological Engineering (2005-present). He is currently a Prof. and Dean of College of Forest and Environmental Sciences, Kangwon National University. He is involved in scientific research for more than 25 years. His research interests are focused on ecological landscape architecture, urban ecosystem planning and greenspace planning.

12

CLIMATE-ORIENTED ASSESSMENT OF MAIN STREET DESIGN AND DEVELOPMENT IN BUDAPEST

Maria CSETE, Attila BUZASI

Department of Environmental Economics, Budapest University of Technology and Economics, Magyar Tudósok körútja 2, 1117 Budapest, Hungary

Abstract. Main streets play pivotal role in urban areas in terms of economic, social, moreover environmental contexts. Such streets are generally situated in densely built-up areas, where the adverse effects of climate change, such as rising temperature and changing precipitation patterns occur emphatically. Increasing urban heat island effect or extreme amount of runoff water during severe storms and floods significantly decrease the adaptive capacity of a city, consequently its residents becoming more vulnerable. Therefore involving climate-oriented design principles into planning and construction phase contributes to reach more sustainable and climate-friendly open spaces what are strongly relevant especially in main streets which are designed for a great amount of people. Present study provides a criteria matrix for assessing the climate-friendly level of recently renewed main streets in Budapest. Due to this assessment tool the adaptation and mitigation performance of the selected projects can be evaluated. For identifying strengths and weaknesses of a given project, a rating scheme has been applied by selecting and using 42 indicators grouped into mitigation, adaptation and awareness raising categories. Thus planners, decision-makers and other stakeholders can easily define future opportunities and challenges, accordingly this study may contribute to take urban street design practices toward climate-friendliness by paying more attention on changing local weather patterns and related consequences.

Keywords: urban sustainability, adaptation, mitigation, assessment, main street, matrix, indicators.

Introduction

As cities becoming more densely populated (UN 2014) and the impacts of climate change are getting more severe (IPCC 2014), there is an urgent need for elaborating effective evaluation systems for assessing our cities' climate-oriented performance. Urban areas are crucial sites due to their concentrated economic, social, environmental and spatial features (Andersson-Sköld *et al.* 2015), consequently reliable assessment systems are inevitable for avoiding maladaptation and for inspiring innovative adaptation actions (Medved 2016). Extreme weather events may increase the losses in urban areas, therefore cities and their population are largely vulnerable regarding climate-related hazards and risks (Birkmann *et al.* 2014; Mitchell *et al.* 2015; Norton *et al.* 2015). It can be declared that urban population has to take adaptation efforts and being more resilient regarding adverse effects and risks caused by climate change (Norton *et al.* 2015; Voskamp, Van de Ven 2015), since economic and social consequences of above mentioned impacts may go beyond the administrative borders of urban zones (Zheng *et al.* 2014). Therefore, integration of sustainability principles into urban development actions is a prerequisite of effective resource management in a given urban area, thus reducing environmental degradation, consequently making cities climate-friendly (Gaitani *et al.* 2014). Nowadays a shift can be seen from mitigation to adaptation in current urban studies and such solutions and questions are raising, which are focusing on middle-, and long term impacts of climate change and paying significant attention to adaptive capacity (Costa *et al.* 2013). It shall be emphasised that uncertainties regarding the possible impacts of climate change are significant, therefore this challenge fundamentally determines the allocation of available resources and knowledge.

As one of the East-Central European metropolises of the former socialist block, the Hungarian capital Budapest has to face a number of challenges concerning urban regeneration, moreover changing climate patterns raising numerous questions about its adaptive capacity as well.

Corresponding author: Maria Csete
E-mail: csete@eik.bme.hu

Due to its unique and specific local weather features and predicted climate patterns in the future (Probáld 2014), urban heat island and extreme floods can be determined as the key climate-related challenges of Budapest (Buzási 2014).

The beginning of the years 2000 found the historic city centre including its public spaces in fairly run-down conditions, but the EU accession of Hungary in 2004 opened new ways for the regeneration of the city. After 2004, with the help of the available EU and national subsidies Budapest got a unique possibility for the regeneration of its run-down historic city centre. By offering fast and spectacular results, public space renewal became a popular tool for local decision-makers and during the last few years all the inner districts realised their main square or high street projects. Being co-financed by the EU these projects were obligated to apply European standards including different sustainability requirements as well. Nevertheless, projects diverge in achieved qualities and are often criticised for not serving effectively the harmonised social, economic and environmental development of the city. Regarding climate aspects there are some promising initiatives emerging recently originating from both public and private actors in order to gain more consciousness in climate friendly urban regeneration and public space renewal[1].

Present study has three different, but relating aims regarding climate-friendly main street design, moreover it can foster to define sustainable and climate-oriented measures and techniques on micro scale. The first aim is focusing on assessment methodology by developing a criteria matrix for evaluating climate-friendliness of main street renewal projects. The matrix is based on CH2MILL evaluation system elaborated by Bevan *et al.* (2007), however modified in climate-friendliness point of view, such as mitigation, adaptation and awareness raising aspects have been involved in it. The system and the evaluation methodology are detailed below. By applying the elaborated criteria matrix for specific sites, the second aim of the study seeks to evaluate recent main street renewal projects of the historic city centre of the Hungarian capital Budapest regarding climate related issues. The elaborated framework represents the different aspects of design, construction and maintenance phase in a given main street project by applying comprehensive indicators and by using an easily understandable three-coloured assessment system. The third aim is to identify present challenges and future opportunities by analysing the strengths and the weaknesses of main street reconstruction projects.

The simplified assessment method can be used not only in case of Budapest but also in other cities with continental climate and common climate risks. Apart from the objective of wider application of this method, it must be emphasised, that meteorological conditions emerging in cities are greatly varied especially with respect to microclimatic features. Underlying this statement, the paper seeking to draw up an assessment method, in which adaptation, mitigation and awareness raising aspects are involved, in addition climate-oriented criteria may take into consideration in renewal projects of urban spaces. This approach has remarkable significance because of the vulnerability of cities concerning predicted extreme weather events.

1. Literature review

1.1. Urban regeneration – macro level

As it was mentioned in the previous paragraphs, changing weather patterns affect more and more significantly the urban areas, whereas urban regeneration and the preparedness for climate hazards are on the focus in recent urban studies (Reckien *et al.* 2014). Urban areas shall pay increasing attention on present and future impacts and risks generated by climate change, since cities and their population are largely exposed to the future impacts. One of the most studied and analysed phenomenon is the urban heat island (UHI) effect, i.e. the larger surface and air temperature in city centres compared to non-urban surrounding areas (Lokoshchenko 2014; Brown *et al.* 2015; Tam *et al.* 2015; van Hove *et al.* 2015). There are several causes what are responsible for increasing the UHI, such as non-reflective and low-albedo materials, human generated heat sources, densely built-up areas and the lack of green spaces and water bodies (Maimaitiyiming *et al.* 2014; Zoras *et al.* 2014). Therefore urban greening, changing surface materials and other techniques in densely populated areas have significant and well-known role in urban regeneration projects for creating sustainable and climate-friendly cities (Csete, Horváth 2012; Jim 2013). Other weather risks which have impact on our cities are changing precipitation patterns: the more frequent storm surges, and regarding to this, the more serious fluvial and pluvial floods (Pathirana *et al.* 2014; Zhang *et al.* 2015).

Climate predictions play highly important role in urban design and regeneration, since according to Costa *et al.* (2013) urbanism may be influenced by them in two manner: firstly, there may be substantial impact in planning and construction phase; secondly the approaches with regards to infrastructure will be changed because of the need for adaptation to changing climate. Similar to these statements, Benkő (2011) in her study noted that one of the most significant debate concerning urban planning is the re-designing and the re-evaluating of urban spaces in order to reach a sustainable condition.

[1] It is worth to mention the Association of Climate Friendly Municipalities, the EU financed Urban Heat Island project or the Green City Hungary movement among others.

Since climate predictions and future climate hazards, moreover related risks are changing from place to place, there is no rigorous guidelines for improving cities' adaptive capacity. It is noted that every city need divergent actions to meet its demand with regards to achieve climate goals and improve the quality of life of its citizen. Nevertheless climate-related urban regeneration techniques can be grouped into three main categories, which are the followings:

- mitigation actions, such as: enhancing green-covered areas; preferring non-motorised ways; reusing of construction materials; using low-maintenance and low-energy devices; improving accessibility by promoting public transport; traffic calming (Csete, Török 2009; Kelemen-Erdős 2012; Juhász 2013; Wamsler *et al.* 2013; Bede, Török 2014; Brown *et al.* 2015; Gaál *et al.* 2015),
- adaptation actions, such as: passive cooling and shading techniques; tackling stormwater issues; promoting integrated water management; paying attention on biodiversity in urban green spaces; designing climate-resistant materials (Gersonius *et al.* 2012; Jim 2013; Gaitani *et al.* 2014; Takebayashi *et al.* 2014; Djukic *et al.* 2016),
- awareness raising actions, such as: collaborating with citizens; using public information about actions; promoting innovative solutions; sustainable designing in planning phase; early warning system regarding heatwaves (Gaitani *et al.* 2014; Takebayashi *et al.* 2014; Groot *et al.* 2015; Lee, Painter 2015).

Costa *et al.* (2013) well summarised the changed demand with respect to urban design what is stimulated by climate change. They determined five different changes in the topic, such as: re-evaluating of present urban infrastructure; investing in improvement of applicability of innovative urban design techniques; reassessing of urban plans with respect to climate resilience; reducing the importance of strategies focusing on short-term goals and objectives; re-evaluating of risk management and civil protection strategies based on newly acceptable climate data.

Present study may reflect to above mentioned challenges and needs in the field of climate-friendly street design by applying criteria matrix with qualitative indicators for assessing climate-related aspects of main street renewals.

1.2. Public spaces – micro level

As it can be seen in the previous sections, city level plays an important role in preparing to changing climate patterns. Since cities consist of public spaces, where regeneration projects are made, micro level has become one of the most relevant place for building cities' adaptive capacity and resilience. As Costa *et al.* (2013) stated, climate change can be understood on global scale, however its impacts have local features, thus demonstrating the obvious importance of micro level in climate-friendly street design. This increasing importance has been recognised in recent urban studies, proved by rising number of publication regarding appropriate street and other public space design from sustainable and climate-friendliness point of view.

Erell (2008) emphasised the importance of application of climatology in urban design processes by distinguishing the challenges and opportunities of design techniques and influencing factors regarding urban microclimate on city and micro level. Jim (2013) provided recommendations concerning urban forestry based on case studies of 100 cities from 25 countries. His recommendations contribute urban spaces to become climate-proof ones concerning ecological design, greenspace geometry, biodiversity, tree and soil issues etc. The statements can be applied not only in design but in construction phase as well.

Zoras *et al.* (2014) performed simulations regarding the use of cool materials in the centre of Florina, Greece. Their results obtained show that surface temperature could be reduced by using cool materials. Rehan (2013) in his comprehensive study conceptualised the sustainability of streetscape and street, as well as determined the main elements to achieve sustainable streetscape. Shashua-Bar *et al.* (2012) analysed the opportunities in amelioration of citizens' thermal comfort in summertime by using climate-oriented urban planning approaches and passive cooling techniques. According to the authors, greening and vegetation issues play more significant role in decrease of temperature compared to actions regarding modification of albedo of walls of buildings. Connecting with latter study, Takebayashi *et al.* (2014) studied technologies concerning decrease of UHI and provided the same consequences, completing this with a statement, that water retention has also important role in decreasing the surface temperature.

Gaitani *et al.* (2014) analysed microclimatic features for urban regeneration project in case of Acharnes. They suggested recommendations with regards to land coverage and building materials in order to ameliorate the thermal comfort in selected streets of Acharnes. Riera Pérez, Rey (2013) focused on neighbourhood scale, between building and city level which aspect provide opportunities to the evaluation of actions and complex urban renewal projects in a highly sensitive way. Beside, on neighbourhood scale both sustainable building and city level exist and can be analysed. Radulovic *et al.* (2015) focused on extreme heat waves and related impacts in case of Serbian cities. They studied compact urban structure, as an opportunity to mitigate urban heat in lights of urban planning whether it has an importance in creating climate-friendly cities.

Klemm *et al.* (2015) emphasised and numerically simulated the potential role of street greenery in moderate climates in case of nine streets of Utrecht, the Netherlands. It can be stated that green facilities in urban streets play significant role in increasing thermal comfort and mitigating heat discomfort. Through above mentioned mechanisms (evapotranspiration, shading, etc.) they contribute to create not only attractive but climate-resilient streets in cities with soft and hard measures. The results of their study pointed out that significant role of trees in urban streets can be observed, moreover taking green aspects into consideration in designing streets is required for achieving global sustainability- and climate-oriented goals.

Based on the reviewed literature it can be stated that street-level refurbishments and renewals plays significant role on city scale, because they are crucial elements of climate-friendliness in a given city.

1.3. Existing evaluation methods

However numerous set of indicators exists regarding urban renewal on international scale, there is no widely accepted method for assessing impacts and contributions of renewal projects in lights of climate-oriented challenges and opportunities.

Norton *et al.* (2015) elaborated a five-step approach for evaluating potential impacts of green infrastructure on micro scale, as a comprehensive urban development tool. The role and the importance of green infrastructure in light of present study is undoubted through both mitigation and adaptation efforts. Therefore green areas have to be treated in the centre of urban renewal projects in developing the most effective urban street renewal strategy. Andersson-Sköld *et al.* (2015) evaluated different scenarios in the field of urban renewal by applying individual rankings for taking environmental and social impacts into consideration, then using these information, recommendations have been drawn up. The main aim of the study was to reduce climate-related risks through a renewal project of Gothenburg with paying more attention on inhabitants' perceptions. A framework for assessing climate-.related risks in the field of extreme heat waves, floods and air pollution has been provided in order to be able to avoid latter incompetent urban development decisions. The potential impacts of the given actions are ranging from −2 (counter-productive effect) to +2 (very effective). Finally, the design alternatives of impact categories (such as flood risk mitigation; urban heat stress; energy and raw material consumption; wellbeing) have been scaling from −3 (very negative contribution) to +3 (very positive contribution). Madureira *et al.* (2015) focusing on social factors by surveying and evaluating urban residents' belief associated with urban green spaces in case of Paris, Angers, Lisbon and Porto. They stated that urban residents rated climate-related aspects at a lower level than well-being and recreation-based benefits provided by urban green spaces. Since urban residents and their needs are in the heart of a given regeneration project, understanding their opinion is crucial in the co-operation phase. Therefore present study provides indicators regarding co-operation between planners, constructers and residents. The Sustainable Pittsburgh released a sustainability assessment tool (Sustainable Pittsburgh 2008) as guidance for decision makers and residents. The brochure introduced different aspects of sustainable urban renewal projects, such as:

- site selection (brownfield or greenfield),
- site preparation (tree removal; water bodies; green areas),
- infrastructure (capacity issues; storm water runoff),
- site design (environmental aspects; bike lanes; accessibility; traffic calming),
- transportation (parking places; promoting non-motorised modes),
- housing (affordability; design),
- buildings (materials; orientation),
- and community character (street furniture; community spaces; public involvement).

Each subcategory have checklists (for example "preservation of flora and fauna habitat and migration corridors" in site preparation category), so the reporter can easily decide and mark in a cell whether a given statement is true or not in case of a given site renewal project.

In summary it can be stated that there are several studies regarding ex-post analysis of a given renewal project, they are mainly focusing on a few aspects of climate change. Involvement of broader scale of subsystems, such as evaluation of both mitigation and adaptation topics – completed by social factors, such as awareness raising – is lacking nowadays in the scientific field, therefore present study can be fill the gap existing in recent urban studies.

2. Methodology

2.1. Criteria matrix for evaluating climate-friendliness

Based on the sustainable urban street options of CH2MHILL (Bevan *et al.* 2007) a simplified criteria matrix for the evaluation of main street renewals' climate friendliness aspects have been elaborated. The evaluation matrix can be divided into three main categories, namely mitigation, adaptation and awareness raising. Each category includes the main aspects from urban regeneration and climate-oriented point of view, by covering the most frequently mentioned topics in present scientific are (see Table 1). Mainly qualitative indicators are involved into each topic which are rated in a three-scale scheme from plus to minus across neutral (minus sign refers to bad performance, zero signs moderate one, and plus ranks well-performed actions). This rating system contributes to straightforward results and can support the evaluation of strengths and weaknesses related to an examined urban main street regeneration projects, moreover it can support planners and decision makers for identifying consequences and future improvement opportunities. Each rating scheme is slightly different from others, in addition to this some indicators have a two-step approach, if an intermediate value could not be identified (see Appendix).

Selection of indicators is based on the set of indicators composed by Bevan *et al.* (2007), however restructured from a climate friendly perspective, modified for local features and data availability. As it was mentioned above, present set of indicators and assessment system are mainly referring to main street renewal projects and their consequences in Budapest, however with a slight modification considering local circumstances it can be applied for assessing other renewal projects in other cities. Mitigation topics reflect to reduction of GHG emissions through site preparation; promoting and preferring non-motorised transportation modes; material issues in maintenance and construction; and green areas. Adaptation ones are based on the improved preparation and adaptive capacity of a given main street for changing weather patterns, such as rising temperature and the risk of more severe floods and storms. As it was declared in the previous section, local climate features and the predicted changes of them largely endanger the adaptive capacity and the resilience of Budapest, moreover this statement is particularly true for densely built-up areas, such as main streets in the downtown. Consequently water management and shading issues are strongly emphasised within adaptation topics, however climate-resistance of materials and green areas have been taken into consideration as well.

Table 1. Evaluated subsystems regarding mitigation, adaptation and awareness raising

Mitigation
Bicycle facilities
Promoting non-motorised transport options
Maintenance
Construction
Public transport facilities
Green areas
Adaptation
Storm water issues
Material and construction elements
Usage of porous pavements
Shading
Water management
Biodiversity issues
Awareness raising
User-oriented solutions
Construction-oriented solutions

Awareness raising aspect acts a slightly different consideration in urban climate studies and often has less attention on it, however stakeholder involvement and information spreading are crucial in developing efficient strategies and creating climate-resilient and sustainable open spaces in urban areas. Therefore the aspects of user-, and construction-oriented solutions have been involved into present analysis by evaluating collaboration between users and planners, by analysing information spreading and the existence of sustainability issues in planning phase and off-site parking.

Data upload of this evaluation matrix is based on divergent methods, such as onsite analysis and planner interviews via phone and email. By performing onsite analysis mainly mitigation- and adaptation-oriented subsystems have been uploaded, however in case of tree removal, construction issues, storm water management actions and water-appropriate plantings there was a need for interviews. Similarly to mentioned subsystems, almost all of the them have been assessed by interviewing planners of the examined projects via phone or email.

2.2. Study area

For the on-site analysis five recently renewed main streets have been chosen, all of them located within the historic inner city districts of Budapest (Fig. 1). The sites correspond in many characteristics that with certain limitations facilitate the comparison of the survey results. These characteristics include:

- high streets: all of them function as the main street of their districts with attractive and intensive retail sector;

Fig. 1. Location of the examined main streets

- rehabilitation sites: all the streets have been totally refurbished during the last two decades, when sustainability and climate-based issues could been taken into consideration;
- many of them are co-financed by the EU;
- traffic calming was a basic objective of the renewal projects.

However, there are some important factors that differ between the sites:

- developers of all the projects are different, most of the projects were coordinated by the local district municipality of each site, except one project that has been realised by the city municipality;
- urban design solutions differ in lack of any kind of common design codes or other design requirements except of basic national regulations, district municipality building codes and utility requirements.

General data about studied projects and areas can be seen in Table 2, that summarises the locations, developers, the date of the given renewal and the length of the streets. It can be stated that the New Main Street project is the biggest action regarding the length of the street with almost 2 kilometres, however the renewed stage of Lövőház Street is only 300 metres. As it was mentioned before, the common primary aim of the projects was traffic calming, therefore mitigation-orientation can be anticipated. This statement shall enhanced by taking the dates of renewals into consideration, since worldwide in the late 90's and the early 2000's mitigation actions outweighed adaptation ones.

3. Results

Table 3 shows the mitigation-oriented features of the renewed streets, where each indicator and the rating scheme can be seen. As it was stated before, traffic calming was distinguished and primary aim of the examined main street projects, therefore transportation issues have

Table 2. General data of the studied areas

Name	(1) New Main Street of the Downtown	(2) Király Street	(3) Baross Street – Krúdy Street axis	(4) Ráday Street	(5) Lövőház Street
Location	District V	District VI and VII	District VIII	District IX	District II
Developer	Local Municipality of District V	Local Municipalities of District VI and VII	Municipality of Budapest	Local Municipality of District IX	Local Municipality of District II
Date of renewal	2010, 2014	2005, 2010	1997, 2002, 2004	1997–2002	2012
Length	1800 m	700 m	500 m	800 m	300 m

Table 3. Mitigation-oriented features of the renewed streets

		Ráday Street	Baross Street	New Main Street	Lövőház Street	Király Street
Bicycle facilities	Bike racks	0	0	+	0	–
	Bicycle route signage	0	+	+	+	–
	Bicycle priority at signals	–	–	–	–	–
Promoting non-motorised transport options	Traffic calming features	+	+	+	+	+
	Remove on-street parking	–	–	–	0	–
Maintenance	Low maintenance pavements	+	+	+	+	0
	Low energy lighting	0	0	0	0	0
	Solar power photovoltaic energy sources	–	0	0	–	–
Construction	Usage of recycled materials	0	+	0	–	–
	Minimal tree removal	NR	+	+	+	NR
	Use of locally obtained materials	0	0	0	N/A	0
Public transport facilities	Enhanced bus zones and shelters	NR	NR	+	NR	NR
	Accessibility by public transport	+	+	+	+	+
Green areas	Change of green-covered area	+	+	+	+	+
	Create pocket green spaces	0	0	+	+	0
	No. Of trees / 100 m	–	+	-	+	0

been emphasised in the set of indicators with five variables. Based on the results of the bicycle issues and of the non-motorised ways it can be stated that in case of Király Street there is a lack of efficient actions in this field, however traffic calming measures (speed humps, chicanes, and chokers) emerged in an appropriate way, similar to the other projects. Removing on-street parking is an other common weak point of the studied renewals by not banning parking places in the streets, though one half of the Lövőház Str. (from Fény Str. to Káplár Str.) is a completely no-traffic zone. The results of maintenance issues is unequal, since low maintenance payments have been applied in case of almost the total stage of the streets, however solar power photovoltaic energy sources emerged only in the Baross and the New Main Street in a small amount (mainly by supplying power to parking meters).

Usage of recycled materials was not a common part of the renewals, it can be seen on Baross Str., however on Lövőház Str. and Király Str. recycled materials have not been used for reducing the footprint of the projects. Minimal tree removal, or more likely additional planting can be observed in the renewed streets in all case, however the original aspect of tree removing was not relevant on Ráday Str. and Király Str. because of the lack of trees previously. The assessment of the accessibility issues is based on the public transport fatalities and stops. Since each of the streets are situated in downtown area, proximity of public transport stops are in 200 metres from the ends of the studied streets. Actions regarding bus stops and related zones are not relevant except in case of New Main Street, because of the lack of public transport route on the other streets.

Green areas are highly relevant actors in offsetting GHG emissions in urban areas, however they have adaptation features, based on the indicators this topic has been involved into the mitigation category. The first evaluated factor was the change of green-covered area, what was found as strength of the projects, since in case of each street, the ration of green-covered area was clearly increased. The fragmentation of green spaces was quite low on New Main Street and Lövőház Str., but moderate on the other three cases as well, what shall endanger ecological connectivity, therefore mitigation capacity shall be reduced in these sites because of the lack of well performed and connected green spaces. The no. of trees / 100 m is the only quantitative indicator in this study, so that the determination of effective value of it is based on literature review (McBride 2008; City of Vancouver 2011; City of Philadelphia 2014). The sustainable and liveable tree spacing is about 8–12 m between trees, so completed by the length of the streets, an effective tree density can be calculated. In case of Ráday Str. and New Main Street, slightly more than 3.5 trees can be observed in average, this value is approximately 7 trees / 100 m on Király Str., 10 for Baross Str. and almost 12 for Lövőház Str. However it must be emphasised that average size of tree canopies was the smallest on Király Str., therefore shading capacity was moderate on this street (see in the next paragraph).

In Table 4, adaptation-centred indicators and related results have been summarised and visualised by the applied three-scored approach. In case of Budapest, increasing temperature and more intense storms and flood risks are the most relevant predicted weather extremes in the future. Therefore water management and shading have been involved into the analysis in a highly emphasised way by using 13 indicators in these fields. It can be stated that storm water management is a totally weak point of the projects, since there is no reusing, treating or reducing actions regarding storm water runoffs on the analysed main streets. However in case of Ráday Str., New Main Street and Király Str., there is an opportunity for infiltrating run off water into a basin, but these technical solutions are out of service. The rating of material and construction elements is basically more positive, since typically modular elements and higher strength pavements have been applied (except on Király Str., where some damages can be observed on the pavement). Usage of climate-resistant design principles and street furniture could not been proved, although each of the furniture have been replaced, but climate-resistance was not a perspective in design and construction phase of the projects regarding street furniture. Opposite of the above mentioned perspectives, usage of porous pavements is widely and properly applied adaptation technique, since impermeable surfaces do not prevent storm water runoff, consequently the risk of storm floods shall increasing. One exception is Király Str., where impermeable surface has been used for the roads, and an other exception is Baross Str., where porous pavements have been partly used for traffic lanes and parking areas. Since intensive warming must be faced by Budapest in the near future, shading of main streets in densely built-up urban areas is crucial in terms of resilience.

In case of three of the studied streets, namely Ráday Str., New Main Street and Király Str., retailers' and restaurant owners' contribution is essential in effective shading, consequently tree canopy layers are not continuous on these streets. However on Ráday Str. there are some shelters when pedestrians can be protected from weather extremes. Existences of public drinking water equipment and water-appropriate plantings have been covered by water management issues in the analysis with moderate or bad results. Firstly, public drinking fountains can be found only on Baross Str. and New Main Street, however in a very small number (1 or 2 fountains per street). Furthermore water-appropriate planting can be applied for designing main streets in a moderate way, since local climate features and predicted changing precipitation patterns have not been into consideration in planning site, although urban-tolerant species have been selected for green spaces. Finally, biodiversity issues are emerging through tree planting and selecting practices and ecological connectivity. With few exceptions, it can be noted that every examined main street has a moderate performance in this field, because of the lack of climate-oriented designing approaches.

Based on Table 5 which represents the results regarding awareness raising aspects it can be stated that these directions of climate-conscious design is in its infancy. This assumption is based on the lack of information concerning collaboration with stakeholders and off-site worker parking. Moreover the streets' performance with regards to information spreading is relatively bad or at least moderate due to the lack of public information boards. Public information boards are crucial in awareness raising by informing broad scale of urban residents about divergent actions and used technology in a given project in terms of sustainability or climate-friendliness. In summary it can be stated that sustainability principles have been partly involved into planning phase by taking technical solutions and opportunities into consideration, however neither project can be seen as sustainable main street renewal one.

Table 4. Adaptation-centred indicators and results

		Ráday Street	Baross Street	New Main Street	Lövőház Street	Király Street
Storm water issues	Reuse of storm water for irrigation	–	–	–	–	–
	Surface conveyance for storm water	–	–	–	–	–
	Reduced storm water vaults through natural drainage systems	–	–	–	–	–
	Storm water infiltration basins in planters	0	–	0	N/A	0
	Rain gardens for storm water infiltration	–	–	–	–	–
	Storm water treatment vaults	–	–	–	–	–
Material and construction elements	Incorporation of recycled vegetative material	0	+	0	+	0
	Precast or modular construction elements	+	+	+	+	0
	Climate-resistant designs and materials	0	0	0	0	0
	Higher strength concrete pavements	+	+	+	+	0
Usage of porous pavements	Porous pavement for traffic lanes	+	0	+	+	–
	Porous pavement in parking areas	+	0	+	+	+
	Porous sidewalk pavements	+	+	+	+	+
Shading	Retention or introduction of tree cover to reduce heat gain on paved areas	0	+	0	+	0
	Shelters for protecting pedestrians from weather	+	+	0	+	0
Water management	Public drinking fountains	–	0	0	–	–
	Water-appropriate plantings	0	0	0	0	0
Biodiversity issues	Diverse plant/tree selections	0	0	0	0	0
	Local climate-oriented plant/tree selections	0	0	0	0	0
	Enable ecological connectivity	0	+	0	+	0

Table 5. Indicators and results regarding awareness raising

		Ráday Street	Baross Street	New Main Street	Lövőház Street	Király Street
User-oriented solutions	Trash and recycling receptacles	0	0	+	0	0
	Collaborate with property owners for mutual benefit	N/A	+	+	N/A	N/A
	Promotional information regarding sustainable elements in the project	–	–	0	–	–
	Public information about actions and used technology	–	–	0	0	–
Constructor-oriented solutions	Off-site construction worker parking	N/A	+	N/A	N/A	N/A
	Sustainable design in planning phase	0	0	0	0	0

4. Discussion

Based on the results of the assessment detailed above, strengths and weaknesses regarding climate-orientation in recent main street design can be defined. First of all it can be stated that there is no clear difference between overall adaptation and mitigation performance of the projects, however as it was mentioned above mitigation-oriented technical solutions was well known on date of the renewals. Traffic calming features, green area-related action can be clearly emphasised as strengths of the examined projects, so applied technologies and design practices shall be continued and spread. On the other hand, bicycle issues, removing on-street parking, using more amount of renewable energy sources and designing with existing and possibly reusable materials are the main opportunities what shall be taken into consideration in future climate-oriented urban main street designs. Involving these missing elements of present street design practices into future projects can enhance sustainability of main streets and contribute to reduce GHG emissions by using innovative and well performed infrastructure solutions, consequently the magnitude of negative effects of climate change shall be reduced.

From adaptation point of view, studied streets are extremely vulnerable to extreme precipitation and related floods because of the lack of effective storm water management. For improving the assessed main streets' adaptive capacity, storm water issues must be taken into consideration during planning and developing phase, since main streets are essentially situated in downtowns. Due to their economic role by attracting retail stores, restaurants and by providing great opportunities for urban residents to meet their recreational needs, enhancing resilience of main streets play pivotal role in the overall cities' adaptive capacity and related vulnerability. After the revision of adaptation-oriented indicators and results, it can be noted that the usage of porous pavements and the strength of used materials are properly performed from climatic point of view, however paying more attention on climate-resistant street furniture is needed. Shading, water management and biodiversity issues are such topics where relevant and greatly important opportunities can be observed by:

- installing public drinking fountains;
- taking water-appropriateness and local climate-orientation into consideration during plant and tree selection;
- paying more attention on enhancing ecological connectivity.

These recommendations completed by total improvement of storm water management and existing good practices contribute to provide more attractive, vibrant and viable open spaces in urban areas.

Based on the results of the awareness raising category, it can be stated that this aspect of climate-friendliness is in its infancy due to the inappropriate information spreading. However collaboration with property owners were effective and proper in case of Baross Str. and New Main Street projects by involving local residents into decision making processes, there is no information regarding this aspect from the other sites. This lack of information makes comprehensive analysis and recommendations more difficult, but the spread of existing good practices with ex-ante stakeholder involvement can be recommended.

Conclusions

Based on the results from the applied criteria matrix, it can be stated that the climate-friendly level of studied main streets in Budapest is moderate. It means that there are good practices observed during on-site analysis, however there are some weaknesses as well, such as bicycle issues, storm water management or information spreading. As improving adaptive capacity of each of cities' subsystem, such as infrastructure elements can contribute to increasing resilience on a larger scale, it can be noted that climate-oriented street design is a crucial part in reducing our cities' and their residents' vulnerability. Therefore adaptation, mitigation and awareness raising aspects must be involved into planning and construction phase of main street design. In terms of climate resilience, this previously mentioned planning approach may contribute to increase adaptive capacity of both infrastructure and urban residents by applying innovative and effective tools for involving climate aspects into street design principles. Since urban main streets have the ability to attract competitive enterprises, consequently they can increase the economic power of a given city. Therefore urban streets with climate-oriented design practices may influence climate resilience and vulnerability on larger scale mainly due to their economic-dependency, however taking climate aspects into urban planning processes may contribute to both economic prosperity and less vulnerability in a given city.

Funding

This study was supported by the János Bolyai Research Scholarship of the Hungarian Academy of Sciences.

References

Andersson-Sköld, Y.; Thorsson, S.; Rayner, D.; Lindberg, F.; Janhäll, S.; Jonsson, A.; Moback, U.; Bergman, R.; Granberg, M. 2015. An integrated method for assessing climate-related risks and adaptation alternatives in urban areas, *Climate Risk Management* 7: 31–50. http://dx.doi.org/10.1016/j.crm.2015.01.003

Bede, Z.; Török, Á. 2014. Theoretical Investigation of Traffic Equilibrium on Bridges, *Transport and Telecommunication Journal* 15(2): 144–150. http://dx.doi.org/10.2478/ttj-2014-0013

Benkő, M. 2011. Budapest Urban Blocks and their Sustainability, *Architektura and Urbanizmus* 45(3–4): 188–201.

Bevan, T. A.; Sklenar, O.; McKenzie, J. A.; Derry, W. E. 2007. Sustainable urban street design and assessment, in *3rd Urban Street Symposium*, 24–27 June 2007, Seattle, Washington, USA.

Birkmann, J.; Garschagen, M.; Setiadi, N. 2014. New challenges for adaptive urban governance in highly dynamic environments: Revisiting planning systems and tools for adaptive and strategic planning, *Urban Climate* 7: 115–133. http://dx.doi.org/10.1016/j.uclim.2014.01.006

Brown, R.; Vanos, J.; Kenny, N.; Lenzholzer, S. 2015. Designing urban parks that ameliorate the effects of climate change, *Landscape and Urban Planning* 138: 118–131. http://dx.doi.org/10.1016/j.landurbplan.2015.02.006

Buzási, A. 2014. Will Budapest be a climate-resilient city? – Adaptation and mitigation challenges and opportunities in development plans of Budapest, *European Journal of Sustainable Development* 3(4): 277–288. http://dx.doi.org/10.14207/ejsd.2014.v3n4p277

City of Philadelphia. 2014. *City of Philadelphia green streets design manual* [online], [cited 2 April 2015]. Available from Internet: http://www.phillywatersheds.org/img/GSDM/GSDM_FINAL_20140211.pdf

City of Vancouver. 2011. *Street tree guidelines for the public realm. Year 2011 Revision* [online], [cited 2 April 2015]. Available from Internet: http://vancouver.ca/files/cov/StreetTreeGuidelines.pdf

Costa, J.; de Sousa, J.; Matos Silva, M.; Santos Nouri, A. 2013. Climate change adaptation and urbanism: a developing agenda for Lisbon within the twenty-first century, *Urban Design International* 19(1): 77–91. http://dx.doi.org/10.1057/udi.2013.15

Csete, M.; Horváth, L. 2012. Sustainability and green development in urban policies and strategies, *Applied Ecology and Environmental Research* 10(2): 185–194. http://dx.doi.org/10.15666/aeer/1002_185194

Csete, M.; Török, Á. 2009. Investment optimisation for the reduction of the emissions of settlements, *Periodica Polytechnica Social and Management Sciences* 17(1): 41–45. http://dx.doi.org/10.3311/pp.so.2009-1.04

Djukic, A.; Vukmirovic, M.; Stankovic, S. 2016. Principles of climate sensitive urban design analysis in identification of suitable urban design proposals. Case study: Central zone of Leskovac competition, *Energy and Buildings* 115: 23–35. http://dx.doi.org/10.1016/j.enbuild.2015.03.057

Erell, E. 2008. The application of urban climate research in the design of cities, *Advances in Building Energy Research* 2(1): 95–121. http://dx.doi.org/10.3763/aber.2008.0204

Gaál, G.; Horváth, E.; Török, Á.; Csete, M. 2015. Analysis of public transport performance in Budapest, Hungary, *Periodica Polytechnica Social and Management Sciences* 23(1): 68–72. http://dx.doi.org/10.3311/PPso.7724

Gaitani, N.; Santamouris, M.; Cartalis, C.; Pappas, I.; Xyrafi, F.; Mastrapostoli, E.; Karahaliou, P.; Efthymiou, C. 2014. Microclimatic analysis as a prerequisite for sustainable urbanisation: Application for an urban regeneration project for a medium size city in the greater urban agglomeration of Athens, Greece, *Sustainable Cities and Society* 13: 230–236. http://dx.doi.org/10.1016/j.scs.2014.02.006

Gersonius, B.; Nasruddin, F.; Ashley, R.; Jeuken, A.; Pathirana, A.; Zevenbergen, C. 2012. Developing the evidence base for mainstreaming adaptation of stormwater systems to climate change, *Water Research* 46(20): 6824–6835. http://dx.doi.org/10.1016/j.watres.2012.03.060

Groot, A.; Bosch, P.; Buijs, S.; Jacobs, C.; Moors, E. 2015. Integration in urban climate adaptation: Lessons from Rotterdam on integration between scientific disciplines and integration between scientific and stakeholder knowledge, *Building and Environment* 83: 177–188. http://dx.doi.org/10.1016/j.buildenv.2014.07.023

IPCC. 2014. *Climate Change 2014: Impacts, adaptation, and vulnerability. Part A: global and sectoral aspects. Contribution of working group ii to the fifth assessment report of the intergovernmental panel on climate change.* Cambridge, United Kingdom and New York, NY, USA: Cambridge University Press.

Jim, C. 2013. Sustainable urban greening strategies for compact cities in developing and developed economies, *Urban Ecosystems* 16(4): 741–761. http://dx.doi.org/10.1007/s11252-012-0268-x

Juhász, M. 2013. Travel Demand Management – Possibilities of influencing travel behaviour, *Periodica Polytechnica Transportation Engineering* 41(1): 45–50. http://dx.doi.org/10.3311/PPtr.7096

Kelemen-Erdős, A. 2012. Sustainable public transport: A Central European Study, *Periodica Polytechnica and Social and Management Sciences* 20(2): 81–90. http://dx.doi.org/10.3311/pp.so.2012-2.03

Klemm, W.; Heusinkveld, B.; Lenzholzer, S.; van Hove, B. 2015. Street greenery and its physical and psychological impact on thermal comfort, *Landscape and Urban Planning* 138: 87–98. http://dx.doi.org/10.1016/j.landurbplan.2015.02.009

Lee, T.; Painter, M. 2015. Comprehensive local climate policy: the role of urban governance, *Urban Climate* 14: 566–577. http://dx.doi.org/10.1016/j.uclim.2015.09.003

Lokoshchenko, M. A. 2014. Urban 'heat island' in Moscow, *Urban Climate* 10(3): 550–562. http://dx.doi.org/10.1016/j.uclim.2014.01.008

Madureira, H.; Nunes, F.; Oliveira, J.; Cormier, L.; Madureira, T. 2015. Urban residents' beliefs concerning green space benefits in four cities in France and Portugal, *Urban Forestry & Urban Greening* 14(1): 56–64. http://dx.doi.org/10.1016/j.ufug.2014.11.008

Maimaitiyiming, M.; Ghulam, A.; Tiyip, T.; Pla, F.; Latorre-Carmona, P.; Halik, Ü.; Sawut, M.; Caetano, M. 2014. Effects of green space spatial pattern on land surface temperature: implications for sustainable urban planning and climate change adaptation, *ISPRS Journal of Photogrammetry and Remote Sensing* 89: 59–66. http://dx.doi.org/10.1016/j.isprsjprs.2013.12.010

McBride, J. R. 2008. A method for characterizing urban forest composition and structure for landscape architects and urban planners, *Arboriculture & Urban Forestry* 34(6): 359–365.

Medved, P. 2016. A contribution to the structural model of autonomous sustainable neighbourhoods: new socio-economical basis for sustainable urban planning, *Journal of Cleaner Production* 120: 21–30. http://dx.doi.org/10.1016/j.jclepro.2016.01.091

Mitchell, D.; Enemark, S.; van der Molen, P. 2015. Climate resilient urban development: why responsible land governance is important, *Land Use Policy* 48: 190–198. http://dx.doi.org/10.1016/j.landusepol.2015.05.026

Norton, B.; Coutts, A.; Livesley, S.; Harris, R.; Hunter, A.; Williams, N. 2015. Planning for cooler cities: a framework to prioritise green infrastructure to mitigate high temperatures in urban landscapes, *Landscape and Urban Planning* 134: 127–138. http://dx.doi.org/10.1016/j.landurbplan.2014.10.018

Pathirana, A.; Denekew, H.; Veerbeek, W.; Zevenbergen, C.; Banda, A. 2014. Impact of urban growth-driven landuse change on microclimate and extreme precipitation – a sensitivity study, *Atmospheric Research* 138: 59–72. http://dx.doi.org/10.1016/j.atmosres.2013.10.005

Probáld, F. 2014. The urban climate of Budapest: past, present and future, *HunGeoBull* 63(1): 69–79. http://dx.doi.org/10.15201/hungeobull.63.1.6

Radulovic, S.; Bobic, A.; Cvejic, J.; Tutundzic, A. 2015. Growing cities in Serbia in the light of projected global warming: the situation in urban morphological zones. *Urban Forestry & Urban Greening* 14(1): 99–106. http://dx.doi.org/10.1016/j.ufug.2014.12.002

Reckien, D.; Flacke, J.; Dawson, R.; Heidrich, O.; Olazabal, M.; Foley, A.; Hamann, J.; Orru, H.; Salvia, M.; De Gregorio Hurtado, S.; Geneletti, D.; Pietrapertosa, F. 2014. Climate change response in Europe: what's the reality? Analysis of adaptation and mitigation plans from 200 urban areas in 11 countries, *Climatic Change* 122(1–2): 331–340.

Rehan, R. 2013. Sustainable streetscape as an effective tool in sustainable urban design, *HBRC Journal* 9(2): 173–186. http://dx.doi.org/10.1016/j.hbrcj.2013.03.001

Riera Pérez, M.; Rey, E. 2013. A multi-criteria approach to compare urban renewal scenarios for an existing neighbourhood. Case study in Lausanne (Switzerland), *Building and Environment* 65: 58–70. http://dx.doi.org/10.1016/j.buildenv.2013.03.017

Shashua-Bar, L.; Tsiros, I.; Hoffman, M. 2012. Passive cooling design options to ameliorate thermal comfort in urban streets of a Mediterranean climate (Athens) under hot summer conditions, *Building and Environment* 57: 110–119. http://dx.doi.org/10.1016/j.buildenv.2012.04.019

Sustainable Pittsburgh. 2008. *Sustainability assessment tool – Southwestern Pennsylvania. Guidance for municipal leaders, developers and concerned citizen* [online], [cited 4 April 2015]. Available from Internet: http://www.sustainablepittsburgh.org/Pub/assessment-tool1.pdf

Takebayashi, H.; Kimura, Y.; Kyogoku, S. 2014. Study on the appropriate selection of urban heat island measure technologies to urban block properties, *Sustainable Cities and Society* 13: 217–222. http://dx.doi.org/10.1016/j.scs.2014.01.008

Tam, B.; Gough, W.; Mohsin, T. 2015. The impact of urbanization and the urban heat island effect on day to day temperature variation, *Urban Climate* 12: 1–10. http://dx.doi.org/10.1016/j.uclim.2014.12.004

UN. 2014. *World urbanization prospects: the 2014 revision.* United Nations Department of Economic and Social Affairs, Population Division, New York.

van Hove, L.; Jacobs, C.; Heusinkveld, B.; Elbers, J.; van Driel, B.; Holtslag, A. 2015. Temporal and spatial variability of urban heat island and thermal comfort within the Rotterdam agglomeration, *Building and Environment* 83: 91–103. http://dx.doi.org/10.1016/j.buildenv.2014.08.029

Voskamp, I.; Van de Ven, F. 2015. Planning support system for climate adaptation: composing effective sets of blue-green measures to reduce urban vulnerability to extreme weather events, *Building and Environment* 83: 159–167. http://dx.doi.org/10.1016/j.buildenv.2014.07.018

Wamsler, C.; Brink, E.; Rivera, C. 2013. Planning for climate change in urban areas: from theory to practice, *Journal of Cleaner Production* 50: 68–81. http://dx.doi.org/10.1016/j.jclepro.2012.12.008

Zhang, B.; Xie, G.; Li, N.; Wang, S. 2015. Effect of urban green space changes on the role of rainwater runoff reduction in Beijing, China, *Landscape and Urban Planning* 140: 8–16. http://dx.doi.org/10.1016/j.landurbplan.2015.03.014

Zheng, S.; Kahn, M.; Sun, W; Luo, D. 2014. Incentives for China's urban mayors to mitigate pollution externalities: the role of the central government and public environmentalism, *Regional Science and Urban Economics* 47: 61–71. http://dx.doi.org/10.1016/j.regsciurbeco.2013.09.003

Zoras, S.; Tsermentselis, A.; Kosmopoulos, P.; Dimoudi, A. 2014. Evaluation of the application of cool materials in urban spaces: a case study in the center of Florina, *Sustainable Cities and Society* 13: 223–229. http://dx.doi.org/10.1016/j.scs.2014.01.007

Maria CSETE. PhD is an Associate Professor, Head of Department at the Budapest University of Technology and Economics, Department of Environmental Economics. She has PhD in Environmental Management Sciences and MSc in Engineering Management at BME; MSc in Regional- and urban development at Sorbonne – Paris IV. Her research interests are in environmental policy and economics, regional and spatial planning, mitigation and adaptation strategies, urban sustainability. Maria Csete has 100 publications, 36 in national and international scientific journals (several indexed in Scopus or Web of Science), 13 book chapters, 1 book and several conference papers.

Attila BUZASI. He is an Assistant Lecturer and PhD Candidate at the Department of Environmental Economics, Budapest University of Technology and Economics, Hungary. His research interests include climate change (mitigation and adaptation), regional and urban sustainability. Attila Buzasi has 12 publications including 2 scientific articles indexed in Scopus.

13

FEATURE SELECTION OF VARIOUS LAND COVER INDICES FOR MONITORING SURFACE HEAT ISLAND IN TEHRAN CITY USING LANDSAT 8 IMAGERY

Nikrouz MOSTOFI[a], Mahdi HASANLOU[b]

[a]*Department of Surveying Engineering, Islamic Azad University, South Tehran Branch, 1777613511 Tehran, Iran*

[b]*School of Surveying and Geospatial Engineering, Norh Karegar, College of Engineering, University of Tehran, 1439957131 Tehran, Iran*

Abstract. Recently, scientists have been taking a great interest in Global warming issue, since the global surface temperature has been significantly increased all through last century. The surface heat island (SHI) refers to an urban area that has higher surface temperatures than its surrounding rural areas due to urbanization. In this paper, Tehran city is used as case study area. This paper tries to employ a quantitative approach to explore the relationship between land surface temperature and the most widespread land cover indices, and select proper (urban and vegetation) indices by incorporating supervised feature selection procedures using Landsat 8 imageries. In this regards, genetic algorithm is incorporated to choose best indices by employing kernel base one, support vector regression and linear regression methods. The proposed method revealed that there is a high degree of consistency between affected information and SHI dataset (*RMSE* = 0.9324, *NRMSE* = 0.2695 and R^2 = 0.9315).

Keywords: Surface heat island, Land use/cover, Support vector regression, linear regression model, Genetic algorithm.

Introduction

The surface temperature is a substantial factor in the study of urban climatology. It changes the air temperature of the lowest layers of the urban area. The surface temperature is also effective in determining the internal climates of buildings and disturbs the energy exchanges that impact the comfort of city life (Voogt, Oke 2003). The urban heat island (UHI) refers to an urban area that has higher surface temperatures than its surrounding rural areas due to urbanization (Xian, Crane 2006). The annual average air temperature of an urban area, with almost one million population, can be one to three degree warmer than its surrounding areas. This phenomenon can affect societies by increasing summertime, air conditioning costs, air pollution, heat related illness, greenhouse gas emissions and water quality. Today, more than fifty percent of the world's population are living in cities (UN DESA 2015), in this regard, urbanization has become a key factor in global warming issue. Tehran, the capital of Iran, one of megacities in the world, is the case study of this research. A megacity is mainly defined as a residential area with a total population in excess of ten million people (Dihkan *et al.* 2015). We have been encountering significant surface heat island (SHI) effect in this area due to rapid urbanization progress and the fact that twenty percent of population in Iran are currently living in Tehran houses.

SHI has been usually monitored and measured by in situ observations acquired from thermometer networks. Recently, remotely sensed observing and monitoring of SHIs has become accessible by incorporating thermal remote sensing technology and satellite data. Satellite thermal imageries, mainly high resolution imageries, have the advantage of providing a repeatable dense grid of temperature data, over a whole urban area, and even distinctive temperatures for individual buildings.

Previous studies of land surface temperatures (LST) and thermal remote sensing of urban and rural areas have been primarily conducted by using AVHRR or MODIS imageries (Streutker 2002; Imhoff *et al.* 2010). Now a days, most of researchers are using high resolution satellite imagery to monitor thermal anomalies in urban areas (Fabrizi *et al.* 2010; Ogashawara, Bastos 2012; Liu *et al.* 2015).

Corresponding author: Mahdi Hasanlou
E-mail: hasanlou@ut.ac.ir

In this study, newly launched Landsat series (Landsat 8) is used to monitor SHI, and retrieve the brightness temperatures and land use/cover types. The Landsat 8 carries two kind of sensors (Landsat 8 2016): The Operational Land Imager (OLI) sensor has former Landsat bands, with three new bands: a deep blue band for aerosol/coastal investigations (band 1), a shortwave infrared band for cirrus detection (band 9), and a Quality Assessment (AQ) band. The Thermal Infrared Sensor (TIRS) provides two high spatial resolution thirty meters thermal bands (band 10 and 11). These sensors both use corrected signal-to-noise ratio (SNR) radiometric performance quantized over a 12-bit dynamic range. Improved SNR performance would cause better determination of land cover type. Details of Landsat 8 band specification is illustrated in Table 1. Furthermore, Landsat 8 imageries incorporate two valuable thermal imagery bands with 10.9 μm and 12.0 μm wavelength. These two thermal bands would improve estimation of SHI by incorporating split-window algorithms, they will also increase the probability of the SHI and urban-modified climates to be monitored. Therefore, it is necessary to design and use new procedures that are able to simultaneously (a) handle the two new high resolution thermal bands of Landsat 8 imagery and (b) incorporate satellite in situ measurement into precise estimation of SHI.

Table 1. Landsat 8 OLI and TIRS bands

Bands	Wavelength (μm)	Res (m)
Band 1 – Coastal aerosol	0.43–0.45	30
Band 2 – BLUE	0.45–0.51	30
Band 3 – GREEN	0.53–0.59	30
Band 4 – RED	0.64–0.67	30
Band 5 - Near Infrared (NIR)	0.85–0.88	30
Band 6 – SWIR 1	1.57–1.65	30
Band 7 – SWIR 2	2.11–2.29	30
Band 8 – Panchromatic	0.50–0.68	15
Band 9 – Cirrus	1.36–1.38	30
Band 10 – Thermal Infrared (TIRS) 1	10.60–11.19	30
Band 11 – Thermal Infrared (TIRS) 2	11.50– 2.51	30

The UHIs can be affected by three main factors (Ogashawara, Bastos 2012): a) reduced evapotranspiration; b) the effects of energy transformation in urban area; and (c) anthropogenic energy production. Also, according to (Actionbioscience 2015), there are three types of UHIs: a) Boundary Layer Heat Island (BLHI); b) Canopy Layer Heat Island (CLHI); and c) Surface Heat Island (SHI). The main difference between BLHI and SHI is that BLHI refers to the warmth of the urban atmosphere while SHI refers to the warmth of the urban surface. Also, the major difference between CLHI and SHI is the place where temperature is appeared and detected. Usually, CLHI is detected by specified air temperature measurement (i.e. in situ data) in the canopy layer, while remotely sensed thermal data observe the spatial patterns and models of upwelling thermal radiance to estimate the LST (Voogt, Oke 2003) of the SHI.

Lately, quantitative algorithms for urban thermal environment and dependent factors have been studied, for example, the relationship of UHI with land cover types and its corresponding regression model (Xian, Crane 2006; Hasanlou, Mostofi 2015; Liu *et al.* 2015; Odindi *et al.* 2015). Similar works have been done and models of the relation between various vegetation indices and the surface temperature have been established (Chen *et al.* 2006; Xiong *et al.* 2012). This paper tried to employ a quantitative approach to track the relationship between SHI and common land cover indices and select proper indices, including the Normalized Difference Vegetation Index (NDVI) (Kriegler *et al.* 1969), Enhanced Vegetation Index (EVI) (Kriegler *et al.* 1969), Soil Adjusted Vegetation Index (SAVI) (Huete 1988), Normalized Difference Water Index (NDWI) (Gao 1996), Normalized Difference Bareness Index (NDBaI) (Zhao, Chen 2005; Chen *et al.* 2006), Normalized Difference Build-up Index (NDBI) (Zha *et al.* 2003), Modified Normalized Difference Water Index (MNDWI) (Xu 2006), Bare soil Index (BI) (Zha *et al.* 2003; Zhao, Chen 2005). Urban Index (UI) (Kawamura *et al.* 1996), Index based Built up Index (IBI) (Xu 2008) and Enhanced Built up and Bareness Index (EBBI) (As-syakur *et al.* 2012). Behind these indices, the tasselled cap transformation (TCT), which is calculated for Landsat 8 imagery, is used to compact spectral data into a few bands associated with physical scene characteristics with minimal information loss (Baig *et al.* 2014). The three TCT components, Brightness, Greenness and Wetness, are computed and incorporated to predict SHI effect. Therefore, the main objectives of this research are to develop a non-linear and kernel base analysis model for urban thermal environment area by incorporating support vector regression (SVR) method (Drucker *et al.* 1997), and also to compare proposed method with linear regression model (LRM) in which linear combination of incorporated land cover indices (features) is used. The primary aim of this paper is to establish a framework producing an optimum SHI by utilizing proper land cover indices form Landsat 8 imagery. In this regard, three scenarios have been implemented: a) incorporating LRM with full feature set without any feature selection; b) incorporating SVR with full feature set without any feature selection; and c) incorporating genetically selected suitable features in SVR method (GA-SVR). The results of this study can be used to increase the output performance of the SHI estimation in urban area using Landsat 8 imagery by adopting the genetically SVR method with (a) an optimal land cover indices/feature space and (b) customized SVR parameters.

1. Material and methods

In this paper, Landsat 8 imagery is used as input imagery data to estimate SHI map, and also various urban and vegetation indices are calculated. Figure 1 shows the flowchart of proposed methods.

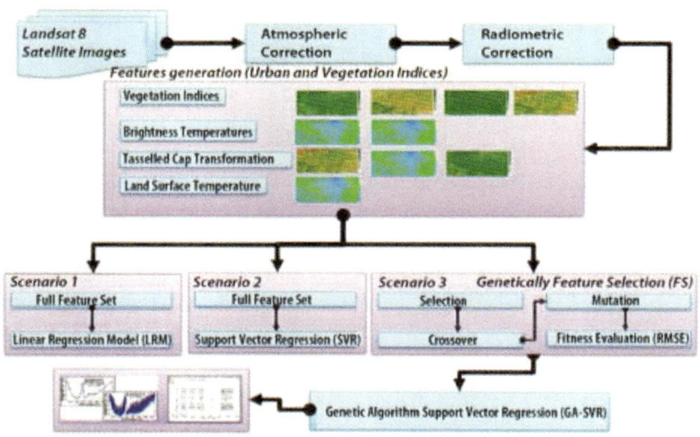

Fig. 1. Flowchart of proposed methods

As shown in Figure 1. atmospheric and radiometric correction is done after importing Landsat 8 images. In utilized procedure, the fast line-of-sight atmospheric analysis of spectral hypercubes (FLAASH) algorithm is used to reduce atmospheric and radiometric effect on incorporated images. The next step starts by producing urban and vegetation indices and also brightness temperature (BT) of thermal bands of Landsat 8 imagery in 10.9 μm and 12.0 μm (band 10, 11) (USGS 2015). Further, in this step, we calculate LST data for incorporated dataset using split window (SW) algorithm (Eq. (1)) introduced in (Jimenez-Munoz *et al.* 2014):

$$LST = c_0 + T_{b10} + c_1\left(T_{b10} - T_{b11}\right) + c_1\left(T_{b10} - T_{b11}\right)^2 + \left(c_3 + c_4\omega\right)\left(1-\varepsilon\right) + \left(c_5 + c_6\omega\right)\Delta\varepsilon, \tag{1}$$

where T_{b10} and T_{b11} are the at-sensor BTs (in kelvins), ε is the mean emissivity, $\varepsilon = 0.5\left(\varepsilon_{b10} + \varepsilon_{b11}\right)$, $\Delta\varepsilon$ is the emissivity difference, $\Delta\varepsilon = \left(\varepsilon_{b10} - \varepsilon_{b11}\right)$, ω is the total atmospheric water vapor content (in g×cm^{-2}) that we set to $\omega = 3$ as mentioned in (Jimenez-Munoz *et al.* 2014), and c_0 to c_6 are the SW coefficients that computed in (Jimenez-Munoz *et al.* 2014). In this study, to estimate ε_{b10} and ε_{b11}, we have simultaneously incorporated MODIS product (MOD11A2 2015) for the same area. As you know, the MOD11A2 is the level-3 MODIS global LST and emissivity, and 8-day data are composed from the daily 1-kilometer LST product. Also this product comprises day time and night time LSTs, quality assessment, observation times, view angles, bits of clear-sky days and nights, and emissivities estimated in bands 31 and 32 from land cover types that are similar to Landsat 8 thermal bands (ε_{b10} and ε_{b11}). Then, the last step is divided in three different scenarios, including LRM implementation (scenario 1), SVR incorporation (scenario 2) and GA-SVR implementation that adopted genetic algorithm (GA) as supervised feature selection (FS) procedure (scenario 3). In the next section, procedure of computing urban and vegetation indices behind utilized algorithm will be discussed.

1.1. Calculating TOA radiance/reflectance and at-sensor BT

The OLI and TIRS bands data can be converted to top of atmosphere (TOA) spectral radiance and planetary reflectance using the reflectance/radiance rescaling parameters provided in the product metadata file (MTL file) (USGS 2015). The following equation (Eq. (2) and Eq. (3)) is used to convert digital number (DN) values to TOA radiance/reflectance for OLI data as follows:

$$L_\lambda = M_L DN + A_L\,; \tag{2}$$

$$\rho_{\lambda'} = M_P DN + A_P\,, \tag{3}$$

where L_λ is TOA spectral radiance (watts/(m^2×srad×μm)), M_L and M_P are band-specific multiplicative rescaling factor extracted from MTL file, A_L and A_P are band-specific additive rescaling factor extracted from MTL file, DN is quantized and calibrated standard digital number values and $\rho_{\lambda'}$ is TOA planetary reflectance, without correction for solar angle. Also, we can correct TOA reflectance by considering the sun angle (Eq. (4)):

$$\rho_\lambda = \frac{\rho_{\lambda'}}{\sin\left(\theta_{SE}\right)}, \tag{4}$$

where ρ_λ is TOA planetary reflectance and θ_{SE} is local sun elevation angle. The scene center sun elevation angle in degrees is extracted from MTL file. For TIRS bands data, we use conversion from spectral radiance to BT using the thermal constants provided in MTL file which derived from Planck's law (Eq. (5)):

$$BT = K_2 / \ln\left(1 + K_1 / L_\lambda\right), \tag{5}$$

where BT is at-sensor BT in kelvin and K_1 / K_2 are band-specific thermal conversion constant extracted from MTL file for each TIRS bands (USGS 2015). By utilizing Eq. (1) to Eq. (5), input features and land cover indices will be produced.

1.2. Urban and vegetation indices

In this study, the most common urban and vegetation indices are used. These indices can be divided to two main type: a) urban indices; and b) vegetation indices. The widespread and common urban indices are shown in the Table 2. The most of these indices would extract urbanization parameters related to spectral difference of near infrared, visible and short wave infrared bands of Landsat 8 Imagery

Table 2. Extracted urban indices form Landsat 8 imagery

Name of urban index	Formulation
Normalized Difference Bareness Index (NDBaI)	$NDBaI = \frac{SWIR1 - TIRS1}{SWIR1 + TIRS1}$
Normalized Difference Build-up Index (NDBI)	$NDBI = \frac{SWIR1 - NIR}{SWIR1 + NIR}$
Bare Soil Index (BI)	$BI = \frac{(SWIR1 + RED) - (NIR + BLUE)}{(SWIR1 + RED) + (NIR + BLUE)}$
Urban Index (UI)	$UI = \frac{SWIR2 - NIR}{SWIR2 + NIR}$
Index-based Built-Up Index (IBI)	$IBI = \frac{\frac{2 \times SWIR1}{SWIR1 + NIR} - \left(\frac{NIR}{NIR + RED} - \frac{GREEN}{GREEN + SWIR1}\right)}{\frac{2 \times SWIR1}{SWIR1 + NIR} + \left(\frac{NIR}{NIR + RED} - \frac{GREEN}{GREEN + SWIR1}\right)}$
Enhanced Built-Up and Bareness Index (EBBI)	$EBBI = \frac{SWIR1 - NIR}{10\sqrt{SWIR1 + TIRS1}}$

Table 3. Extracted vegetation indices form Landsat 8 imagery

Name of vegetation index	Formulation
Normalized Difference Vegetation Index (NDVI)	$NDVI = \frac{NIR - RED}{NIR + RED}$
Enhanced Vegetation Index (EVI)	$EVI = G\frac{NIR - RED}{NIR + C_1 RED - C_2 BLUE + L}$ $L = 1; C_1 = 6; C_2 = 7.5; G = 2.5$
Soil Adjusted Vegetation Index (SAVI)	$SAVI = \frac{NIR - RED}{NIR + RED + L}(L + 1)$ $0 < L < 1 \Rightarrow L = 0.5$
Normalized Difference Water Index (NDWI)	$NDWI = \frac{NIR - SWIR1}{NIR - SWIR1}$
Modified Normalized Difference Water Index (MNDWI)	$MNDWI = \frac{GREEN - NIR}{GREEN + NIR}$
Tasselled Cap Transformation (TCT)	Brightness
TCT	Greenness
TCT	Wetness

(Table 1). All indices from Table 2 are calculated based on incorporating digital number (DN) of Landsat 8 bands.

As before, in Table 3 computable vegetation indices extracted from spectral bands like, near infrared, visible and short wave infrared bands of incorporated dataset are illustrated. All these indices are calculated based on incorporating reflectance/radiance of related Landsat 8 bands by using procedure introduced in (USGS 2015).

1.3. Linear regression model

Given a data set $\{y_i, x_{i1}, \ldots, x_{iq}\}_{i=1}^{n}$ of n statistical units, a linear regression model (LRM) assumes that the relationship between the dependent parameter y_i and the q vector of regressors x_i is linear. This relationship is modeled through a disturbance term or error parameter ε_i – an unobserved random parameter that adds noise to the linear relationship between the dependent parameter (in this study SHI data) and regressors (in this study land cover indices) (Kutner *et al.* 2004). Thus the model takes the form (Eq. (6)):

$$y_i = \beta_i x_{i1} + \ldots + \beta_q x_{iq} + \varepsilon_i = X_i^T \beta + \varepsilon_i, \quad i = 1, \ldots, n. \tag{6}$$

Where T defines the transpose, therefore $X_i^T \beta$ is the inner product between vectors x_i and β. Often these n equations are stacked together and written in vector form as (Eq. (7)):

$$y = X\beta + \varepsilon, \tag{7}$$

$x_{i1}, \ldots, x_{iq}$, are called regressors or independent variables (in this study land cover indices). The matrix X is sometimes called the design matrix. y_i, is called the measured variable or dependent variable (in this study SHI data). The criteria as to which variable in a dataset is modeled as the dependent parameter and which is modeled as the independent parameter may be based on an assumption that the value of one of the parameters is caused by, or directly influenced by the other variables. As an alternative, there may be a reason to model one of the parameters in terms of the others, in which case there would be no need for presumption of any causes. β, is a q dimensional vector. Its elements are also called regression coefficients. Statistical prediction and conclusion in linear regression focuses on β. The elements of this parameter vector are perceived as the partial derivatives of the dependent parameter with respect to the various independent parameters.

1.4. Support vector regression

The SVR is a supervised learning method which emerged in late 1970s (Drucker *et al.* 1997). SVR allows computing a strong nonparametric model of the relationship between urban/vegetation indices and SHI change. This method is also widely used in most remote sensing applications like SST and LST estimation (Moser, Serpico 2009), biophysical parameter estimation and other vegetation index monitoring from multispectral satellite images (Durbha *et al.* 2007). Consider a set of training points, $\{(x_1,z_1),\ldots,(x_l,z_l)\}$, where $x_i \in \mathbb{R}^n$ is a feature vector and $z_i \in \mathbb{R}^1$ is the target output. Under given parameters $C>0$ and $\in>0$ the standard form of support vector regression (Eq. (8)) (Drucker *et al.* 1997; Smola, Schölkopf 2004) is:

$$\min_{\omega,b\xi,\xi^*} \frac{1}{2}\omega^T\omega + C\sum_{i=1}^{l}\left(\xi_i+\xi_i^*\right)$$

$$\text{subject to } \omega^T\phi(x_i)+b-z_i \le \in + \xi_i$$

$$z_i-\omega^T\phi(x_i)-b \le \in + \xi_i^*;$$

$$\xi_i,\xi_i^* \ge 0,\ i=1,\ldots,l. \tag{8}$$

By introducing Lagrange multipliers and exploiting the optimality constraints, the decision function has the following explicit form (Eq. (9)):

$$\sum_{i=1}^{l}\left(\alpha_i+\alpha_i^*\right)K(x_i,x)+b,$$
$$0\le\alpha_i\le C,\ 0\le\alpha_i^*\le C. \tag{9}$$

Where l is the number of support vectors (SVs) and the kernel function (Eq. (10)):

$$K(x_i,x)=\sum_{j=1}^{m}\phi_j(x)\phi_j(x_i) \tag{10}$$

and α_i^* are Lagrange multipliers. In order to run the SVR method with high efficiency, some constraint must be considered, (a) tuning the SVR parameters including $C, \in$ and kernel parameters, (b) optimizing input space (selecting suitable features). In this paper, we are focusing on both constraints by utilizing search procedure to tune the parameters and feature selection method, and to optimize input space (Genetic Algorithm).

1.5. Genetic algorithm

Genetic algorithm (GA) is the most widespread technique among evolutionary algorithms. This method allows us to search potential solutions to optimize problems in reasonable time, particularly when the search space area is very extensive (Goldberg, Holland 1988). This method is heuristic, based on population of individuals (e.g., chromosomes), that each individual performs a candidate solution (Goldberg, Holland 1988) and can be illustrated as a bit in string mode. Each individual is evaluated by fitness function. This function measures the quality of an individual. In GA method, the population commences randomly, or by different strategy based on the problem in question. The population answers some number of evolutions. During GA's process individuals are evolved and reproduced using GA's operations such as: mutation, crossover, and selection. Its main goal is to select and find the individual with the best fitness value (Goldberg, Holland 1988).

We modeled the problem of feature/index selection as follows: each individual has a size of N features/genes, as shown in Figure 2, and each gene represents a binary random vector number including 0s and 1s, associated with a features/indices. Bit strings of 0s and 1s are chosen for coding. Since N features form one combination, chromosome is arranged as comprised of N individual feature sequence numbers, which are arranged in a serial mode. The length of each parameter is automatically detected according to the number of features/indices in the data set (Fig. 2). As previously stated, the quality of each candidate solution is evaluated according to a fitness function. Our fitness function here is the root mean square error (RMSE) of a regression performed by a SVR or LRM method. Stochastic two-point was used as the selection operator. Additionally, Gaussian mutation, as well as uniform crossover were used as GA operators. Finally, migration direction was set to forward mode in addition to the elitism mechanism.

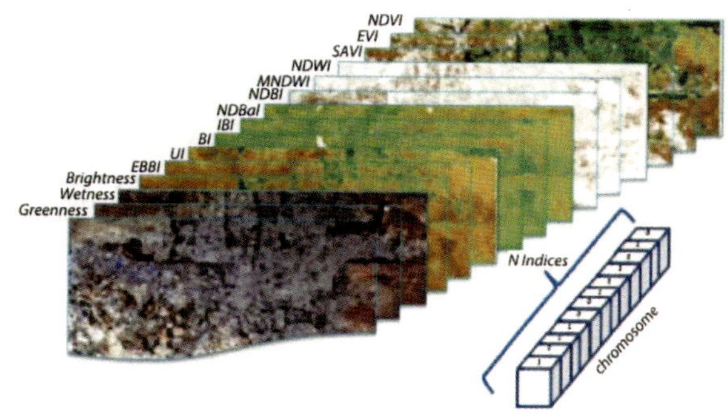

Fig. 2. Chromosome format in genetically feature selection

2. Results

In this study, Landsat 8 imagery acquired from Tehran city is used as dataset. This dataset belongs to summer time (June 15, 2014), which air temperature was nearly 40 °C (Fig. 3). As mentioned in previous section, by incorporating Landsat 8 LST retrieval algorithm (Eq. 1) and contemporary MODIS product estimating emissivity of two thermal bands (i.e. ε_{b10} and ε_{b11}), SHI is estimated. Urban, vegetation and TCT (Brightness, Greenness and Wetness) indices from DN/Ref Landsat 8 images are calculated as well. Calculated indices and information using this dataset is shown in Figure 3.

To establish models in all three scenarios, 2400 points of in situ data were extracted from Tehran urban region. Then, 720 (30% of data) random points were selected as training data and 1680 (70% of data) random points as testing data. Also, some common criteria like, mean square error (MSE), root mean square error (RMSE), Normalized root mean square error (NRMSE) and R-squared

(R^2) were used to examine the output result of each method and scenario.

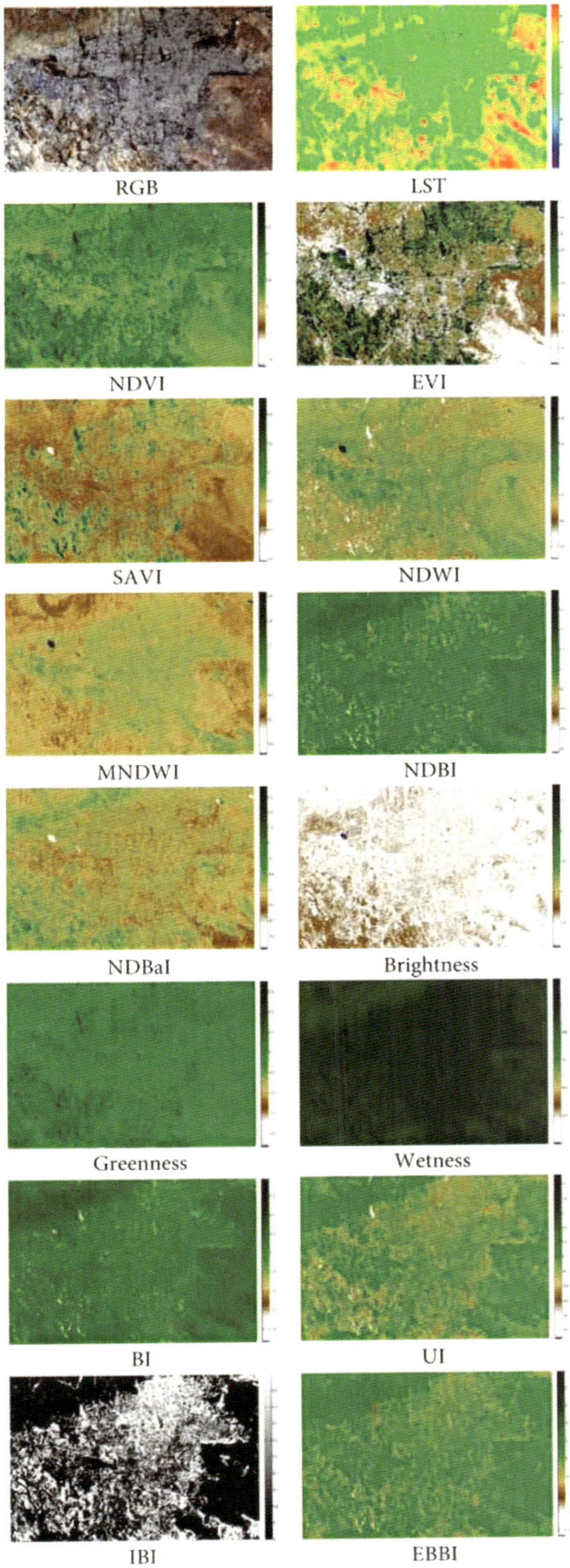

Fig. 3. Urban, vegetation and TCT indices for Tehran dataset

2.1. Scenario 1

In this scenario, we used and implemented LRM method by using linear combination of fourteen features including EVI, NDVI, NDWI, NDBI, NDBaI, MNDWI, BI, UI, IBI, EBBI, SAVI, Wetness, Greenness and Brightness as formulized in Eq. (11). Then, by incorporating training data, the estimation of unknown variable is possible. The output result of LRM method is presented in Table 4.

$$SHI = \beta_1 + \beta_2 EVI + \beta_3 NDVI + \beta_4 NDWI + \beta_5 NDBI + \beta_6 NDBaI + \beta_7 MNDWI + \beta_8 BI + \beta_9 UI + \beta_{10} IBI + \beta_{11} EBBI + \beta_{12} SAVI + \beta_{13} Wetness + \beta_{14} Greenness + \beta_{15} Brightness. \quad (11)$$

Table 4. The output result of LRM method using full fourteen input dataset (scenario 1)

	Estimate	SE	tStat	pValue
Intercept	−54.4486	13.8228	−3.939	0.00010
EVI	−8.2797	1.5004	−5.5183	0.00000
NDVI	−29.6874	3.9611	−7.4947	0.00000
NDWI	−15.9532	5.7056	−2.7961	0.00530
NDBI	−23.3553	11.5275	−2.0261	0.04310
NDBaI	−55.8113	0.9728	−57.3746	0.00000
MNDWI	−9.0744	6.219	−1.4591	0.14500
BI	0.7867	2.0305	0.3874	0.69860
UI	22.0356	4.4001	5.008	0.00000
IBI	−0.0522	0.0991	−0.5267	0.59860
EBBI	79.4409	13.1277	6.0514	0.00000
SAVI	54.911	5.6451	9.7272	0.00000
Wetness	56.0247	8.491	6.5981	0.00000
Greenness	0.6995	6.0152	0.1163	0.90750
Brightness	46.4579	3.0543	15.2106	0.00000

As shown in Table 4, estimated coefficient value for each feature (β_i) is illustrated in first column, the second column contains standard error (SE) of the estimation, the third column shows t statistic (tStat) for a test in which the coefficient is equal to zero and the last column contains p-value for the t statistic. We can also examine our model by incorporating estimated coefficient and using testing data. The result of LRM method is presented in Table 5.

Table 5. The performance of LRM method (scenario 1)

	MSE	RMSE	NRMSE	R^2
Training	0.5672	0.7531	0.3483	0.8762
Testing	0.5486	0.7407	0.3419	0.8785

From Table 5, it is obvious that, there is a good degree of consistency between SHI data and estimated parameters with testing RMSE around 0.74 °C and high compatibility with R^2 around 0.88.

On the other hand, a normal probability illustration of the residuals of a fitted linear model is shown in Figure 4a and added variable plot for whole model is shown in Figure 4b. Figure 4b illustrates the progressive effect on the reflex of specified terms by omitting the effects of all other terms. The slope of the estimated line is the coefficient of the linear compound of the determinate terms projected onto the best fitting direction. Also, from Figure 4b a horizontal line does not fit between the confidence boundaries, but referring to result extracted from Table 5, it is an acceptable result.

2.2. Scenario 2

As before, in this scenario, we used SVR method by using all feature set (fourteen features), as mentioned in flowchart (Fig. 1), next step is performed by SVR technique to relate extracted urban, vegetation and TCT indices to SHI data (Eq. 12).

$$SHI = f(NDVI, EVI, SAVI, NDWI, MNDWI, Brightness, Greenness, Wetness, NDBaI, NDBI, BI, UI, IBI, EBBI). \quad (12)$$

In this regards, as mentioned in previous section we adopted SVR technique. In SVR, the parameter C computes the trade-off between the flatness and the degree to which deviations larger than ε are tolerated in the optimization formulation. In this manner, if C is too large, then the objective is to minimize the empirical risk without regard to flatness part in the optimization formulation. The bigger the ε is, the fewer support vectors will be included. Therefore, more "flat" estimation is a consequence of bigger ε values. In fact, both C and ε values affect the flatness (model complexity). In this paper, C value is computed by (Eq. (13)) base on (Cherkassky, Ma 2004).

$$C = \max(Training\ data) - \min(Training\ data). \quad (13)$$

Also, a Gaussian radial basis function (RBF) kernel (Eq. (10)) is used; this function is widely used in remote sensing algorithms (Hasanlou *et al.* 2013). Before the regression estimating stage, simple normalizing must be applied to the training dataset. The main advantage of normalizing is to avoid attributes in greater numeric ranges dominating those in smaller numeric ranges. Another advantage is to reduce numerical complexity during the computation. The next step is the training procedure, during which some critical SVR parameters, ε, and in the RBF kernel, γ, must be specified.

A simple tool to check a grid of parameters is provided by cross-validation (CV) error (i.e. RMSE as fitness function) with 5-fold. Range of grid search method for estimating ε parameter is [0,5] and for γ RBF parameter is $[2^{-7}, 2^{7}]$. In this manner, Table 6 shows the optimum SVR parameters estimation for Tehran Landsat 8 imagery. It is obvious from Table 6 that the optimum SVR parameters for Tehran scene are $\in = 0$, $\gamma = 2$ and C = 22.40, which fulfils minimum RMSE.

By incorporating the optimum estimated parameters ($\in, \gamma$ and C) with minimum validation error (RMSE), and training dataset, the SVR model would be generated. Then, the performance of the selected final SVR model is computed for Tehran scene using testing data (Table 7).

a)

b)

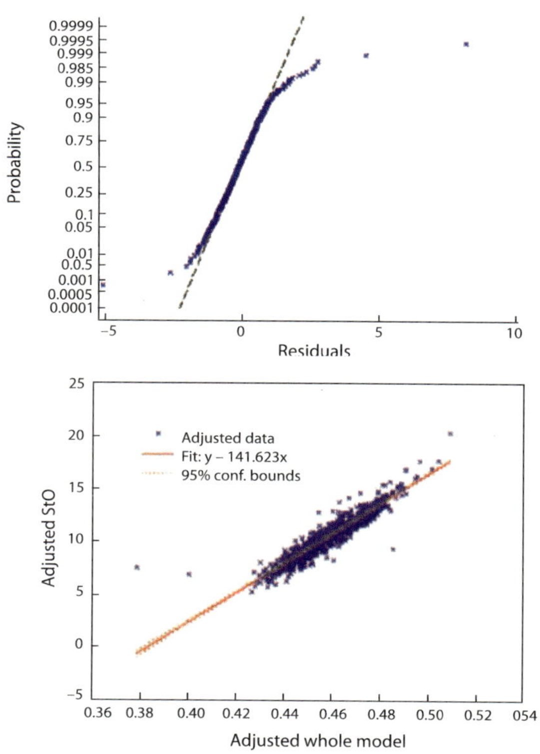

Fig. 4. a) Normal probability plot of residuals and b) Added variable plot for whole model

Table 6. Optimum SVR parameters estimation for Landsat 8 Tehran scene with C = 22.4013

	$\in = 0$	1	2	3	4	5
$\gamma = 2^{-7}$	8.9248	8.964	9.044	9.261	9.491	9.882
2^{-6}	8.1931	8.302	8.436	8.825	9.095	9.660
2^{-5}	7.2267	7.327	7.672	8.162	8.583	9.270
2^{-4}	5.8522	5.994	6.619	7.266	7.977	8.760
2^{-3}	3.9949	4.453	5.289	6.239	7.285	8.105
2^{-2}	2.372	2.939	3.974	5.242	6.397	7.409
2^{-1}	1.5473	2.110	3.174	4.467	5.729	6.894
2^{0}	1.4013	1.801	2.843	4.050	5.457	6.598
2^{1}	**1.3833**	1.683	2.552	3.770	5.179	6.578
2^{2}	1.5092	1.620	2.468	3.711	5.203	6.759
2^{3}	1.7264	1.825	2.685	3.973	5.360	7.122
2^{4}	1.9631	2.233	3.288	4.468	5.929	7.605
2^{5}	2.4885	2.957	4.228	5.561	6.974	8.220
2^{6}	3.554	4.100	5.487	6.811	8.160	9.216
2^{7}	5.1897	5.94	7.131	8.257	9.371	10.385

Table 7. The performance of final SVR model (scenario2)

	MSE	RMSE	NRMSE	R^2
Training	0.7507	0.8664	0.2424	0.9442
Testing	1.1155	1. 0562	0.3053	0.9100

As it is clear from Table 7, there is a high degree of consistency between incorporated information for each feature in kernel method and SHI data. For example, correlation coefficient in training data is $R^2 = 0.94$ and in testing data is $R^2 = 0.91$ and improvement of NRMSE comparing to scenario 2 (NRMSE = 0.3023).

2.3. Scenario 3

In this scenario, we applied GA-SVR method to optimize input space of SVR method. This means that by utilizing all feature set (fourteen features) and the GA procedure as feature selection optimizing tool, the suitable and appropriate features (including urban, vegetation and TCT indices) are selected as input space of SVR technique. Then reduced features would estimate SHI data. An individual's chromosome, i.e., the features present in an individual, was initialized in a random way and the parameters were set according to results of preliminary experiments. Table 8 presents all parameters set in GA. The samples were randomly chosen; however, the total number of samples has an important impact on the performance. The higher the number of samples are; the more time would be consumed calculating the fitness for each individual. In order to ensure high reliability of results, 10 runs of GA for each dataset were performed. Then those features that appeared more frequently were selected.

Table 8. GA's Parameters

Parameters	Value
Population size	20
Crossover rate	0.7
Elitism Ration	1
Mutation Ratio	0.05
Crossover Method	Two point
Max Iterations	50
Elite count	1

Since we have decided to let GA find the optimal features, it is also important to note that result of this scenario is optimal features (including urban, vegetation and TCT indices) with minimum RMSE validation error and maximum R^2. Figure 5 shows the best and the mean of fitness values (RMSE) in each generation for a single run of GA.

Five indices are selected as the optimal and best features among fourteen features, including, NDBI, NDBaI, BI, Greenness and Brightness indices and the rest of the features are omitted from SVR input dataset (Eq. (14)).

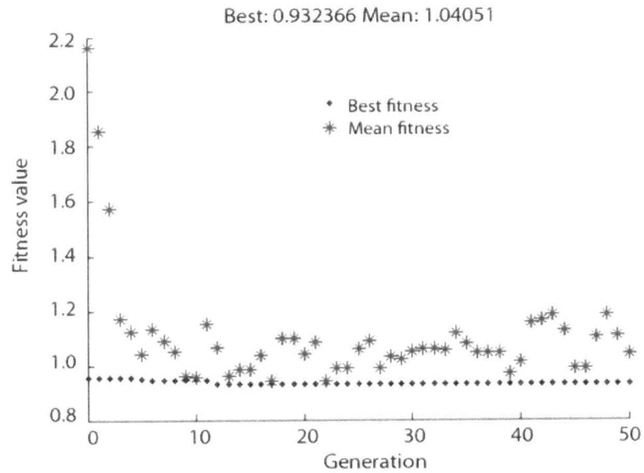

Fig. 5. Fitness values achieved by the proposed GA-SVR (scenario 3)

$$SHI = f\left(NDBI, NDBaI, BI, Greenness, Brightness\right) \quad (14)$$

By incorporating these selected features, procedure of training and testing in SVR technique is commenced. Result and performance of this scenario (GA-SVR) is computed for Tehran Landsat 8 scene (Table 9). As it is clear from Table 9, the minimum NRMSE (0.26) has been achieved by this scenario. NRMSE facilitates the comparison between datasets or models with different scales and it is often expressed as a percentage, where lower values indicate less residual variance. Also, in this scenario, R-squared coefficient for testing data is $R^2 = 0.93$, calculated by using GA and SHI data, revealing high compatibility between selected features.

Table 9. The performance of GA-SVR model (scenario 3)

	MSE	RMSE	NRMSE	R^2
Training	0.8852	0.9409	0.2633	0.9338
Testing	0.8693	0.9324	0.2695	0.9315

2.4. Discussion

To sum up, utilizing three different scenarios with various regression methods would enable us to evaluate our estimation procedures better. Further, adopting NRMSE criteria for comparing purpose, enables the evaluation of training and testing result extracted from three scenarios. In this regards, scenario 3 has represented high performance in relating the input space features (input indices) to SHI data. Also, incorporating genetic FS method has some advantages comparing to previous scenarios, a) reduced dimensionality of input space increasing reliability and reducing computation complexity; and b) revealed the best and proper indices that used to connect input space and SHI data. Using genetically FS enables us to monitor more affective indices that influence estimating SHI data (i.e. NDBI, NDBaI, BI, Greenness and Brightness indices).

Conclusions

All range of Landsat 8 spectral bands, particularly thermal bands, have been used to estimate SHI of Tehran city. In this study, urban indices including NDBaI, NDBI, BI, UI, IBI and EBBI have been calculated using recent urban parameters and factors. Further, to investigate vegetation factors better, more common vegetation and water indices including NDVI, EVI, SAVI, NDWI, MNDWI and TCT information including Brightness, Greenness and Wetness have been used. By utilizing these information and indices, the modeling and monitoring process of SHI is more practical. Also as a part of this study, three scenarios were implemented to compare the performance of each scenario. In scenario one, all calculated features/indices (full features) were used as input space of linear regression model. Result of scenario one is illustrated in Table 5. Then, in scenario two, same as previous scenario, full features were used as input space but with kernel base method (i.e. support vector regression). This scenario is more complicated than scenario one but it can handle high dimensional data and has better performance result (Table 7). Finally, in scenario three, we used supervised feature selection procedures (genetic algorithm) to select proper and affective features (indices). Estimated result (Table 9) revealed that scenario three has more reliable performance using NRMSE criteria comparing to two others (NRMSE = 0.2695 for scenario one, NRMSE = 0.3053 for scenario two and NRMSE = 0.3419 for scenario three). Also, incorporating genetic FS in these three scenarios indicated that, to estimate SHI data using Landsat 8 images, it is better to use more affective and optimum indices like NDBI, NDBaI, BI, Greenness and Brightness indices. This study would be completed by incorporating supervised feature extraction (FE) method to select suitable transform features from urban and vegetation information.

Acknowledgements

The authors would like to thank the Islamic Azad University's (South Tehran Branch) research deputy office for supporting this work. Also, we would like to thank the Earth Resources Observation and Science (EROS) – U.S. Geological Survey and the Land Remote Sensing Program of the U.S. Geological Surveying collaboration with NASA for generously availing the MODISLST and Landsat 8 data.

References

Actionbioscience. 2015. *Urban heat islands: hotter cities* [online], [cited 20 May 2015]. Available from Internet: http://www.actionbioscience.org/environment/voogt.html

As-syakur, A. R.; Adnyana, I. W. S.; Arthana, I. W.; Nuarsa, I. W. 2012. Enhanced built-up and bareness index (EBBI) for mapping built-up and bare land in an urban area, *Remote Sensing* 4(10): 2957–2970. https://doi.org/10.3390/rs4102957

Baig, M. H. A.; Zhang, L.; Shuai, T.; Tong, Q. 2014. Derivation of a tasselled cap transformation based on Landsat 8 at-satellite reflectance, *Remote Sensing Letters* 5(5): 423–431. https://doi.org/10.1080/2150704X.2014.915434

Chen, X.-L.; Zhao, H.-M.; Li, P.-X.; Yin, Z.-Y. 2006. Remote sensing image-based analysis of the relationship between urban heat island and land use/cover changes, *Remote Sensing of Environment* 104(2): 133–146. https://doi.org/10.1016/j.rse.2005.11.016

Cherkassky, V.; Ma, Y. 2004. Practical selection of SVM parameters and noise estimation for SVM regression, *Neural Networks* 17(1): 113–126. https://doi.org/10.1016/S0893-6080(03)00169-2

Dihkan, M.; Karsli, F.; Guneroglu, A.; Guneroglu, N. 2015. Evaluation of surface urban heat island (SUHI) effect on coastal zone: the case of Istanbul megacity, *Ocean & Coastal Management* 118(Part B): 309–316.

Drucker, H.; Burges, C. J.; Kaufman, L.; Smola, A.; Vapnik, V. 1997. Support vector regression machines, *Advances in Neural Information Processing Systems* 9: 155–161.

Durbha, S. S.; King, R. L.; Younan, N. H. 2007. Support vector machines regression for retrieval of leaf area index from multi angle imaging spectro radiometer, *Remote Sensing of Environment* 107(1–2): 348–361. https://doi.org/10.1016/j.rse.2006.09.031

Fabrizi, R.; Bonafoni, S.; Biondi, R. 2010. Satellite and ground-based sensors for the urban heat island analysis in the city of Rome, *Remote Sensing* 2(5): 1400–1415. https://doi.org/10.3390/rs2051400

Gao, B. 1996. NDWI–A normalized difference water index for remote sensing of vegetation liquid water from space, *Remote Sensing of Environment* 58(3): 257–266. https://doi.org/10.1016/S0034-4257(96)00067-3

Goldberg, D. E.; Holland, J. H. 1988. Genetic algorithms and machine learning, *Machine learning* 3(2): 95–99. https://doi.org/10.1023/A:1022602019183

Hasanlou, M.; Mostofi, N. 2015. *Investigating urban heat island effects and relation between various land cover indices in Tehran City using Landsat 8 Imagery*. MDPI, f004.

Hasanlou, M.; Samadzadegan, F.; Homayouni, S. 2013. SVM-based hyperspectral image classification using intrinsic dimension, *Arabian Journal of Geosciences* 8(1): 477–487. https://doi.org/10.1007/s12517-013-1141-9

Huete, A. R. 1988. A soil-adjusted vegetation index (SAVI), *Remote Sensing of Environment* 25(3): 295–309. http://dx.doi.org/10.1016/0034-4257(88)90106-X

Imhoff, M. L.; Zhang, P.; Wolfe, R. E.; Bounoua, L. 2010. Remote sensing of the urban heat island effect across biomes in the continental USA, *Remote Sensing of Environment* 114(3): 504–513. https://doi.org/10.1016/j.rse.2009.10.008

Jimenez-Munoz, J. C.; Sobrino, J. A.; Skokovic, D.; Mattar, C.; Cristobal, J. 2014. land surface temperature retrieval methods from Landsat-8 thermal infrared sensor data, *IEEE Geoscience and Remote Sensing Letters* 11(10): 1840–1843. https://doi.org/10.1109/LGRS.2014.2312032

Kawamura, M.; Jayamana, S.; Tsujiko, Y. 1996. Relation between social and environmental conditions in Colombo Sri Lanka and the urban index estimated by satellite remote sensing data. The International Archives of the Photogrammetry, *Remote Sensing and Spatial Information Sciences* 31: 321–326.

Kriegler, F. J.; Malila, W. A.; Nalepka, R. F.; Richardson, W. 1969. Preprocessing transformations and their effects on multispectral recognition, in *Remote Sensing of Environment*, VI. 97.

Kutner, M.; Nachtsheim, C.; Neter, J.; Li, W. 2004. *Applied linear statistical models*. 5th ed. Boston: McGraw-Hill/Irwin.

Landsat 8. 2016. *Landsat 8* [online], [cited 21 May 2016]. Available from Internet: http://landsat.usgs.gov/landsat8.php

Liu, K.; Su, H.; Zhang, L.; Yang, H.; Zhang, R.; Li, X. 2015. Analysis of the urban heat island effect in Shijiazhuang, China using satellite and airborne data, *Remote Sensing* 7(4): 4804–4833. https://doi.org/10.3390/rs70404804

MOD11A2. 2015. *NASA land data products and services* [online], [cited 22 May 2015]. Available from Internet: https://lpdaac.usgs.gov/products/modis_products_table/mod11a2

Moser, G.; Serpico, S. B. 2009. Automatic parameter optimization for support vector regression for land and sea surface temperature estimation from remote sensing data, *IEEE Transactions on Geoscience and Remote Sensing* 47(3): 909–921. https://doi.org/10.1109/TGRS.2008.2005993

Odindi, J. O.; Bangamwabo, V.; Mutanga, O. 2015. Assessing the value of urban green spaces in mitigating multi-seasonal urban heat using MODIS land surface temperature (LST) and Landsat 8 data, *International Journal of Environmental Research* 9(1): 9–18.

Ogashawara, I.; Bastos, V. da S. B. 2012. A quantitative approach for analyzing the relationship between urban heat islands and land cover, *Remote Sensing* 4(11): 3596–3618. https://doi.org/10.3390/rs4113596

Smola, A. J.; Schölkopf, B. 2004. A tutorial on support vector regression, *Statistics and Computing* 14(3): 199–222. https://doi.org/10.1023/B:STCO.0000035301.49549.88

Streutker, D. R. 2002. A remote sensing study of the urban heat island of Houston, Texas, *International Journal of Remote Sensing* 23(13): 2595–2608. https://doi.org/10.1080/01431160110115023

UN DESA. 2015. *World's population increasingly urban with more than half living in urban areas* [online], [cited 20 May 2015]. Available from Internet: http://www.un.org/en/development/desa/news/population/world-urbanization-prospects-2014.html

USGS. 2015. *Using the USGS Landsat 8 product* [online], [cited 20 May 2015]. Available from Internet: http://landsat.usgs.gov/Landsat8_Using_Product.php

Voogt, J. A.; Oke, T. R. 2003. Thermal remote sensing of urban climates, *Remote Sensing of Environment* 86(3): 370–384. https://doi.org/10.1016/S0034-4257(03)00079-8

Xian, G.; Crane, M. 2006. An analysis of urban thermal characteristics and associated land cover in Tampa Bay and Las Vegas using Landsat satellite data, *Remote Sensing of Environment* 104(2): 147–156. https://doi.org/10.1016/j.rse.2005.09.023

Xiong, Y.; Huang, S.; Chen, F.; Ye, H.; Wang, C.; Zhu, C. 2012. The impacts of rapid urbanization on the thermal environment: a remote sensing study of Guangzhou, South China, *Remote Sensing* 4(7): 2033–2056. https://doi.org/10.3390/rs4072033

Xu, H. 2006. Modification of normalised difference water index (NDWI) to enhance open water features in remotely sensed imagery, *International Journal of Remote Sensing* 27(14): 3025–3033. https://doi.org/10.1080/01431160600589179

Xu, H. 2008. A new index for delineating built-up land features in satellite imagery, *International Journal of Remote Sensing* 29(14): 4269–4276. https://doi.org/10.1080/01431160802039957

Zha, Y.; Gao, J.; Ni, S. 2003. Use of normalized difference built-up index in automatically mapping urban areas from TM imagery, *International Journal of Remote Sensing* 24(3): 583–594. https://doi.org/10.1080/01431160304987

Zhao, H.; Chen, X. 2005. Use of normalized difference bareness index in quickly mapping bare areas from TM/ETM+, in *Geoscience and Remote Sensing Symposium, 2005. IGARSS '05*, 29 July 2005, Seoul, South Corea. IEEE International, 1666–1668.

Nikrouz MOSTOFI. He received the BSc degree in Civil Eng.-Surveying & Geomatics Eng. from industrial K. N. Toosi University, Tehran, Iran in 2001, and the MSc degree in photogrammetry engineering from the Tehran University, Tehran, Iran in 2005. Since 2014 he has been PhD in Remote Sensing and Geospatial Information System in Islamic Azad University, Sciences and Researchs Branch. Also, he is a member of Islamic Azad University, South Tehran Branch, and Department of Surveying Engineering, Tehran, Iran. His current research interests are using urban heat island monitoring, close range photogrammetry and machine learning algorithms in photogrammetry and remote sensing.

Mahdi HASANLOU. He received the BSc degree in Surveying and Geomatics Eng. from the University of Tehran, Tehran, Iran, in 2003, the M.Sc. degree in Remote Sensing from University of Tehran, Tehran, Iran, in 2006, and the PhD degree in Remote Sensing from University of Tehran, Tehran, Iran, in 2013. Since 2013, he has been as an assistant professor in the School of Surveying and Geospatial Engineering, College of Engineering, University of Tehran, Iran, His research activities are mainly focused on Thermal, optical and SAR remote sensing for urban and agro-environmental applications.

14

DEVELOPMENT OF A PERFORMANCE THRESHOLD APPROACH FOR IDENTIFYING THE MANAGEMENT OPTIONS FOR STABILISATION/SOLIDIFICATION OF LEAD POLLUTED SOILS

Pietro Paolo Falciglia[a], Abir Al-Tabbaa[b], Federico G. A. Vagliasindi[a]

[a]Department of Civil and Environmental Engineering, University of Catania, Viale A. Doria, 6-95125 Catania, Italy

[b]Department of Engineering, University of Cambridge, Trumpington Street, Cambridge CB2 1PZ, UK

Abstract. Two soils spiked with lead at different rates were stabilised/solidified using Portland cement and fly ash at different soil:binder ratios, and tested for their setting time, unconfined compressive strength, leachability and durability. A performance threshold approach was used in order to identify optimal management options for the products of the S/S treatment. Results show that soil texture, percentage of binders and lead concentration play an important part in the treatment, significantly influencing the performance of the resulting products in terms of curing, compressive strength and durability. Pb soil concentrations higher than 15000 mg kg^{-1} were found to heavily reduce the applicability of the treatment requiring the maximum amount of binder in order to satisfy the performance criteria. The performance of sandy soils was shown to be limited by setting time and UCS features due to the retardation of the hydration reactions and also by its leaching behaviour, whereas for silt-clayey soils the critical parameter is the mechanical resistance.

Keywords: cement, fly ash, lead (Pb), management options, soil contamination, stabilisation/solidification (S/S).

Reference to this paper should be made as follows: Falciglia, P. P.; Al-Tabbaa, A.; Vagliasindi, F. G. A. 2014. Development of a performance threshold approach for identifying the management options for stabilisation/solidification of lead polluted soils, *Journal of Environmental Engineering and Landscape Management* 22(02): 85–95.
http://dx.doi.org/10.3846/16486897.2013.821070

Introduction

Lead (Pb)-based compounds have been a major source of environmental contamination in the past few decades where lead has been reported to affect human health and it is considered as a possible cause of human cancer (Tang, Yang 2012).

Treatment options for Pb-polluted soils have been reported to be either expensive (soil washing, Torres *et al.* 2012), require intensive energy consumption (electrokinetic remediation, Ryu *et al.* 2011), or time consuming, as in the case of biological methods such as bioleaching (Cheng *et al.* 2009) or phytoremediation (Bech *et al.* 2012), or environmentally unsustainable, as in the case of disposal in landfills (Harbottle *et al.* 2007).

Recent studies (Leonard, Stegemann 2010) have also shown the possibility of using stabilization/solidification (S/S) with hydraulic binders as a possible treatment option, with the potential of using the S/S product for useful purposes. S/S has been widely used due to its versatility, efficiency, time and costs to dispose of low-level radioactive and hazardous wastes, as well as to remedy metal or radionuclides contaminated soils (Falciglia *et al.* 2012).

Stabilisation/solidification (S/S) is a treatment process by which contaminated soils, sediments or waste materials are mixed with a binder and specific additives with the aims of reducing the mobility of the toxic contaminants by increasing the pH and fully or partially binding the contaminants in the solid matrix (stabilisation), and of improving the physical properties (strength, compressibility, permeability and durability) of the final treatment products (solidification) (Antemir *et al.* 2010; Polettini *et al.* 2001).

Corresponding author: Pietro P. Falciglia
E-mail: ppfalci@dica.unict.it

Binder-based stabilisation/solidification (S/S) techniques are used for Pb-polluted soil treatment because they ensure physical and chemical stabilisation by significantly reducing the mobility and the solubility of Pb in soil (Yin *et al.* 2006), requiring minimal input of energy, resulting in potential applications in concrete works (Al-Ansary, Al-Tabbaa 2007; Harbottle *et al.* 2007).

For Pb-polluted soil treatment, most applications of S/S are cement-based, and rely on Portland cement (PC) as the primary binder (Svensson and Allard 2008). PC is a heterogeneous mixture of five mineral phases: 50–70% alite (C_3S), 20–30% belite (C_2S), 5–12% alluminate (C_3A), 5–12% ferrite (C_4AF) and 2% gypsum. More recently PC has been combined in blends with other minerals such as lime, blast-furnace slag, clays and fly ash. Fly ash (FA), also known as pulverised fuel ash (PFA), is generated from coal fired power plants and, due to its low cost, has been widely used in PC manufacturing and as a partial substitute for PC as stabilising agent for Pb in S/S treatments. If FA contains more than 20% CaO, it is classified as self-cementing Class C FA, while Class F FA generally contains less than 10% CaO. FA presents mechanical features slightly lower than PC (especially Class F), influencing the performance of the S/S treated matrices if used instead of PC as a binder, but it can be successfully used for soil stabilisation or in other civil construction applications (Moon, Dermatas 2007).

The hydration of PC is a sequence of overlapping chemical reactions between dry binder compounds and water, leading to continuous cement paste stiffening and hardening. The early behaviour of hydrating PC is governed by reactions of aluminate phases while the setting and the early strength development behaviour is mostly dependent on the hydration of silicates, particularly alite. The formation of hydration products and the development of micro-structural features depend on solution processes and interfacial and solid-state reactions. The hydration products of PC are mainly made up by 20–25% $Ca(OH)_2$ (CH), 60–70% calcium silicate hydrate gel (CSH) and 5–15% other phases including grains of still-unhydrated cement (Chen *et al.* 2009).

In particular, using Class F FA, even if it does not readily exhibit self cementing characteristics, upon PC or lime addition, pozzolanic reactions take place leading to the formation of calcium silicate hydrate gel (CSH) (Dermatas, Meng 2003).

In Pb polluted soils treated with PC and/or FA, two possible mechanisms may be responsible for the immobilisation of the contaminant during the hydration phases (Moon, Dermatas 2007). One mechanism may be precipitation resulting from the formation of lead silicate oxide. Another may be inclusion, either by physical encapsulation and or by chemical inclusion. Physical encapsulation can be achieved by creating a solidified monolith, while chemical inclusion can be achieved through the incorporation of lead in binder hydration products, such as CSH gel phases, which play such an important role in the retention of metal. The main mechanisms that determine Pb immobilization in the solid matrix, similarly to other heavy metals, are: sorption on clay and pozzolanic reaction products (Moon, Dermatas 2007) and addition (Eq. (1)) or substitution (Eq. (2)) reactions with CSH (Ouki, Hills 2002):

$$\text{CSH} + \text{Pb} \rightarrow \text{Pb} - \text{CSH}; \qquad (1)$$

$$\text{CSH} + \text{Pb} \rightarrow \text{Pb} - \text{CSH} + \text{Ca}^{2+}. \qquad (2)$$

Several studies on S/S treatment have been performed to better understand the fundamentals of heavy metal immobilisation and leaching (Kundu, Gupta 2008) or to investigate the performance of PC or PC-FA-based S/S treatments of soils polluted by lead. Work carried out by Jing *et al.* (2004) on leachability of sandy soil contaminated by Pb at rate of 3800 mg kg^{-1}, treated using PC, lime and lime-FA mixtures, showed that S/S treatments significantly reduced Pb leaching (below the regulatory concentration) in all performed tests and that Pb concentration in the leachate is mainly controlled by the leachate pH. Moon and Darmatas (2007) investigated the effectiveness of a FA-based S/S treatment of two soils polluted by Pb at concentration of 459 and 3530 mg kg^{-1} by means of semi-dynamic leaching tests. Results showed a reduction of up to 98.5% in the Pb release upon addition of 25% FA and that soil contaminant concentration influenced Pb level in leachate. Moreover, other literature findings (Dermatas, Meng 2003; Qian *et al.* 2008; Yin *et al.* 2006) reported that, in a PC and/or FA based treatment, the percentage of binder used, curing age and Pb contamination level influenced the setting time, compressive strength, leachability and chemical and crystalline structure of the treated soils.

Indeed, it was found that the hydration of cement, and materials such as FA, can be highly modified by heavy metal concentration due to coating around binder grains (Malviya, Chaudhary 2006) and that altered features of hydration products can result in a change of characteristics of the matrices treated by S/S in terms of mechanical strength, durability, curing time, permeability and leachability (Jing *et al.* 2004; Ouki, Hills 2002). In particular, in the presence of PC and FA, Pb, if present at high concentrations, may form, depending on the pH value, hydroxide and silicate phases characterized by a low solubility such as leadhillite (lead carbonate sulfate hydroxide, $Pb_4SO_4(CO_3)_2(OH)_2$), lead carbonate hydroxide hydrate ($3PbCO_3{\cdot}2Pb(OH)_2{\cdot}H_2O$) (Lee 2007) and lead silicate (Pb_2SiO_4, Pb_3SiO_5) (Moon, Dermatas 2007) that reduce the permeability of the matrices and increase the setting times.

Therefore, PC and FA based S/S techniques have been documented as appropriate for Pb-contaminated soil treatment but the Pb concentration level has been reported to possibly significantly influence the effectiveness of the treatment. However the correlation between metal contamination level variation and the performance of S/S treated matrices has not been investigated in any detail and in particular, the limits for the application of S/S techniques to treat high Pb concentration polluted soils are not clear.

Hence, the general objective of this work was to better understand the potential of S/S for the treatment of Pb polluted soils and in particular the effects of Pb concentration and soil:binder ratio on the physical and mechanical properties of soil treated with PC and FA which were assessed analysing the setting time, unconfined compressive strength (UCS), leachability and durability of the treatment products.

1. Materials and methods

1.1. Soil, contaminants and binders

A sandy soil (soil A) and a silty-clay soil (soil B), with properties shown in Table 1, were used for the experiments. Selected soils were spiked with Pb at different rates (C) (1000, 2000, 4000, 8000, 15000 and 25000 mg kg^{-1}), by adding a known quantity of a contaminant solution containing deionized water and reagent grade lead (II) nitrate, purchased from Merck KGaA (Darmstadt, Germany). This nitrate form of Pb was chosen due to its high solubility, representing a "worst-case" scenario (Jing *et al.* 2004) and because it is a main pollutant contained in industrial wastes (Gervais, Ouki 2002).

Table 1. Characteristics of the soils

Parameter	soil A	soil B
Texture	sandy	silty-clay
Sand (silica s. 75–350 μm) [%]	80	20
Silt (silica flour 10–75 μm) [%]	10	56
Clay (kaolin <75 μm) [%]	10	24
pH (L:S of 10)	8.73	8.39
Organic matter [%]	2.79	2.98
Total Organic Carbon [%]	1.67	1.75
Bulk density [g cm^{-3}]	1.42	1.31
Surface area [m^2 g^{-1}]	3.33	14.1
Moisture content [%]	14	25.5

After the contamination procedure, the soils were kept in a closed vessel and stored in a dark room at 4 °C for 1 month then five samples for each contamination level were collected and analysed for Pb content before S/S treatments. Pb content was performed by ICP-OES (Perkin Elmer Optima 4300 with Dual View).

The cement used was purchased as type CEM I Portland cement (PC) CEM 11/B-LL 32.5R from Italcementi S.p.A. (Italy). Class F fly ash (FA) was obtained from Buzzi Unicem S.p.A. (Italy). The chemical and physical properties of PC and FA are presented in Table 2.

Table 2. Characteristics of the binders

Parameter	PC (type I)	FA
Properties		
Moisture content [%]	0.1	0.4
Bulk density [g cm^{-3}]	1.2	1.5
pH (L/S of 10)	13.1	12.4
Surface area [m^2 g^{-1}]	7.8	30.0
Chemical composition		
SiO_2 [% w/w dry]	23.5	42.4
Fe_2O_3 [% w/w dry]	3.7	2.4
Al_2O_3 [% w/w dry]	4.9	37.4
CaO [% w/w dry]	66.3	8.3
MgO [% w/w dry]	1.6	1.9

1.2. Binder systems and S/S sample production

The S/S treatment was performed by mixing control or spiked soil (S) samples with a binder mixture (B) of PC and FA (PC:FA 1:1) at three different S:B ratios (3.3:1, 4.0:1, 5.0:1), applied wet using a water (W) to dry binder (DB) ratio of 0.42:1 as summarised in Table 3.

Table 3. Details of the soil-binder mixes

PC:FA	W:DB	S:B
1:1	0.42:1	5.0:1
1:1	0.42:1	4.0:1
1:1	0.42:1	3.3:1

The mixing was performed by means of a food mixer for 15 min to a homogeneous consistency and the treated soil samples were then cast and compacted into cylindrical moulds (100.0 mm in height and 50.0 mm in diameter) in accordance with the ASTM D1557-91 standard. After 1 day, the samples were demoulded then cured for 28 days in sealed sample bags at a temperature of 20±2 °C and a relative humidity of 95±3% prior to UCS and durability testing.

1.3. Testing protocol

To verify the effectiveness of the S/S treatment, it is necessary to assess the characteristics of the treatment products and compare them with specific performance criteria. It is appropriate to establish a testing regime that addresses the relevant issues for the management scenario of the treatment products (e.g. disposal or utilisation) being considered (Perera *et al.* 2004). The testing protocol on control and soils contaminated at different C included: (1) setting time; (2) UCS; (3) leaching; and (4) wet-dry and freeze-thaw durability values.

The initial and final setting times of the mixtures were determined by using the ASTM C191-82 method. The setting time of a cementitious mixture is referred to as the period from which water is introduced into the mixture system to the onset of hardening. The initial setting time occurred when the Vicat needle 1.00 mm in diameter penetrated the mortar mixture to a point of 25±1 mm, while final setting time occurred when the needle did not visibly sink into the paste.

UCS test relates to the mechanical resistance of the S/S products. UCS values were measured according to ASTM test method D1633 by applying a vertical load axially at a constant strain rate of 0.5 MPa s^{-1} using a Laumas Electronics CTS compressive strength testing apparatus until failure of the cylindrical specimen.

Durability test methods are applied to analyse the long-term performance of the S/S products and in particular the resistance of the material to repeated cycles of weathering. Cured test specimen were subjected to twelve wet/dry (W/D) and freeze/thaw (F/T) cycles according to ASTM D4843 and ASTM D4842 methods, respectively. Specifically, for W/D test each cycle consisted of a period of 5 hours submerged under water and 42 hours in an oven under low-temperature drying condition (71 °C), while for the F/T test, each cycle consisted of a 24-hours freezing period at –20 °C and a 24-hours thawing period in water.

Pb^{2+} leaching behaviour of the products was investigated applying the EN 12457-2 test and the results were compared with E.U. landfill acceptance criteria (Council Decision 2003/33/EC). For the EN 12457-2 test, after the UCS test, coarse particles were separated using a 4.0 mm sieve. Ninety grams of particle samples were placed into a 1000 mL polypropylene plastic bottle containing 900 mL of deionised water (pH 6.81) (L:S weight ratio 10:1). The suspension was shaken in a rotary shaker for a period of 24 hours at 10 rpm and 20 °C. After extraction, the final pH of the leachate was measured and the liquid was separated from the solids by filtration through a 0.45 μm glass fibre filter. The filtered leachates were then preserved for Pb content, which was performed by ICP-OES (Perkin Elmer Optima 4300 with Dual View).

All tests were carried out in triplicate and mean values are shown.

1.4. Quality criteria

In preparation for full-scale treatment, and to evaluate the effectiveness of the S/S treatment and the degree to which the S/S objectives were met, some specific quality performance criteria were defined. Performance criteria are also usually developed in conjunction with the objectives of the treatment and the management scenario of the end material (Perera *et al.* 2004).

The criteria refer to some physical and chemical properties of the S/S solids, measured at 28 days, considering the specific methods adopted for the experiments. The quality criteria were extracted from regulatory limits proposed by US Environmental Protection Agency (US EPA) and United Kingdom Environmental Agency (UK EA) (Stegemann, Cote 1990) and are summarized in Table 4.

Table 4. Quality criteria for S/S materials

		Range/ value	Regulatory
Physical properties			
Setting time (h)		5–72	US EPA (Perera *et al.* 2004)
Initial setting time (h)		2–8	BS EN 196-3:2005 US EPA (Perera *et al.* 2004)
Final setting time (h)		<24	BS EN 196-3:2005 US EPA (Perera *et al.* 2004)
UCS (MPa)	Landfill disposal	0.35	USEPA/530-SW-016 US EPA (Perera *et al.* 2004)
UCS (MPa)	Landfill disposal	1.00	BS EN 196-3:2005
UCS (MPa)	Sanitary landfill disposal or construction application	3.45	USEPA/530-SW-016; WTC (1991) US EPA (Perera *et al.* 2004)
UCS (MPa)	Mortar manu-facturing	20	UK waste disposal regulatory (Yin *et al.* 2006)
Weight loss after 12 cycles of durability test (%)		<30	WTC (1991); US EPA (Perera *et al.* 2004), Stegemann and Cote (1990)
Chemical properties			
Leachate Pb (mg kg^{-1})	Inert waste landfill disposal	0.5	EU landfill acceptance criteria (2003/33/EC) US EPA (Perera *et al.* 2004)
Leachate Pb (mg kg^{-1})	Non hazardous waste landfill disposal	10	EU landfill acceptance criteria (2003/33/EC) US EPA (Perera *et al.* 2004)
Leachate Pb (mg kg^{-1})	Hazardous waste landfill disposal	50	EU landfill acceptance criteria (2003/33/EC) US EPA (Perera *et al.* 2004)

2. Results

2.1. Testing protocol

2.1.1. Setting times

Figures 1 (a) and 1 (b) show the effects of Pb soil concentration (C) on the initial and final setting times of the S/S treated soils for a S:B ratio of 3.3:1, 4.0:1 and 5.0:1 and for soils A and B, respectively. Results indicate that, especially for soil A, C and S:B ratio significantly influenced the setting times of the S/S treated soils. In particular, setting times increased for both soils, with increasing C and decreased with decreasing the S:B ratio. Higher setting times were observed for soil A, where the presence of Pb at a S:B ratio of 5.0 strongly delayed the hydration reactions and significantly lengthened setting time values (up to 240 h) for C higher than 15000 mg kg^{-1}. This specific behaviour suggests that when a Pb:B threshold ratio is exceeded, a significant increase in the setting time occurs.

These results are consistent with the literature findings where Pb has been reported to suppress cement hydration and lengthen the setting times due to the precipitation of protective coatings of gelatinous hydroxide around the cement grain surface (Chen *et al.* 2009). Moreover, results are in agreement with experimental findings obtained by other authors: Gervais and Ouki (2002) found that the initial setting time for a PC system doped with Pb as nitrate salt at a Pb:B ratio of 0.01 and a W:B ratio of 0.45 is delayed to 30 h, whereas, for a PC S/S treated soil with an S:B ratio of 1, Yin *et al.* (2006) reported final setting times of 4, 8, 20 and 22 h for a C of 500, 5000, 25000 and 50000 mg kg^{-1}, respectively.

Considering the less rigorous US EPA acceptance quality criteria of 72 h for final setting time, inadequate values were obtained for soil A at a S:B ratio of 5.0 for any Pb contamination level and at a S:B ratio of 4.0 only for a C of 25000 mg kg^{-1}. Adequate values were obtained for soil B for all the experimental conditions tested. If the BS EN criteria is considered (initial setting time <8 h, final setting time <24 h), adequate values were obtained only for soil B contaminated at the maximum level of 2000 mg kg^{-1} for all the S:B ratios investigated.

2.1.2. UCS

Results of compressive strength (UCS) tested at 28 days of curing for control and treated polluted soils (soil A and B) are presented as a function of C in Fig. 2.

Results were observed to have an average margin of error of ±7%. Results showed that C highly influenced UCS and the same trend between UCS and C was observed for both tested soils. Specifically, for all the S:B ratios tested, an increase of UCS was observed for a C = 1000 mg kg^{-1} (up to 13000 KPa) respect to the control samples, followed by a decrease with increasing C for the highest values. As expected, UCS of the samples increased with decreasing S:B ratio, and highest values were measured for sandy soil. Overall, excluding the data referred to a C = 25000 mg kg^{-1}, a difference of between 3000 and 6000 KPa was recorded between UCS of soil A and B for all the S:B ratios. For the samples spiked at a C = 25000 mg kg^{-1}, a drastic UCS decrease was observed only for the sandy soil. Therefore, it is clear that the presence of low C (<4000 mg kg^{-1}) results in an improvement of the mechanic characteristics of the S/S treated soils respect to

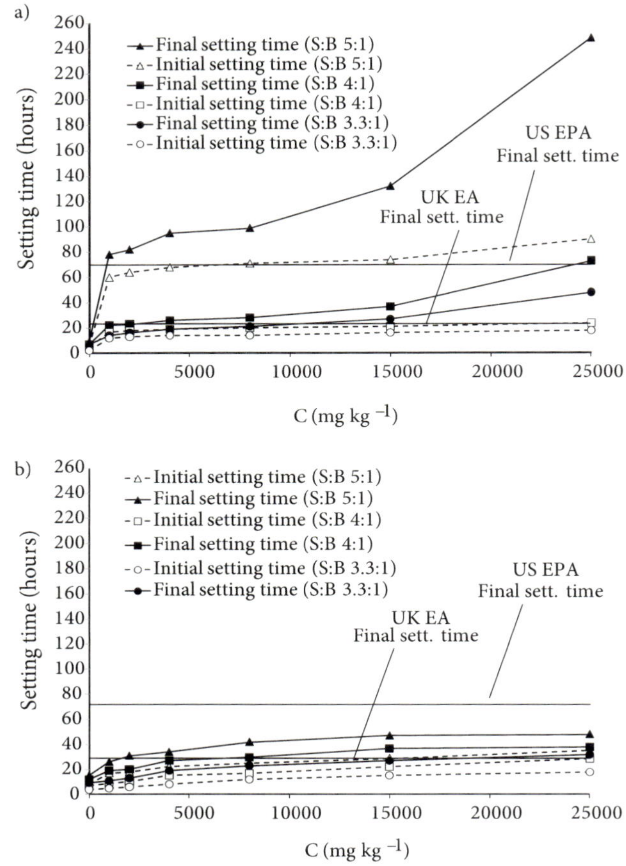

Fig. 1. Setting times of S/S treated soils vs Pb soil concentration for soil A (a) and soil B (b) (S:B ratio of 3.3, 4.0 and 5.0)

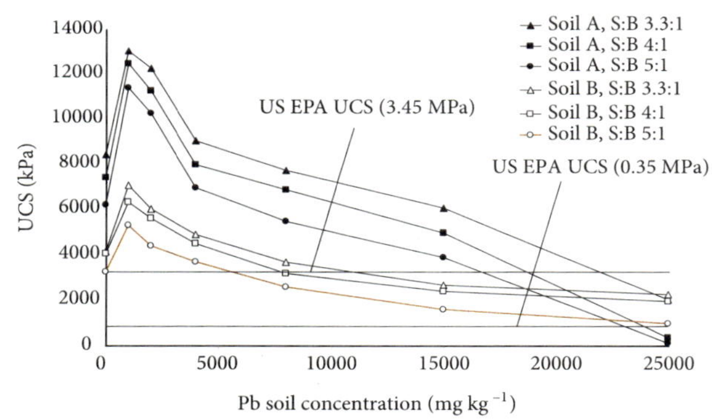

Fig. 2. Effect of Pb soil concentration on the UCS development of S/S soils for soil A and B (S:B ratio of 3.3, 4.0 and 5.0)

the control samples, whereas for higher C, UCS decreased, doing so more rapidly for sandy soil up to values lower than 100 KPa.

This specific behaviour is probably due to the lengthening of the hydration reaction that was observed to be more consistent for the sandy soils. As a matter of fact, the better strength performances of the sandy soils are hindered by the higher setting times recorded, especially for high values of C. But, for the lowest C, it seems that the presence of Pb in the structure of the S/S treated soils gave them an improvement in the mechanical features that are not weakened by a significant retardation of the hydration reactions. A previous study (Yin *et al.* 2006) on S/S treated soils reported an increase of UCS with increasing the Pb contamination level in soil. Authors observed for a matrix of PC and soil (S:B 2:1) cured at 28 days an UCS of 34, 37 and 47 MPa for a contamination level of 0, 500 and 25000 mg kg^{-1}, respectively. UCS of 34 MPa was also found by Gervais and Ouki (2002) but for a 100% PC solidified previously doped with Pb at a 0.01 Pb:B ratio.

In terms of quality acceptance, considering the US EPA criteria, insufficient strength values were obtained only for soil A considering a S:B ratio equal to or higher than 4.0 and the maximum C (25000 mg kg^{-1}) in the cases

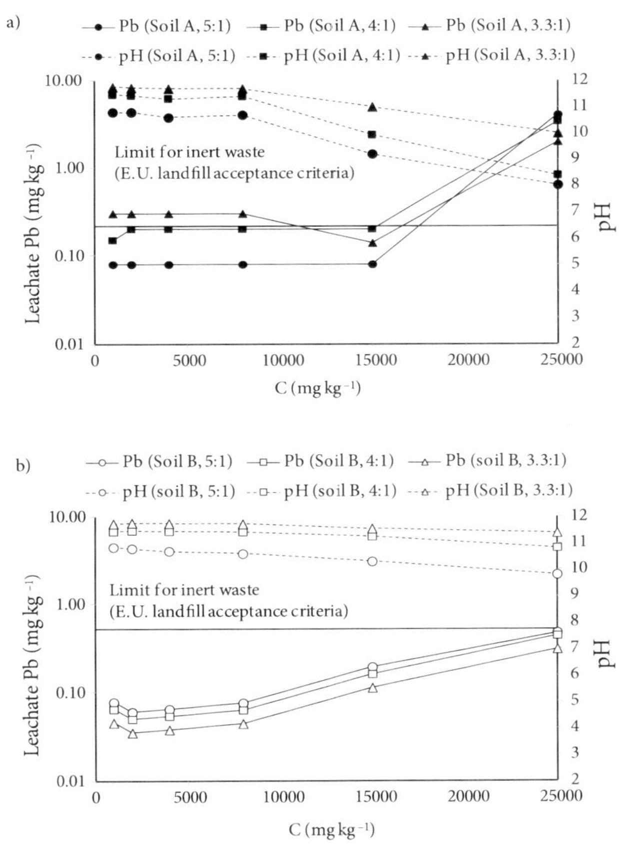

Fig. 3. Effect of Pb soil concentration on the leachability of lead and pH of S/S treated soils for soil A (a) and soil B (b) (S:B ratio of 3.3, 4.0 and 5.0)

of S/S treated soils landfill disposal (UCS > 0.35 MPa). If the more restrictive limit for construction application (UCS > 3.45 MPa) is considered, sufficient UCS values were reached for C equal to or lower than 15000 mg kg^{-1} for sandy soil and 4000 mg kg^{-1} for silty-clay soil. Higher C silty-clay soils (up to 8000 mg kg^{-1}) could be successfully treated using a S:B ratio of 3.3 or changing the composition of the binder (i.e. increasing the PC content).

2.1.3. Leachability

The effect of C on the Pb leachability expressed as mg kg^{-1} of solid material and on pH values of leachates resulting from EN 12457-2 test are shown in Fig. 3.

Results were observed to have an average margin of error of ±4%. For all the single batch extractions low values of leached Pb were observed for C values up to 8000 mg kg^{-1}, while for higher C a significant increase was observed in soil A and a slight increase in soil B. This specific behaviour corresponded to pH values which were above 9.8 for soil B but decreased down to a value of 8.0 for soil A at the highest C, indicating the higher buffering capacity of the treated samples of soil B compared to those of soil A. As expected, an increase of leached Pb and a decrease of pH values were observed with increasing S:B ratio. Generally, for most alkaline materials the leaching concentration of Pb decreases for a pH reduction from 10 to 8, but in this case a different trend was observed, probably due to the higher concentration tested that determined an increase in the leaching value even though the reduction in pH would have yielded a lower concentration.

Furthermore, the theory for which leaching of contaminants such as Pb is reduced with increasing the percentage of fine texture soil and/or humidity (i.e. soil B) seems to be confirmed. Indeed, an increase of the specific surface area of the S/S matrices and, therefore an increased adsorption of contaminant onto the clay fraction of the soil may produce a decrease in the amount of coating on the cement grains allowing the cement hydration. Moreover, for soil B, the presence of kaolin, that is known to be a good Pb^{2+} adsorbent (Jiang *et al.* 2009), especially for high pH values, improved the adsorption phenomena reducing Pb leaching. This phenomenon is more relevant for Pb concentration equal to or higher than 8000 mg kg^{-1}. For both types of soil, the good results obtained, in terms of leachability, are also probably due the presence of a low concentration of organic matter. This condition is in fact known to play an important role in the increase of the immobilization phenomena of Pb (Janoš *et al.* 2010).

Comparing the obtained results with E.U. landfill acceptance criteria (Council Decision 2003/33/EC), the leached Pb for soil B (all treatments) was below the limit of 0.5 mg kg^{-1} for inert waste landfill disposal. For soil A, the limit of 0.5 mg kg^{-1} was respected only for C equal to or lower than 8000 mg kg^{-1} or for C equal to or lower than 15000 mg kg^{-1} using a S:B ratio of 4.0 or 3.3. For the other treatments, leached Pb was in each case below the limit of 10 mg kg^{-1} for non-hazardous waste landfill disposal. Results confirm that the main factor controlling the Pb concentration in the leachate is the final pH, and they are in agreement with those obtained by Jing *et al.* (2004). This finding on Pb leachability highlights the possibility of successfully treating also heavy Pb soil contamination using minimal binder percentage (i.e. S:B = 5.0:1) for silty-clayey soils or using higher binder percentages (depending on the objectives of the treatment) for sandy soils.

2.1.4. Durability

For S/S treated soils long-term performance assessment durability tests were performed and results are shown in Figures 4 (a) and 4 (b).

Results were observed to have an average margin of error of ±9%. For the F/T test a slight increase of the weight loss was observed with increasing C for C ≤ 15000 mg kg^{-1}. A significant average variation (about 15%) in terms of mass loss was observed between both tested soils. C > 15000 mg kg^{-1} significantly worsened the performance of the S/S matrices, shown by an increase of the mass loss during the durability cycles. Specifically, for the samples where C was 25000 mg kg^{-1} a weight loss of 21, 24 and 49% was achieved respectively for the soil A (S:B 3.3:1), soil B (S:B 3.3:1) and soil B (4.0:1). For the other experimental conditions, all the tested samples were disintegrated after 2 cycles, for soils A (S:B 5.0:1) and B (S:B 5.0:1) and after 8 cycles for soil A (S:B 4.0:1). These results clearly show that, despite the overall best durability performance of the sandy soils, at high Pb contamination level, they could be more vulnerable than fine texture soils such as soil B. This confirms that in sandy soil, the large retardation of the hydration reactions observed, due to the high Pb level, may significantly worsen the characteristics of the S/S treated soils and consequently their performance.

For the W/D test a small increase in the weight loss was observed with an increasing of C for C equal to or lower than 8000 mg kg^{-1}. For all the samples no difference was observed between C = 15000 and 25000 mg kg^{-1}, whereas an increase of about 10% was observed for soil B in comparison with soil A. The observed decrease in weight loss with an increasing Pb level in soil may be attributed to the water saturation phenomena regarding the samples during their immersion in water for 5 hours at each cycle of W/D. The presence of a higher amount of water in the sample matrix could play an important role in the improvement of the hydration kinetics that gave the best performance in terms of W/D durability to the treated soils.

Based on the durability acceptance criteria proposed by WTC 1991 and reported by Stegemann and Cote (1990) (weight loss < 30%), a minimal S:B ratio is sufficient to successfully treat a Pb contamination level equal to or lower than 15000 mg kg^{-1}, whereas a S:B ratio of 3.3 is needed to treat contamination levels up to 25000 mg kg^{-1}.

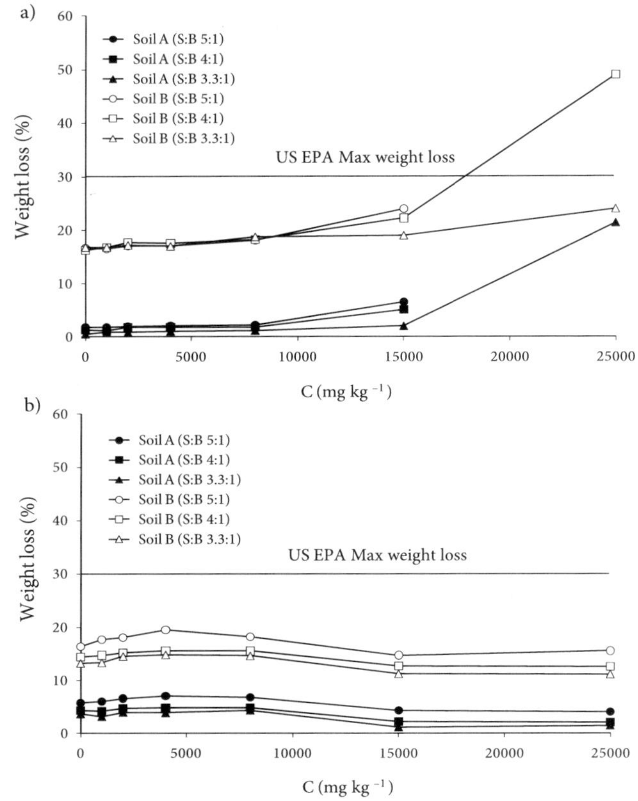

Fig. 4. Effect of Pb soil concentration on the weight loss of S/S treated soils during freeze-thaw (a) and wet-dry (b) durability test for soil A and B (S:B ratio of 3.3, 4.0 and 5.0)

2.2. Assessment of S/S management options

Based on the results obtained by the experimental phase of this work, a performance threshold approach was used to assess potential management options for the S/S treatments. Four management options, namely (1) in situ S/S treatment; (2) hazardous waste landfill disposal; (3) non-hazardous waste landfill disposal; and (4) inert waste landfill disposal or reuse of S/S products for construction materials were chosen for the two investigated soils (sandy and silty-clay soils). Performance thresholds were taken from the performance acceptance criteria reported in Table 4 and management options were represented as minimal percentage of binder to use (S:B ratio) versus the Pb contamination level of the soil (C) for which the selected performance thresholds are all satisfied. The ranges of the values shown on the x and y axes represent the limits of the experimental conditions investigated. The Pb soil concentration limit which can be treated and the related limiting factor and maximum S:B ratio required for sandy and silt-clay soils is reported in Table 5.

Table 5. Pb soil concentration limit and limiting factor for sandy and silt-clay soils as a function of the scenario adopted

Scenario – S:B ratio	Sand soil		Silt-clay soil	
	Pb concentration limit (mg kg^{-1}) (S:B ratio required)	Limiting factor	Pb concentration limit (mg kg^{-1}) (S:B ratio required)	Limiting factor
1	8000 (3.3)	Setting time	4000 (4.0)	UCS
2	–	–	–	–
3	15000 (4.0)	UCS	4000 (5.0)	UCS
4	15000 (4.0)	UCS-Leachability	4000 (5.0)	UCS

2.2.1. In situ S/S treatment

The in situ S/S soil mixing treatment is carried out using mixing augers through which a grout is introduced and mixed with the soil resulting in stabilized/solidified overlapping soil-grout columns (Fleri, Whetstone 2007). Compared to the ex-situ treatments, they could be more hazardous due to the potential risk of a contamination propagation during the realization of the interventions; consequently, high performance acceptance criteria limits were adopted for the purpose.

A fundamental property for the S/S soil-grout columns is the final setting time, since a delay in the S/S mass curing could significantly enhance the leachability of the contaminants. A downward contaminant migration could take place during the in situ mixing processes due to the transitory increase of water amount in soil. For this reason the BS EN limit of 24 h was adopted. Moreover, to ensure a high mechanical resistance another US EPA restrictive limit of 3.45 MPa was applied for UCS. EU landfill acceptance criteria (2003/33/EC) limit (0.5 mg kg^{-1}) and WTC limit (30% weight loss) were adopted for leaching and durability tests, respectively. Figures 5 (a) and 5 (b) show the minimal percentage of binder to use (S:B ratio) as a function of C for which the selected limits are satisfied in the case of an in situ treatment, for a sandy and a silty clay soil, respectively. Results highlight that due to the strict limits adopted, the in situ treatment may be successfully adopted only for not heavy Pb contamination levels. Specifically, a maximum level of 8000 mg kg^{-1} and 4000 mg kg^{-1} is treatable for a sandy and a silty clay soil, respectively. Moreover, a S:B ratio of 4.0 is adoptable for both the soils up to 2000 mg kg^{-1}, whereas a S:B ratio of 3.3 is necessary for higher contaminations.

2.2.2. Hazardous waste landfill disposal

A maximum final setting time of 72 h and an UCS of 350 kPa at 28 days are suggested by USEPA guidelines for materials that are to be disposed to landfill (Perera *et al.* 2004) which takes into consideration events such as weight of overburden and land moving equipment. For leaching performance and durability, EU landfill acceptance criteria limit for hazardous waste disposal (50 mg kg^{-1}) and WTC limit (30% weight loss) were used. Results (Figs 5 (c) and 5 (d)) show that a maximum C of 15000 mg kg^{-1} may be treated using a S:B of 4.0 and 5.0 for sandy and silty clay soil, respectively, whereas a maximum percentage of binder must be adopted for higher contamination levels. Specifically, for silty clay soil this limitation is due to the durability features, whereas for sandy soil it is due to all the adopted limits except for the leaching limit that is always satisfied. In addition, the worst results, in terms of setting time, were observed for sandy soil, resulting in an S:B values equal to or lower than 4.0 being required for its treatment.

2.2.3. Non-hazardous waste landfill disposal

Compared to the previous scenario, a higher UCS value of 3.5 MPa has been suggested by USEPA and WTC (Perera *et al.* 2004) for disposal to non-hazardous waste (sanitary) landfill because compaction of municipal waste might subject the S/S material to higher stresses because handling, placement and covering operations are not tailored for S/S material. Moreover, the EU landfill acceptance criteria limit for non-hazardous waste disposal (10 mg kg^{-1}) was used. The use of 3.5 MPa limit for UCS results in a strong limitation of the applicability field (Figs 5 (e) and 5 (f)) especially for silty clay soil for which a maximum C of 4000 mg kg^{-1} may be treated using a S:B from 3.3 to 5.0. For sandy soil a higher C up to 15000 mg kg^{-1} may be treated but using a S:B from 3.3 to 4.0.

2.2.4. Inert waste landfill disposal or reuse for construction

Compared to the sanitary landfill disposal scenario, a lower limit of 0.5 mg kg^{-1} for leaching is required by the EU for disposal in inert waste landfill. For both soils, the use of this severe limit for leaching results in the same limitation as the previous scenario (Figs 5 (g) and 5 (h)).

Conclusions

A sandy and a silty-clay soil, spiked with lead at different rates, were S/S treated using a binder mixture of PC and FA (PC:FA 1:1) at different soil:binder ratios, and a testing protocol included setting time, UCS, leaching and durability test were performed to assess the effects of lead and binder level on physical properties of S/S matrices. Based on the experimental results and defining specific quality performance criteria, a performance threshold approach was used to assess potential management options for the S/S treatments. The following conclusions have been drawn according to the results presented above:

1. Soil texture, the percentage of binders used and lead concentration in soil significantly influence

the performance of the S/S treated soils in terms of curing, compressive and weather cycling strength;

2. The observed influence of lead content on studied parameters may be useful in predicting setting time, UCS, leaching and durability of soils treated by PC and FA at any contamination level;
3. Lead soil concentration higher than 15000 mg kg^{-1} heavily reduces the applicability of the S/S techniques requiring a large amount of binder to satisfy the

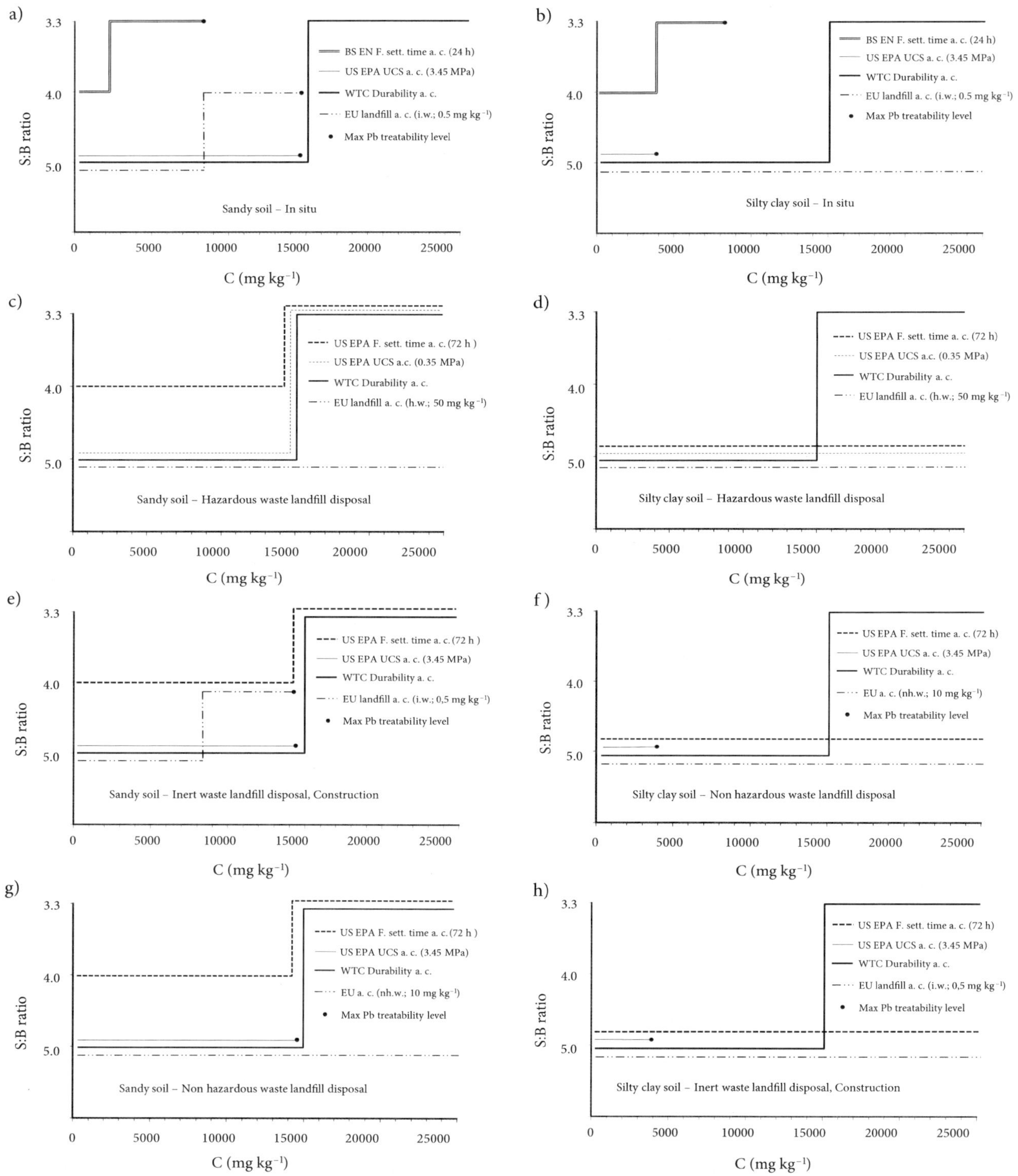

Fig. 5. Percentage of binder to use (S:B ratio) versus C for which the quality performance criteria thresholds are satisfied for sandy and silty clay soils and different scenarios

selected performance criteria; this could make the treatment very expensive. Specifically, despite the best compressive strength observed at the lowest lead concentration values, soils performance was shown to be limited by setting time and UCS features due to the lengthening of the hydration reaction;

4. In terms of leachability, lead soil concentration does not represent a limitation, indeed also heavy contamination may be successfully treatable also using minimal binder percentage for any type of soil, except for the in situ and inert waste landfill disposal scenarios, for which, due to the severe limits adopted, a higher soil:binder ratio of 4.0 is required;
5. A strong limitation of the applicability field was observed for an in situ treatment, S/S products reuse or disposal to a landfill, especially for silty clay soils for which only lead concentration lower than 4000 mg kg^{-1} may be treated in the case of landfill disposal or reuse, while a maximum contamination of 8000 mg kg^{-1} must be treated using a soil:binder ratio of 3.3:1 in the case of sandy soil in situ treatment;
6. The obtained results are of practical interest and may be used for preliminary cost assessment of full scale remediation activities of lead polluted soils.

References

Al-Ansary, M. S.; Al-Tabbaa, A. 2007. Stabilisation/solidification of synthetic petroleum drill cuttings, *Journal of Hazardous Materials* 141(2): 410–421. http://dx.doi.org/10.1016/j.jhazmat.2006.05.079

Antemir, A.; Hills, C. D.; Carey, P. J.; Magnié, M. C.; Polettini, A. 2010. Investigation of 4-year-old stabilised/solidified and accelerated carbonated contaminated soil, *Journal of Hazardous Materials* 181(1–3): 543–555. http://dx.doi.org/10.1016/j.jhazmat.2010.05.048

Bech, J.; Duran, P.; Roca, N.; Poma, W.; Sánchez, I.; Roca-Pérez, L.; Boluda, R.; Barceló, J.; Poschenrieder, C. 2012. Accumulation of Pb and Zn in *Bidens triplinervia* and *Senecio sp.* spontaneous species from mine spoils in Peru and their potential use in phytoremediation, *Journal of Geochemical Exploration* 123: 109–113. http://dx.doi.org/10.1016/j.gexplo.2012.06.021

Chen, Q. Y.; Tyrer, M.; Hills, C. D.; Yang, X. M.; Carey, P. 2009. Immobilisation of heavy metal in cement-based solidification/stabilisation: a review, *Waste Management* 29(1): 390–403. http://dx.doi.org/10.1016/j.wasman.2008.01.019

Cheng, Y.; Guo, Z.; Liu, X.; Yin, H.; Qiu, G.; Pan, F.; Liu, H. 2009. The bioleaching feasibility for Pb/Zn smelting slag and community characteristics of indigenous moderate-thermophilic bacteria, *Bioresource Technology* 100(10): 2737–2740. http://dx.doi.org/10.1016/j.biortech.2008.12.038

Dermatas, D.; Meng, X. 2003. Utilization of fly ash for stabilization/solidification of heavy metal contaminated soils, *Engineering Geology* 70(3–4): 377–394. http://dx.doi.org/10.1016/S0013-7952(03)00105-4

European Communities Council Decision 2003/33/EC. Establishing criteria and procedures for the acceptance of waste at landfill.

Falciglia, P. P.; Cannata, S.; Romano, S.; Vagliasindi, F. G. A. 2012. Assessment of mechanical resistance, g-radiation shielding and leachate g-radiation of stabilised/solidified radionuclides polluted soils: preliminary results, *Chemical Engineering Transactions* 28: 127–132.

Fleri, M. A.; Whetstone, G. T. 2007. In situ stabilisation/solidification: project lifecycle, *Journal of Hazardous Materials* 141(2): 441–456. http://dx.doi.org/10.1016/j.jhazmat.2006.05.096

Gervais, C.; Ouki, S. K. 2002. Performance study of cementitious systems containing zeolite and silica fume: effects of four metal nitrates on the setting time, strength and leaching characteristics, *Journal of Hazardous Materials* 93(2): 187–200. http://dx.doi.org/10.1016/S0304-3894(02)00005-5

Harbottle, M. J.; Al-Tabbaa, A.; Evans, C. W. 2007. A comparison of the technical sustainability of in situ stabilisation/solidification with disposal to landfill, *Journal of Hazardous Materials* 141(2): 430–440. http://dx.doi.org/10.1016/j.jhazmat.2006.05.084

Janoš, P.; Vávrová, J.; Herzogová, L.; Pilařová, V. 2010. Effects of inorganic and organic amendments on the mobility (leachability) of heavy metals in contaminated soil: a sequential extraction study, *Geoderma* 159(3–4): 335–341. http://dx.doi.org/10.1016/j.geoderma.2010.08.009

Jiang, M.; Wang, Q.; Jin, X.; Chen, Z. 2009. Removal of Pb(II) from aqueous solution using modified and unmodified kaolinite clay, *Journal of Hazardous Materials* 170(1): 332–339. http://dx.doi.org/10.1016/j.jhazmat.2009.04.092

Jing, C.; Meng, X.; Korfiatis, G. P. 2004. Lead leachability in stabilized/solidified soil samples evaluated with different leaching tests, *Journal of Hazardous Materials* 114(1–3): 101–110. http://dx.doi.org/10.1016/j.jhazmat.2004.07.017

Kundu, S.; Gupta, A. K. 2008. Immobilization and leaching characteristics of arsenic from cement and/or lime solidified/stabilized spent adsorbent containing arsenic, *Journal of Hazardous Materials* 153(1–2): 434–443. http://dx.doi.org/10.1016/j.jhazmat.2007.08.073

Lee, D. 2007. Formation of leadhillite and calcium lead silicate hydrate (C-Pb-S-H) in the solidification/stabilization of lead contaminants, *Chemosphere* 66(9): 1727–1733. http://dx.doi.org/10.1016/j.chemosphere.2006.07.012

Leonard, S. A.; Stegemann, J. A. 2010. Stabilization/solidification of petroleum drill cuttings: leaching studies, *Journal of Hazardous Materials* 174(1–3): 484–491. http://dx.doi.org/10.1016/j.jhazmat.2009.09.078

Malviya, R.; Chaudhary, R. 2006. Factors affecting hazardous waste solidification/stabilization: a review, *Journal of Hazardous Materials* 137(1): 267–276. http://dx.doi.org/10.1016/j.jhazmat.2006.01.065

Moon, D. H.; Dermatas, D. 2007. Arsenic and lead release from fly ash stabilized/solidified soils under modified semi-dynamic leaching conditions, *Journal of Hazardous Materials* 141(2): 388–394. http://dx.doi.org/10.1016/j.jhazmat.2006.05.085

Ouki, S. K.; Hills, C. D. 2002. Microstructure of Portland cement pastes containing metal nitrate salts, *Waste Management* 22(2): 147–151. http://dx.doi.org/10.1016/S0956-053X(01)00063-0

Perera, A. S. R.; Al-Tabbaa, A.; Reid, J. M.; Stegemann, J. A. 2004. *State of practice reports*, UK stabilization/solidification treatment and remediation – testing and performance criteria, in Al-Tabbaa, A.; Stegemann, J. A. (Eds.). *Stabilization/Solidifi-*

cation Treatment and Remediation, Advances in S/S for Waste and Contaminated Land. London: A. A. Balkema Publishers.

Polettini, A.; Pomi, R.; Sirini, P.; Testa, F. 2001. Properties of Portland cement – stabilised MSWI fly ashes, *Journal of Hazardous Materials* 88(1): 123–138. http://dx.doi.org/10.1016/S0304-3894(01)00292-8

Qian, G. R.; Shi, J.; Cao, Y. L.; Xu, Y. F.; Chui, P. C. 2008. Properties of MSW fly ash–calcium sulfoaluminate cement matrix and stabilization/solidification on heavy metals, *Journal of Hazardous Materials* 152(1): 196–203. http://dx.doi.org/10.1016/j.jhazmat.2007.06.118

Ryu, B. G.; Park, G. Y.; Yang, J. W.; Baek, K. 2011. Electrolyte conditioning for electrokinetic remediation of As, Cu, and Pb-contaminated soil, *Separation and Purification Technology* 79(2): 170–176. http://dx.doi.org/10.1016/j.seppur.2011.02.025

Stegemann, J. A.; Cote, P. L.; 1990. Summary of an investigation of test methods for solidified waste evaluation, *Waste Management* 10(1): 41–52. http://dx.doi.org/10.1016/0956-053X(90)90068-V

Svensson, M.; Allard, B. 2008. Leaching of mercury-containing cement monoliths aged for one year, *Waste Management* 28(3): 597–603. http://dx.doi.org/10.1016/j.wasman.2007.02.031

Tang, X.; Yang, J. 2012. Long-term stability and risk assessment of lead in mill waste treated by soluble phosphate, *Science of the Total Environment* 438: 299–303. http://dx.doi.org/10.1016/j.scitotenv.2012.08.054

Torres, L. G.; Lopez, R. B.; Beltran, M. 2012. Removal of As, Cd, Cu, Ni, Pb, and Zn from a highly contaminated industrial soil using surfactant enhanced soil washing, *Physics and Chemistry of the Earth* 37–39: 30–36.

Yin, C. Y.; Mahmud, H. B.; Shaaban, M. G. 2006. Stabilization/solidification of lead-contaminated soil using cement and rice husk ash, *Journal of Hazardous Materials* 137(3): 1758–1764. http://dx.doi.org/10.1016/j.jhazmat.2006.05.013

Pietro Paolo FALCIGLIA. A graduate of University of Catania, Italy, received a Master of Science from Superior School of Catania in 2004, and a PhD in Civil and Environmental Engineering from University of Salerno (Italy) in 2008. Currently he is a Research Associate and Contract Professor of Remediation and Reclamation of Polluted Sites at the Civil and Environmental Engineering Department of the University of Catania. Publications: author/co-author of over 40 scientific papers. Research interests include waste management and treatment, remediation of polluted sites, wastewater treatment.

Abir AL-TABBAA. She graduated from Bristol University in 1983 in Civil Engineering and obtained an MPhil and PhD degrees in Soil Mechanics in 1984 and 1987 respectively from Cambridge University. In 1991 she took up a lectureship at the University of Birmingham and in 1997 returned to Cambridge University as a University Lecturer, now Reader in Geotechnical Engineering. She is a fellow of Sidney Sussex College. She is a Chartered Engineer and a Member of the Institution of Civil Engineers. She was a member of the British Geotechnical Association Executive Committee (2001–2004), a member of the Institution of Civil Engineers Geotechnical Engineering Advisory Panel (2003–2005), Ground Improvement Advisory Panel (2007–2012) and the UK Representative on the International Society of Soil Mechanics and Geotechnical Engineering Technical Committee TC211 on Ground Improvement. She was awarded the Institution of Civil Engineers Reed and Mallik Medal in 2003 for her paper on the five-year soil mixing treatment work at the West Drayton site near Heathrow Airport. Research interests include soil mix technology, field characterisation of soils, waste, waste management and contaminated land, groundwater, seepage and permeability, geomaterials and constitutive models, materials for extreme environments.

Federico G. A. VAGLIASINDI. A graduate of University of Catania, Italy, received a Master of Science from the Civil Engineering Department at Colorado State University, Fort Collins (CO) in 1991, and a Full Professor of Environmental and Sanitary Engineering at University of Catania, Italy, since March 2002. Has participated to research projects financed by the European Commission, the Italian National Research Council, MIUR (PRIN 1999, 2001, 2004, 2006; Project Aquatec PON 2002), AWWARF (American Water Works Association Research Foundation, USA), EPRI (Energy and Power Research Institute, USA). He has consulted several public administrations involved with the ongoing emergency in the solid waste management and contaminated site remediation in Sicily. He is a member of several professional association, has been a component of the Environmental and Sanitary Section of the Italian Great Risks Commission. Since April 2008, he has been a President of the Italian Association of Sanitary and Environmental Engineering Professors. His research interests include water and wastewater treatment and reuse, integrated solid waste management, contaminated site remediation.

15

LANDSCAPE AND TRAFFIC FACTORS AFFECTING ANIMAL ROAD MORTALITY

Shyh-Chyang LIN

Department of Civil Engineering and Engineering Management, National Quemoy University
Jinning, Kinmen, Taiwan

Abstract. Roadkill is a significant indicator in realizing the impacts of roads on adjacent ecosystems. This study undertakes a comprehensive survey of roadkills in Kinmen (Taiwan) and analyzes their causes. Two models, Traffic Flow Model and Geometric Model, combined with animal road-crossing behaviors, are used to derive survival probability. Survey results and model predictions yield similar results for moderate traffic flow and agree in bird and small mammal roadkill frequency prediction. It is found that traffic volume, adjacent landscape and road condition are the major contributing factors related to roadkills. Higher traffic volume near habitats always augments the probability of roadkill; however, roadside trees, adjacent landscapes, and road longitudinal slope also affect the probability of successful crossing by small animals, especially birds. This study also proposes measures, to be applied to future road planning and design, aimed to lower roadkill probability.

Keywords: roadkill, ecological conservation, traffic flow theory, Kinmen, geometric model.

Introduction

The globalization of regional development has been dramatically destroying ecological habitats and decreasing global biodiversity. Harmony and coexistence between nature and human beings are in jeopardy. The impact of human development on the global ecosystem is unprecedented. Wilson (1989) estimates that 10,000 to 15,000 species become extinct worldwide each year; 34 species disappear on the earth each day due to the destruction of forests and animal habitats. The most severe impact by humans on ecosystems is the construction of road networks. Road construction directly affects the adjacent habitat and leads to increased human invasion and development. Roads impact ecosystems through: destruction of habitat, disturbance to animals, roadkills, and barrier effects (Seiler 2001). Although some positives, such as providing corridors for animal movement (Adams, Geis 1983; Seiler 2001), can be derived from the presence of roads, they cause much more harm to ecosystems than good. Five major factors to be considered when planning ecologically friendly roads are: the road itself (width, pavement, mitigation measures), traffic (vehicle speed, traffic volume), road verge (width, shape, vegetation), adjacent land use (forest, community, agriculture), and species (size, weight, food web, territory, migration characteristics, dispersion ability, and rarity) (Jaarsuma, Willems 2002).

The patterns of avian-vehicle collisions are difficult to assess because they are three dimensional and the birds' crossing methods vary. Granivorous and omnivorous birds are more likely to collide with cars as they forage on the roads and are attracted by the heat of roads (Dhindsa *et al.* 1988). Avian roadkills happen more frequently when birds are chasing each other and flying low over roads (Eritzoe 2002). Some research suggests that accidents involving birds are rare at vehicle speeds of less than 80 km/hr (Nankinov, Todorov 1983) and can occur only at vehicle speeds over 100 km/hr (Dhindsa *et al.* 1988). Based on the results of this survey, this fact is questionable and needs further investigation.

Though it is believed that small animal collisions have a minor effect on human safety, drivers can be seriously injured due to avoiding hitting animals. One surprising fact is that non-forest areas have more roadkills than forested or reserved areas (Gunther *et al.* 1998). The ecotones between natural and developed areas are the places with higher probability of roadkills. Refuse and grains dropped

Corresponding author: Shyh-Chyang Lin
E-mail: sclin@nqu.edu.tw

from vehicles are fatal attractions (Slater 1994) to wildlife as they are often hit by oncoming traffic (Dhindsa *et al.* 1988). The impacts of roadkills upon animal populations vary depending on the sensibility of the species. A survey in Yellow Stone National Park shows that only 2% of the estimated population of wolves and mule deer, and 1% of large mammals were killed by vehicles each year (Gunther *et al.* 1998). However, the impact of roadkill on the population of rare or endangered species is still unknown but may be an important issue in conserving those species.

It is the responsibility of transport planning engineers to reduce the ecological impact roads have and use mitigation measures as tools in ecological conservation. The basic, necessary information for road-related ecological data is the locations of roadkill hot spots and their causes. With that ecological information, implementation of ecological friendly transportation systems can be achieved. Most factors in animal casualties are biological aspects; however, certain engineering designs, improper construction practices, or conflicts between engineering codes and conservation policies may potentially lead to a massive number of roadkills. Therefore, some roadkills can be prevented simply by changing engineering designs and methods. This, however, is usually overlooked by engineers because few studies regarding roadkill prevention have been undertaken by engineering researchers or scholars. This study models roadkill probability by applying traffic flow theory and analyzes the effects of road attributes, roadside trees, traffic volume, and adjacent landscapes on the frequency of roadkill. Then, field investigation data is used to test the models. Finally, the research results and models of this study are used to propose measures to mitigate the probability of vehicular-wildlife accidents from an engineering point of view.

1. Study site and methods of surveying roadkill

In this study, roadkill research is carried out on Kinmen Island, Taiwan: an island county near Mainland China with abundant ecological features. With a diverse landscape, high road density, and the abundance of birds and small mammals, roadkills in Kinmen happen with great frequency. This makes it an ideal place for roadkill investigation. Therefore, the results of the research show the biological correlation between habitats and roadkill hot spots as well as the influence human development has on the probability of animal casualties. For a more comprehensive survey, 8 major roads are selected for roadkill investigation: Central Road (CR, 11.86 km), Wuandaou E. Road (WDE, 3.25 km), Wuandaou W. Road (WDW, 12 km), Wuandaou N. Road (WDN, 12.45 km), Jiochan Road (JR, 1.1 km), Banlin Road (BL, 0.74 km), University Road (UR, 0.5 km), and Gaoyaung Road (GY, 4.1 km), totally 46 km as shown in Figure 1. The roads cover the four townships in Kinmen; each with distinct adjacent landscapes, traffic flows, and speed limits are distinct.

The field investigation was carried out November 1, 2005 to October 31, 2006. Researchers rode motorbikes at speeds of about 40 km/hr along the 8 roads once a day, every day. Once an animal body was spotted, they: took a picture of the body, initially identified the species, used a laser distance measurer to check the location, removed the body, reconfirmed the initial identification of the species, and buried it on the road side to avoid double counting. After 12 months of investigation, roadkill hot spots were identified and their associated factors, such as road pattern, adjacent landscapes, traffic flow, road slopes and vehicle speeds, are also studied. 466 roadkills were discovered. Their locations are shown in Figure 1. The results of the survey are summarized in Table 1 and roadkills on each investigated road are shown in Table 2. The majority of identified animal bodies are birds (206), rodents (143), and amphibians (mostly toad, 55).

Central Road, with the highest traffic volume and speed limit, has a substantially lower animal accident rate. Therefore, traffic volume and vehicle speed may not be the only factors that affect roadkill frequency. They may act as ecological filter (McDonald, St. Clair 2004) or absolute barriers (Bennett 1992; Forman 1995) to animals

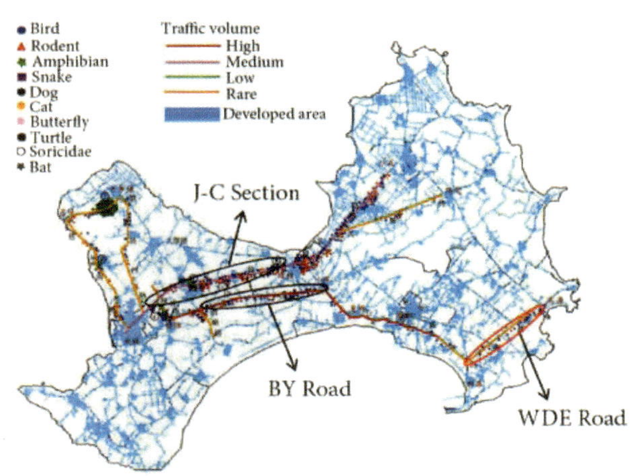

Fig. 1. The surveyed 8 roads in Kinmen and the locations of hot spots

Table 1. Summary of roadkills

Species	Bird	Rodent	Amphibian	Insectivora	Snake	Cat	Turtle	Dog	Butterfly	Bat	Total
Roadkills	206	143	55	27	12	12	4	3	3	1	466

Table 2. Rates of roadkill for each investigated road

Road	Length (km)	Peak traffic volume (veh/hr)	No. of roadkills	Average rate of roadkills (No./km/ month)
BL*	0.74	283	18	324
WDN	12.45	151	250	1.67
JR	1.10	371	9	06$
WDW	12.0	118	97	0.68
UR	0.50	42	3	0.50
WDE	3.25	80	18	0.46
CR	11.86	802	61	0.43
GY	4.10	216	7	0.14

*The survey period of BL is only 7.5 months (1st, November, 2005 to 14th, June, 2006). Widening BL started on 14th, June, 2006 and since then no roadkills have been spotted.

approaching. On the other hand, pollution, disturbance, road width, roadside trees, and animal populations are other important factors.

BL has the highest rate of roadkill even though its traffic flow and vehicle speeds are not high and its adjacent landscape is similar to that of other roads. BL road's slope is a special feature and an important ecological factor in increasing animal accidents. Two major wetlands, which are habitats of Kinmen amphibians, are located along a portion of WDW. Thus, during toad migration season, numerous Spectacled Toads (*Bufo melanostictus*) are run over by vehicles. It is evident that roadkills are related to habitat and season for most species. Tall and dense roadside trees along WDN and WDE roads attract wildlife to the roadside and, thereby, increase collision probability. The study of roadkill hot spot locations can help identify the dominant factors which affect animal accidents from a biological and, more importantly, an engineering point of view.

As researchers did not see animal accidents in person, the patterns of roadkill were unidentifiable. Since the patterns are significant in this study for modelling and analysis, we interviewed local taxi drivers and categorized four types of avian-vehicle collisions.

(I) Birds run over by vehicles as they stand or walk on the road.

(II) Birds hit by oncoming vehicles as they fly up from the road surface.

(III) Birds hit by vehicles as they fly across the road directly.

(IV) Birds hitting the side of vehicles as they fly into the road.

The interviews and observations did not find head-on collisions between birds and vehicles. This suggests that birds can sense directly approaching objects but may only vaguely detect vehicles approaching at right angles. Surveys found that categories (III) and (IV) account for 72% of bird accidents recorded. This suggests that birds have difficulty avoiding vehicles when flying across roads.

The four scenarios of avian-vehicle collision can also be applied to small mammals. Despite crossing at much lower speeds than birds, small mammals will not be able to survive if they collide with the side of a vehicle. Therefore, category (IV) still applies. With the knowledge of animal accident patterns, we can model roadkill probability.

2. Modeling roadkill probability

2.1. Geometric model

According to numerous researchers, traffic volume and vehicle speed are the major causes of roadkills (Forman *et al.* 2003; Reijnen *et al.* 1995). The higher the traffic volume or vehicle speed, the higher the roadkill probability. However, other factors, such as animal crossing speed and vehicle characteristics, are significant in assessing and predicting the probability. Based upon the investigation results and animal-vehicle collision patterns, two models are proposed in this study in order to predict the number of roadkills. First of all, we denote:

- λ (no./sec) : Traffic volume in one direction.
- V_v (m/sec) : Average speed of vehicles.
- V_a (m/sec) : Animal crossing speed.
- W_v (m) : Average vehicle width.
- L_v (m) : Average vehicle length

The time, t_a, needed for an animal to cross a road is.

$$t_a = \frac{W_V}{V_a}. \quad (1)$$

As an animal enters into the "Impact Width" zone, L_i, defined as $L_i = L_x + L_v$ (Fig. 2), an accident will occur. L_x is the distance ahead of vehicle that would cause category III accidents (hit by vehicle). L_v is the range that causes category IV accidents (hit side of vehicle). L_x can be obtained by the relationship

$$L_x = V_v \cdot t_a = W_v \frac{V_V}{V_a}. \quad (2)$$

Based on Eq. (2), the impact width is then

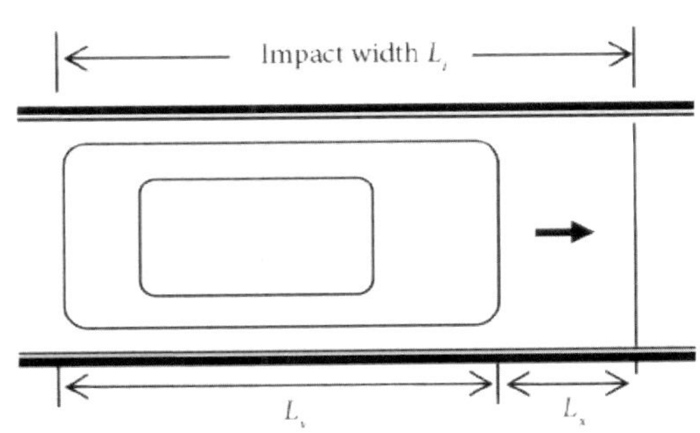

Fig. 2. The range of Impact Width consisting L_x and L_v. L_x may cause roadkill scenario (III) and L_v may cause roadkill scenario (IV)

$$L_i=L_v+L_x=L_v+W_v\frac{V_v}{V_a}. \quad (3)$$

The relationship between traffic volume λ and average spacing of oncoming vehicles (L_λ in Fig. 3) is

$$\frac{L_\lambda}{V_v}=\frac{1}{\lambda}. \quad (4)$$

Such that

$$L_\lambda=\frac{V_v}{\lambda}. \quad (5)$$

Figure 3 shows that the probability of animal accident is a geometric ratio of L_i/L_λ that is defined as Geometric Model of roadkill probability p as

$$p=\frac{L_i}{L_\lambda}. \quad (6)$$

Substituting (3) and (5) into (6) yields

$$p_L=\frac{L_v+W_v\frac{V_v}{V_a}}{\frac{V_v}{\lambda}}=\lambda\left(\frac{L_v}{V_v}+\frac{W_v}{V_a}\right). \quad (7)$$

Eq. (7) is a simple linear probability for animal accidents that may occur when crossing a single lane road. Here, we apply a geometric relationship to the distance between cars. The derivation is based on three assumptions:

(a) Road crossing behaviors of animals are stochastic and blind. Since it is not possible to include an animal's sense of danger or ability to avoid oncoming cars in a mathematic model, these factors were excluded from the derivative.

(b) For the same reason, the driver's reaction to crossing animals is also neglected.

(c) The case of extremely high traffic volume is excluded from the derivations since high traffic volume would make $L_x > L_\lambda$ (Overlap of impact width) and $p > 1$ (Roadkill probability is greater than 1).

2.2. Traffic model

Traffic Theory (Matson *et al.* 1971; May 1990) can also be used to predict the probability of accidents. Traffic Theory is used to derive the probability (p_x) number of cars (x) passing a certain checkpoint in time interval t as a Poisson's distribution,

$$p_x=\frac{e^{-\lambda t}(\lambda t)^x}{x!}. \quad (8)$$

Probability of an animal successfully crossing a single lane road p_o (no car passes within time interval $t : x = 0$) can be obtained based on Eq. (8) as,

$$p_0=e^{-\lambda t}. \quad (9)$$

Van Langeveld and Jaarsma (2004) utilized this formula to discuss roadkill probability for mammals which led to:

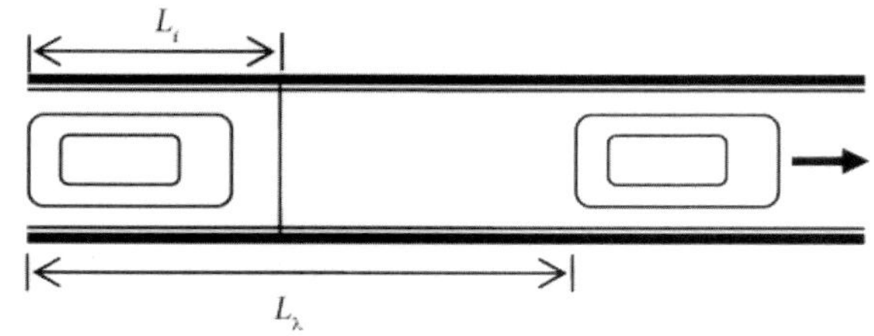

Fig. 3. Relationship of space between cars (L_λ) and impact width (L_i). The safety zone is $L_\lambda - L_i$ within which animals can cross the road safely

$$p_0=e^{-\lambda\left(\frac{B+W_r}{V_v}\right)}, \quad (10)$$

where B denotes body length of the animal and W_r is road width. Since this study focuses on birds and small mammals, B can be neglected and W_r should be replaced by W_v. Because Eq. (10) considers L_x as the entire impact range but does not include L_v (Fig. 2). If the time of vehicle passing L_v is taken into account, then the probability of a small animal successfully crossing a single lane road would be modified as

$$p_0=e^{-\lambda\left(\frac{L_v}{V_v}+\frac{W_v}{V_a}\right)}. \quad (11)$$

Hence, roadkill probability yields

$$p_T=1-p_0=1-e^{-\lambda\left(\frac{L_v}{V_v}+\frac{W_v}{V_a}\right)}. \quad (12)$$

Observing Eq. (12), we find that even though traffic volume (λ) and vehicle speed (V_v) are independent variables, there is an implicit correlation between them. In the case of constant traffic volume, the relationship can be clarified. In the case of high average vehicle speed: the ratio L_v/V_v is small, the probability of an animal hitting the side of a vehicle is small. The second term of the exponent dominates the probability which implies that in this type of collision, the animal would be hit by the front of the vehicle. In the case of low average vehicle speed: the ratio L_v/V_v becomes dominant which means that the space between vehicles is reduced and the probability of an animal hitting the side of the car increases. In this case, both vehicle length and speed are significant. When the average vehicle speed is extremely low, it resembles a traffic jam that leads to an accident probability approaching 100%. To animals, this is like a wall on the road: each animal crossing will hit the side of a vehicle.

With low traffic volume, the difference between the Geometric Model and Traffic Model is insignificant, but traffic volume and vehicle speed are included in the models. While vehicle speed reflects the situation of side collisions, Eq. (10) (van Langevelde, Jaarsma 2004) does not consider it. Therefore, Eqs. (7) and (12) are more effective when considering small animal roadkill probability.

2.3. Modification of the models

Still, the two equations cannot precisely predict roadkill probability since an animal's ability to dodge oncoming objects and the driver's reaction are neglected in the mathematical derivations. Pure mathematical models cannot predict or reflect complicated animal behaviors; therefore, more studies have to be conducted to understand animal conceptions of moving objects. However, our survey shows that adjacent landscape is also an important factor. If all the independent factors-landscape, driver's reaction, and animal behaviors-are included in the derivation, the final equations will be much more precise in predicting the probability of an accident. The three independent factors are denoted as:

(a) Landscape Factor F_L (Including topography, road condition, adjacent vegetation and habitat).
(b) Driver's Factor F_D (The reactions of a driver to the appearance of animals on the road).
(c) Animal Factor F_A (The animal's ability to dodge oncoming vehicles).

The numerical value of each factor is between 0 and 1 but needs to be determined by experiment or survey. Including the three factors in the derivation, Eqs. (7) and (12) are modified and yield

$$\bar{p}_L = p_L \cdot F_L \cdot F_D \cdot F_A = \lambda\left(\frac{L_v}{V_v} + \frac{W_v}{V_a}\right) \cdot F_L \cdot F_D \cdot F_A \quad (13)$$

(Geometric Model);

$$\bar{p}_T = p_T \cdot F_L \cdot F_D \cdot F_A = \left[1 - e^{-\lambda\left(\frac{L_v}{V_v} + \frac{W_v}{V_a}\right)}\right] \cdot F_L \cdot F_D \cdot F_A \quad (14)$$

(Traffic Flow Model).

These equations provide more representative models and can be used as tools in planning or designing roads.

3. Testing roadkill models

To test the models, two roads with similar surrounding landscapes are selected to compute the roadkill probabilities. The WDN road J-C section and WDE road have a total length of 5.44 km and 3.25 km respectively (landscapes shown in Figs 4 and 5). J-C section of WDN Road is with dense roadside trees of beef wood (*Casuarina equisetifolia*) and its both sides of the road are mostly sorghum farmlands and are birds' favorite foraging places. Landscape of WDE Road has dense roadside trees. Both sides of the road are mostly natural areas and habitats of various kinds of wildlife. The models are also applied to examine the ecological characteristics of BY Road (part of CR Road with a total length of 5.85 km and landscape shown in Fig. 6). BY Road is the widest road in Kinmen with the highest traffic volume. The original roadside trees (beef wood) were replaced by camphor trees (*Cinnamomum camphora*) after a typhoon hit Kinmen in 2000. It can be seen that the landscape of BY Road is quite distinct from WDN and WDE Roads and this may imply some different ecological features of the roads. The traffic data of the three roads are summarized in Table 3.

Table 3. Traffic information for J-C section, WDE and BY Roads

Road	Length (km)	Average vehicle speed (km/hr)	Penk traffic volume (No./hr)	Roadkills (bird rodent)	Total roadkills
J-C of WDN	5.44	60	221 (East bound) / 150 (West bound)	50/63	113
WDE	3.25	60	45 (North bound) / 35 (South bound)	10/5	15
BY	5.85	70	369 (East bound) / 433 (West bound)	36/4	40

Since the roadside trees and adjacent landscapes of the two roads are similar, we assume that the F_L, F_D, and F_A of J-C are equal to those of WDE. Bird flight speeds vary greatly depending on the species. Bramblings fly 24~88 km/hr, Crows 80~90 km/hr, Heron 29~46 km/hr, and Osprey 66 km/hr (Tyne, Berger 1959). Furthermore, speeds during road-crossing or take off tend to be significantly lower than normal flight. It is, therefore, very difficult to ascertain the flight speed. However, flight speed V_a does not significantly affect the results

Fig. 4. Landscape of J-C section of WDN Road

Fig. 5. Landscape of WDE Road

Fig. 6. Landscape of BY Road

in the models; therefore, the average speed is assumed to be V_a = 40 km/hr. In addition, the crossing speed of small mammals (mostly rodents) is assumed to be V_a = 10 km/hr. The average crossing speed for animals is then V_a = 25 km/hr. To test the models and carry out the computation, some assumptions are made as follows: the average vehicle length is L_v = 5 m, vehicle width is W_v = 2.5 m, then taking double lanes into consideration. Animal roadkill probability p_L, by Geometric Model, gives us:

$$p_L = 1 - \left[1 - \lambda_1\left(\frac{L_v}{V_v} + \frac{W_v}{V_a}\right)\right]\left[1 - \lambda_2\left(\frac{L_v}{V_v} + \frac{W_v}{V_a}\right)\right]. \quad (15)$$

Then, the modified Geometric Model probability is:

$$\bar{p}_L = p_L \cdot F_L \cdot F_D \cdot F_A = \left\{1 - \left[1 - \lambda_1\left(\frac{L_v}{V_v} + \frac{W_v}{V_a}\right)\right] \cdot \left[1 - \lambda_2\left(\frac{L_v}{V_v} + \frac{W_v}{V_a}\right)\right]\right\} \cdot F_L \cdot F_D \cdot F_A. \quad (16)$$

While the Traffic Model, p_T, yields:

$$p_T = 1 - \left[e^{-\lambda_1\left(\frac{L_v}{V_v} + \frac{W_v}{V_a}\right)}\right]\left[e^{-\lambda_2\left(\frac{L_v}{V_v} + \frac{W_v}{V_a}\right)}\right]. \quad (17)$$

The modified Traffic Flow Model probability is then:

$$\bar{p}_T = p_T \cdot F_L \cdot F_D \cdot F_A = \left\{1 - \left[e^{-\lambda_1\left(\frac{L_v}{V_v} + \frac{W_v}{V_a}\right)}\right]\left[e^{-\lambda_2\left(\frac{L_v}{V_v} + \frac{W_v}{V_a}\right)}\right]\right\} \cdot F_L \cdot F_D \cdot F_A. \quad (18)$$

3.1. Comparing roadkill probability for the J-C section and WDE road

Substituting the data into Eqs (16) and (18), the ratio of roadkill probability $r_{(\text{J-C/WDE})}$ can be obtained as

$$r_L = 4.52 \text{ (Geometric Model);} \quad (19)$$

$$r_T = 4.46 \text{ (Traffic Flow Model).} \quad (20)$$

The difference between the results of the two models is insignificant. To predict the number of roadkills (X) along WDE we utilize the number of roadkills from the J-C section, and establish the relationship:

$$\frac{X}{3.25} \cdot r_L = \frac{113}{5.44}. \quad (21)$$

Then, we have

X =14.9 (Geometric Model), and X = 15.1 (Traffic Model), (22)

which is surprisingly consistent with our observations (15 roadkills on WDE). The result confirms the previous assumption that F_L, F_D, and F_A are identical for J-C and WDE roads. Since unknown factors are eliminated, the roadkill numbers for WDE can be predicted. If we take V_a = 40 km/hr and V_a = 10 km/hr and apply Traffic Flow Model, X = 15.0 and X = 15.4 are obtained. This implies that the animal crossing speeds are insignificant in this case. As adjacent landscapes are similar, the models can be applied to predict the number of roadkills.

3.2. Probability ratio comparison between J-C section and BY roads

With regards to landscape, J-C section has high and dense beef woods while BY Road has low and dispersed camphor trees. Also, BY Road, the major commuting conduit in Kinmen, is wide and experiences high traffic volume. The distinct difference between these two roads should be a significant factor in the determination of roadkill probability. Substituting associated data into the models, $r_{(J\text{-}C/BY)}$ is computed as:

$$(23)\ r_L = 0.504\frac{(F_L)_{J-C}}{(F_L)_{BY}} \text{ (Geometric Model);} \quad (23)$$

$$(24)\ r_T = 0.512\frac{(F_L)_{J-C}}{(F_L)_{BY}} \text{ (Traffic Flow Model).} \quad (24)$$

Knowing the number of the roadkills on J-C section, the number of animal casualties of BY (X) is calculated with the relationship:

$$\frac{X}{5.85}\cdot r = \frac{113}{5.44}. \quad (25)$$

Assuming that the landscape factors of the two roads are equal, $(F_L)_{J\text{-}C} = (F_L)_{BY}$, the roadkills of BY road are:

$$X = 241 \text{ (Geometric Model) and}$$
$$X = 237 \text{ (Traffic Flow Model),} \quad (26)$$

respectively. The results differ greatly from our observations (40 roadkills on BY) and the assumptions about landscape factors are not confirmed. Instead of equal landscape factors $(F_L)_{J\text{-}C} = (F_L)_{BY}$ for the two roads, $(F_L)_{J\text{-}C} = 6\times(F_L)_{BY}$ would produce $X \cong 40$ which means that the landscapes of the two roads are significantly distinct and are reflected in the calculation. Since J-C section and BY road are parallel and only about 1 km apart, we can assume that the adjacent ecosystems, and animal population and composition are similar for the two roads. Expressions of (26) reveal that the probability of roadkill occurring on J-C section is 6 times higher than on BY road and implies that the landscape of J-C section is prone to causing more animal accidents. We can then conclude that higher traffic volume increases roadkill probability; but, the traffic may also become a barrier blocking animal crossing. Noise, pollution, and disturbance may cause birds to fly across roads at higher altitudes or avoid mammals approaching the roads. Thus, fewer bird collisions take place along BY road. In addition, increased road width and traffic prevent birds from foraging on the road; therefore, reducing avian roadkill probability. This assessment may be an important tool in planning, designing and constructing an ecologically friendly road.

3.3. Inclined roads (BL road)

Among the surveyed roads, BL is the most intriguing because it has the highest roadkill probability among the roads surveyed. Fifteen roadkills (9 birds, 2 rodents, 1 frog, 1 snake, 1 unidentified) occurred along BL road as vehicles descended the slope but only four (2 birds, 1 rodent, 1 cat) as they ascended. The ratio of bird roadkills is, therefore, 9/2 = 4.5 (down/up slope). Since traffic volume is low, Eqs (13) and (14) can be utilized. The average vehicle speed on decline and incline are measured as 55.7 km/hr and 49.8 km/hr respectively. The road width is 6m, and its peak traffic volume is 140 vehicles per hour while off peak is 75 vehicles per hour. By applying the models to avian roadkill to find roadkill ratio r (down/up), with the assumption that $L_v = 5$ m, $W_v = 2.5$ m, and $V_a = 40$ km/hr, we find that $r = 1.045$ for the Geometric Model and $r = 1.045$ for the Traffic Flow Model. This means that the number of roadkills that occur over the course of the downward gradient is only 1.045 times that of the upward. The minor difference between the values suggests that vehicle speed is not the major cause of bird roadkill on BL road.

The difference in values indicates that slope may be a factor. As a vehicle travels up the gradient, it is more visible to birds crossing the road. When it travels down, the vehicle visually blends into the road surface. Many varieties of birds view objects using rapid spiral scan (saccades) with a magnitude of 10°~12° (Wallman, Letelier 1993). With BL road slope of 2.04° (or 4.2%), the bird's visual range up the gradient is only (assuming a scan magnitude 11°) 11° /2–2.04° = 2.96°, while down the slope it is 11° /2 + 2.04° =7 .54°. This difference contributes to the varying results of opposing gradients. Also, the accidents that took place on this road were most likely birds run over by vehicles as they were walking or foraging on the road. Only birds walking or standing on the road have an obstructed view, hence flying over can be excluded as factor in the accidents.

We can confirm that the slope of the road is the major factor of roadkills on BL road by observing the changes of road conditions. Since 1.5 months after survey commenced, BL road has been in the process of being widened. In the first 3 months of construction, only roadside grading was undertaken and traffic remained normal. More importantly, birds could fly over the road during this period of time; however, no bird casualties were discovered after the commencement of construction. If the avian casualties found before the start of construction were the result of normal avian-vehicle collisions, further fatalities should have been recorded during construction as well. This may confirm the speculation that most of the avian deaths occurred as the birds walked or foraged on the road and, due to road gradient, had an obstructed view of oncoming traffic. Therefore, this unusual roadkill phenomena can be studied by applying our models.

3.4. The effects of roadside trees on roadkill

Roadside trees influence the occurrence of avian-vehicle collisions. Attributes such as density, height, leaves and branches are all factors. Dense leaves and branches can

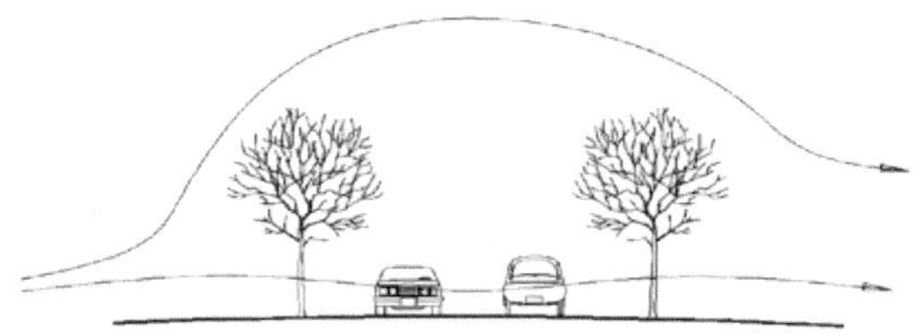

Fig. 7. Leaves and branches of roadside trees block part of flying route of birds.

block part of flying route of birds. Most birds, especially understory species, would fly crossing the road between trunks of trees whose height is almost coincident with vehicles'. Therefore this type of roadside tree arrangement would increase avian roadkill probability (Fig. 7). Observations along WDN Road confirm this scenario.

To estimate the impact roadside trees have on the occurrence of avian roadkill, four roads in Kinmen were selected to observe, record, and assess:

(a) WDN Road near Shiao-Jin Road: The south side of the road is mainly farmland and the north side is about 1 hectare of woodland bird habitat.

(b) CR Road (between Cheng-Kong and Shia-Shin): This part of CR runs east-west. There is a small forest on the north side of the road and another forest at a lower altitude on the south side.

(c) WDN Road (between Chiong-Lin and Ho-Pan): Both sides of the road are farmland. The roadside trees are high and dense beef woods.

(d) WDW Road: The adjacent landscapes of both side of the road are farmland. Similar to WDN Road, the roadside trees are mainly beef woods.

The observations of the four roads are shown in Table 4.

It is obvious that the probability of birds flying into the "Impact Zone" will be much higher for (c) and (d) (high and dense beef woods) than the other two roads. We can conclude that dense roadside trees cause birds to cross the road at a lower altitude which leads to more avian casualties and a greater landscape factor (F_L). In fact, the observation of avian roadkills along all eight roads also confirms this phenomenon.

4. Suggestions

4.1. Mitigation measures for avian roadkills

4.1.1. Topography

Flying birds' migration patterns generally follow certain terrain and vegetation criteria (Bevanger, Brøseth 2004). The height of flight is also dependent upon the height of vegetation (Ramp *et al.* 2006). Consequently, fewer birds are killed on raised roads as opposed to roads on flat terrain (Clevenger *et al.* 2003) because the topography forces birds to fly high over the roads and prevents collisions with oncoming cars. Therefore, a raised road bed or altered surrounding vegetation will reduce avian-vehicle collisions.

4.1.2. Vegetation

High and dense roadside trees increase the probability of accidents involving avian species. In order to decrease this probability, vegetation can be designed to influence bird flight patterns and direct them over the road safely as shown in Figure 8. Certain aspects (such as trash, temperature, and water) may attract birds to forage on the road surface; however, with this type of vegetation, the chance of birds foraging on the road is lower, thus, reducing collision probability.

Table 4. Flight patterns over roads with various forms of roadside trees

Road	Attributes of roadside trees	Observation time	Effective observation range	Number of birds flying into Impact Zone*	Number of birds flying away from Impact Zones	Total number of birds observed
(a)	Low and dispersed	17:15~18:15	100 m	5	71	76
(b)	Low and dispersed	17:05~18:35	200 m	10	75	15
(c)**	High and dense	17:05~18:40	300 m	95 (15 birds walking on the road)	41	135
(c)***	High and dense	16:55~18:15	900 m	155 (24 birds walking on the road)	125	210
(d)	High and dense	16:00~18:00	100 m	49 (1 bird walking on the road)	37	16

Note: *The Impact Zone is defined as the range between road surface and 2 m below; ** Observed by bare eyes; *** Observed by telescope.

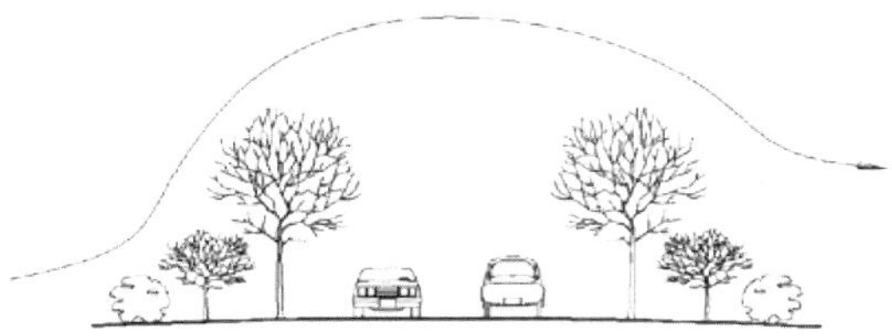

Fig. 8. Reducing avian roadkills by vegetation arrangement which is to plant trees and shrubs in upward trend to guide birds fly higher over a road

4.1.3. Spacing of vegetation

Gaps in linear vegetation (such as roadside trees) become avian migration conduits and, in turn, roadkill hot spots. Therefore, vegetation should not have gaps in the vicinity of high traffic volume sections of a road.

4.1.4. Road attributes

4.1.4.1. Sloping road hot spots

As observed on BL road, the probability of avian roadkill is greater on negatively sloping sections then positive due to higher vehicle speeds and lower visibility for birds at risk. At high speeds, drivers have a limited time to respond to emerging birds. The same is true for birds attempting to avoid oncoming vehicles. Lowering vehicle speeds can reduce collisions. Flattening slopes and installing warning signs and speed bumps will reduce vehicle speeds and, in turn, give drivers and animals more time to respond.

4.1.4.2. Curving section hot spots

Limited vision at sharp corners causes insufficient response time for both drivers and animals and increases the chance of collision. Straightening abrupt turns or clearing nearby obstacles can improve the visibility and lower roadkill probability. Where these two methods can not be implemented, installing warning signs and lowering the speed limit are also effective measures.

4.1.4.3. Flat or depressed surface hot spots

Birds tend to follow the terrain and fly into flat or depressed road surfaces and, therefore, cause accidents. Raising the road bed to guide birds over the road is another way to reduce avian roadkill. However, the barrier effect of this engineering method has to be considered before designing the road.

4.1.5. Traffic volume and speeds

4.1.5.1. Traffic volume

According to our models, traffic volume is obviously the most influential factor in determining the probability of roadkill. Lower speed limits and reduced traffic volume are the most effective measures to implement. If hot spots are located in high traffic areas, diverting traffic to alternative routes is an option. Other measures to lower traffic volume can be considered: building underpasses or overpasses at hot spots, temporarily closing roads in high casualty seasons, encouraging public transportation, and subsidizing car pooling.

4.1.5.2. Vehicle speed

For drivers, speed reduces the visual field, restricts peripheral vision and limits the time available to receive and process information (AASHT 2001). Low vehicle speed can not only reduce roadkills but also lower noise intensity and prevent barrier effects. Reducing speed limits, installing speed bumps and radar, putting up traffic signals, and establishing warning signs in routes are methods that can be implemented to lower vehicle speeds.

4.2. Mitigation measures for terrestrial animals

On Kinmen, small mammals, amphibians and reptiles are the most common terrestrial animals; among them, rodents and frogs are the most common victims of roadkill. On April (157 days after survey commenced), on WDW road, 29 Spectacled toads were killed by vehicles on roads adjacent to Suan-Li wetland. This suggests that some amphibians migrate at certain times during mating season. For this specific species, roadkill may be mitigated by temporarily closing the road or installing temporary underpasses (or overpasses) at the hot spots.

Mitigation measures for terrestrial animals are dependent upon species. To prevent terrestrial animal roadkill, passages are the most common. However, vegetation and fences have to be considered as well in order to achieve best results. Guidelines for installing passages are: (a) close to animal habitat, (b) close to animal migration routes, (c) close to hot spots or locations where animals frequently appear, and (d) connect animal habitats. The location and adjacent landscape are also important factors for the success of the passages. Therefore, the surroundings of the passages should adhere to the following: (a) a small wetland near passage to attract animals, (b) regular checks to ensure a clear entrance to the passage, (c) fences to lead animals to the entrance while maintaining metapopulation dynamics, (d) guiding facilities such as slopes, curves, ditches, etc. that allow animals to access the passage, and (e) reduced human access or disturbances (Lin 2006).

Conclusions

Based on our research, three aspects influence the probability of roadkill. Firstly, the comparison of J-C section and WDE road proves that if surrounding landscapes are similar there is a positive correlation between traffic volume and roadkill probability thus giving us the 'Road-Traffic aspect'. In this case, traffic volume is the most significant factor affecting the probability. In addition, vehicle speed, car length, and animal crossing speed can also determine the frequency of animal accidents. If the traffic volume is high but vehicle speed low, the traffic will be congested and most accidents will consist of an animal hitting the side of a vehicle. Thus, roadkill probability increases. On the other hand, with a constant traffic volume but high vehicle speed, the length of vehicle becomes unimportant and most accidents will be head-on collisions.

Secondly, roadkill probably is high when roads pass through natural habitats or bisect animal migration routes. This situation is referred to as, the 'Landscape aspect'. It involves the direct engagement of humans and animals. Tall and dense roadside trees cause higher animal casualties, especially among understory avian species, by attracting animals and changing the flight patterns of avian species. While, low roadside trees, wide road surfaces, or high traffic flow (see BY road) form an ecological barrier, restrict the movement of some species, and, therefore, reduce gene exchange.

Finally, the 'Biological aspect' refers to the fact that, in general, most drivers will slow down to avoid hitting animals and, therefore, lower the accident rate. Many animals also possess the ability to avoid collisions. For species that are accomplished fliers (such as the Barn Swallow and Chestnut bat), their agility greatly reduces the risk of accident even when they congregate in great numbers at the roadside. In contrast, slow moving animals such as reptiles and amphibians cross the road with great risk. These biological factors cannot be precisely determined and more research on these subjects needs to be undertaken. Nevertheless, applying models that incorporate similar landscapes around different roads, accident probabilities may be effectively obtained without knowing the influence of biological factors.

Acknowledgements

This work was supported by National Science Council of ROC in Taiwan grant no. NSC 103-2621-M-507-002.

References

Adams, L. W.; Geis, A. D. 1983. Effects of roads on small mammals, *Journal of Applied Ecology* 20: 403–415. http://dx.doi.org/10.2307/2403516

Bennett, A. F. 1992. Roads, roadsides, and wildlife conservation: a review, in D. A. Saunders, R. J. Hobbs (Eds.). *Nature conservation 2: the role of corridors*. Surrey Beatty and Sons, 99–118.

Bevanger, K.; Brøseth, H. 2004. Impact of power lines on bird mortality in a subalpine area, *Animal Biodiversity and Conservation* 27(2): 67–77.

Clevenger, A. P.; Chruszcz, R.; Gunson, K. E. 2003. Spatial patterns and factors influencing small vertebrate fauna road-kill aggregations, *Biological Conservation* 109: 15–26. http://dx.doi.org/10.1016/S0006-3207(02)00127-1

Dhindsa, M. S.; Sandhu, J. S.; Sandhu, P. S.; Toor, H. S. 1988. Roadside birds in Punjah (India): Relation to mortality from vehicles, *Environmental Conservation* 15(4): 303–310. http://dx.doi.org/10.1017/S0376892900029799

Forman, R. T. T. 1995. *Land mosaics: the ecology of landscapes and regions*. Cambridge University Press.

Forman, R. T. T.; Sperling, D.; Bissonette, J. A.; Clevenger, A. P.; Cutshall, C. D.; Dale, V. H.; Fahrig, L.; France, R.; Goldman, C. R.; Heanue, K.; Jones, J. A.; Swanson, F. J.; Turrentine, T.; Winter, T. C. 2003. *Road ecology, science and solutions*. Washington: Island Press.

Gunther, K. A.; Biel, M. J.; Robison, H. L. 1998. Factors influencing the frequency of road-killed wildlife in Yellowstone National Park, in *Proceedings of the International Conference on Wildlife Ecology and Transportation*, 10–12 February 1998, Fort Myer, Florida, 32–36.

Jaarsuma, C. F.; Willems, G. P. A. 2002. Reducing habitat fragmentation by minor rural roads through traffic calming, *Landscape and Urban Planning* 58: 125–135. http://dx.doi.org/10.1016/S0169-2046(01)00215-8

Lin, S. C. 2006. A review for the types of wildlife passages, *Journal of National Park* 16(1): 83–98 (in Chinese).

Matson, T. M.; Smith, W. S.; Hurd, F. W. 1971. *Traffic engineering*. New York: McGraw-Hill.

May, A. D. 1990. *Traffic flow fundamentals*. Englewood Cliffs: Prentice-Hall, Inc.

McDonald, W. R.; St. Clair, C. C. 2004. The effects of artificial and natural barriers on the movement of small mammals in Banff National Park, Canada, *Oikos* 105: 39–407.

Nankinov, D. N.; Todorov, N. M. 1983. Bird casualties on highway, *Soviet Journal of Ecology* 42: 594–596.

Ramp, D.; Wilson, V. K.; Crof, D. B. 2006. Assessing the impacts of roads in peri-urban reserves: Road-based fatalities and road usage by wildlife in the Royal National Park, New South Wales, Australia, *Biological Conservation* 129: 348–359. http://dx.doi.org/10.1016/j.biocon.2005.11.002

Reijnen, R.; Foppen, R.; ter Braak, C.; Thissen, J. 1995. The effects of car traffic on breeding bird populations in woodland III, Reduction of density in relation to the proximity of main roads, *Journal of Applied Ecology* 32: 187–202. http://dx.doi.org/10.2307/2404428

Seiler, A. 2001. *Ecological effects of roads, a review*. Riddarhytan: University of Agriculture Sciences.

Slater, P. 1994. Wildlife road casualties, *British Wildlife* 5: 214–221.

Tyne, J. V.; Berger, A. J. 1959. *Fundamentals of ornithology*. New York: John Wiley and Sons, Inc.

van Langevelde, F.; Jaarsma, C. F. 2004. Using traffic flow theory to model traffic mortality in mammals, *Landscape Ecology* 19: 895–907. http://dx.doi.org/10.1007/s10980-004-0464-z

Wallman, J.; Letelier, J-C. 1993. Eye movements, head movements, and gaze stability in bird, in H. P. Zeigler and H-J. Bischof (Eds.). *Vision, brain and behaviour in birds*. Cambridge: The MIT Press.

Wilson, E. O. 1989. Threats to biodiversity, *Scientific American* 261(3): 108–116. http://dx.doi.org/10.1038/scientificamerican0989-108

Shyh-Chyang LIN is an associate professor at the Department of Civil Engineering and Engineering Management at National Quemoy University in Taiwan. His research has been evolving from applied mechanics, engineering materials, to ecological engineering, island studies, and finally to social networking. The broad spectrum of his research interests allows him to explore deeper ecological, social, economical, and political issues. Currently, his research focuses on applications of social network technology, and scale issues of small islands. He was a recipient of Presidential Award for Environmental Conservation in 2005.

ASSESSMENT OF FUTURE URBAN GROWTH IMPACT ON LANDSCAPE PATTERN USING CELLULAR AUTOMATA MODEL

Cheng LI[a], Jie ZHAO[b]

[a] *Department of Geo-Information Science, School of Resources and Geosciences, China University of Mining and Technology, Daxue road 1, 221116, Xuzhou, Jiangsu, China*

[b] *Huaihai Development Research Institute, Jiangsu Normal University, Heping road 57, 221009, Xuzhou, Jiangsu, China*

Abstract. Understanding and predicting of the urban growth process and its impact have become increasingly important for decision making toward sustainable development. In this paper, we presented a cellular automata model to assess the consequence of future urban growth. The hybrid calibration method combining logistic regression with trial-and-error was applied to estimate the parameters. The study proposed the integration method of Multi-Criteria Evaluation and Analytic Hierarchy Process that can be utilized to effectively translate the qualitative descriptions of scenarios into quantitative spatial analysis. Finally, the comparison of the different scenarios provided an insight into the impacts of different urban development strategies on landscape patterns. The result indicates that CA model can be effectively connected with the urban decision making processes. The moderate development scenario could be considered as the best one in achieving the objectives of compact urban form, good residential environment, as well as environmentally and economically efficient development.

Keywords: cellular automata model, urban growth, calibration, scenario, landscape metrics, landscape pattern.

Introduction

In recent decades, urbanization, the most extreme anthropogenic land cover/use transformation has been a universal and important socioeconomic phenomenon around the world. Urban growth has been accelerating with the significant increase in urban population. The world urban population was only about 3% of the global population in the 1800s, but increased to nearly 30% in 1950. Currently, over half of the world population live in urban areas, and the figure is projected to reach 67.1% (6.25 billion) by 2050 (United Nations 2012). Although urban areas cover a very small percentage of the world's land surface in comparison with other land cover types, their rapid expansion has marked effects on environment and socio-economy, such as loss of natural vegetation and farmland (Tan *et al.* 2005), local and regional climate change (Kaufmann *et al.* 2007), decline in biodiversity (Zimmermann *et al.* 2010), hydrological circle alternation (Barron *et al.* 2013), etc. Without effective planning, there is no doubt that the pressure for sustainable development will continue to increase (Dewan, Yamaguchi 2009; Lambin *et al.* 2001).

Modeling of the urban growth process is an important technique to provide a better understanding of causes and mechanisms governing urban growth; to analyze alternative urban growth consequences, therefore, to support the appropriate urban planning and decision making responses to urban growth (Berling-wolff, Wu 2004). Cellular automata (CA) models are among the most popular ways to simulate the evolution of urban growth. They are able to handle large amounts of data and many fields of studies, such as population, land use, socioeconomic activity (Batty 2005). CA models are a powerful tool for representing and simulating spatial processes underlying the spatial decisions due to their simplicity, flexibility, and intuitiveness (Munshi *et al.* 2014; Santé *et al.* 2010). Temporal and spatial complexity of urban growth process can also be well modelled using CA based models (Barredo *et al.* 2003). Additionally, CA models have become an experimental tool for urban planning by producing different scenarios under various urban planning policies (Fuglsang *et al.* 2013).

Calibration and validation are the basis of their successful implementation because they provide a tool to

Corresponding author: Cheng Li
E-mail: cheng.li@cumt.edu.cn

ensure that models can conduct accurate and reasonable simulation regarding current and future urban growth scenarios (Wu 2002). Various calibration methods have been developed to generate more precise parameters, such as logistic regression (Sui, Zeng 2001), multicriteria evaluation (Wu 1998), support vector machine (Yang *et al.* 2008), artificial neural networks (Li, Yeh 2002), genetic algorithm (Li *et al.* 2008), kernel-function (Liu *et al.* 2008), etc. Although many methods have been explored for calibration, there is not general method to calibrate urban CA models because the objectives and structure of these models are different. Wu (2002) argued that the calibration is dependent on the objective of the simulation. For validation, the cell by cell comparison method is the most commonly used for measuring goodness-of-fit between simulated results and observed one. This assessment, however, cannot consider the urban growth patterns which are important in analyzing the impact of the variables on landscape patterns. Therefore, the pattern similarity should also be involved to evaluate the performance of a CA model.

The integration of CA model and GIS has a potential to explore different urban development scenarios under various policies. The simulation serves as not only a matter of visualization but also a bridge between urban growth patterns and decision making (Wu 1998). Many studies have investigated different aspects of scenarios which are designed under the consideration of different urban development policies in specific areas (He *et al.* 2006; Shen *et al.* 2009; Zhang *et al.* 2011). Furthermore many researchers argued that scenario evaluation should be a key aspect of land use modeling in order to test and compare different land-use planning policies (Thapa, Murayama 2012; Zhang *et al.* 2011). Various landscape metrics have been widely used to evaluate and compare the scenarios, such as edge density, number of patches, area weighted mean shape index, etc (Aguilera *et al.* 2011; Mitsova *et al.* 2011; Petrov *et al.* 2009; Zhang *et al.* 2011). This can be attributed to their usefulness for quantification and interpretation of land use patterns. They make the processes and patterns of urban development more prominent. Besides this general analysis, only a few researches have been done on the evaluation of scenarios by other more detailed analysis, such as those that use landscape metrics at local scales to better localize changes in land occupation patterns (Aguilera *et al.* 2011; Thapa, Murayama 2012).

This study aims at enhancing the understanding of urban growth process by simulating future scenarios using CA model, with a case study of Xuzhou city in China. The CA model is developed to simulate the urban growth by integrating various factors. In order to provide support for decision making process, the future development scenarios are designed. According to the qualitative description of each scenario and the objective of development strategies, the parameters in CA model are quantitatively identified. Furthermore, a set of landscape metrics are applied to quantitatively compare the urban growth patterns of different scenarios.

1. Material and methods

1.1. Study area and data

Xuzhou city is situated in the plains of the Yellow River and the Huaihe River. It has a total administrative area of approximately 11,258 km^2, with 1,160 km^2 as the city proper area. It is regarded as a medium-sized metropolitan area in comparison to other cities in China. Xuzhou city is composed of ten county-level divisions, five counties

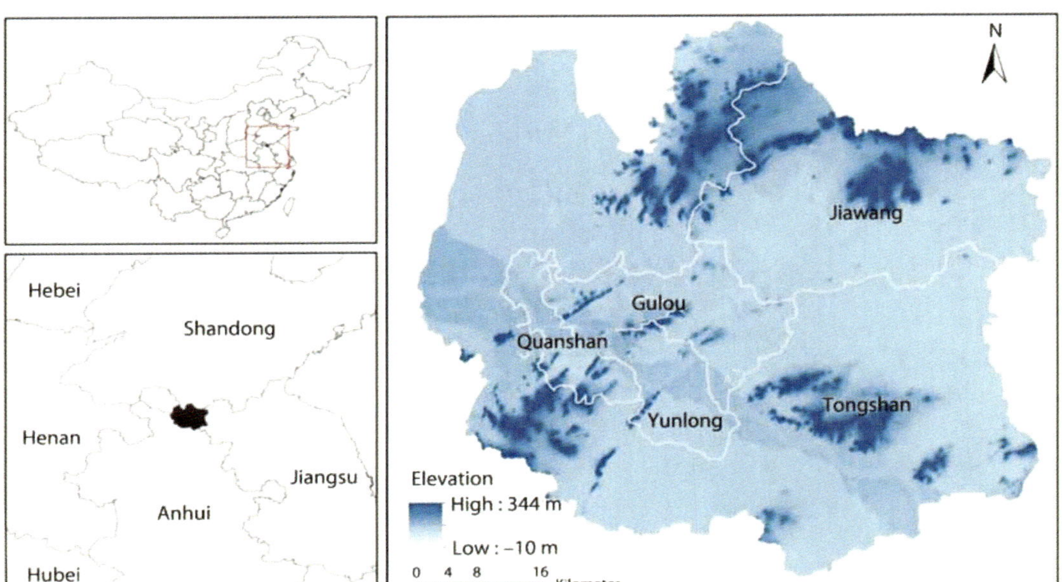

Fig. 1. Location of study area (Xuzhou) and its topography

(Peixian, Fengxian, Suining, Pizhou, and Xinyi), and five municipal districts (Quanshan, Gulou, Yunlong, Jiawang, and Tongshan). As shown in Figure 1, the five municipal districts are identified as the study area. Traditionally, they are viewed as the central city, in which Quanshan, Gulou and Yunlong are composed of city proper area. Jiawang and Tongshan are composed of fringe and rural areas. Mining and industrial manufacturing have been the source of the strong economic activity of the region.

In this research, Landsat images from the years of 1990, 2001, 2005 and 2010, were used to obtain multi-temporal land cover data for Xuzhou city. The land cover classification data was produced through the interpretation from Landsat data (Li, Thinh 2013). Figure 2 shows the multiple temporal land cover maps at the extent of the study area with the overall accuracy of more than 85%.

The urban growth is a complex process which involves the interaction influence of various factors (Li *et al.* 2014). Some variables need to be incorporated into the model, which include: (1) global suitability factors, including Distance to central business district (Dis2CBD), Distance to district centers (Dis2Cen), distance to major roads (Dis2MajR), distance to minor roads (Dis2MinR), slope, and population density (PopDen); (2) neighborhood variable; (3) Spatial policy factor, including conservation zones, construction restriction, master plan 2010–2020, layout of potential subsidence areas (Subsidence), layout of environmental protection areas (Environment). However the estimated coefficients of these variables could be misleading in analysis of urbanization process when the variables are measured in different units. Therefore, all variables should be standardized into the range from 0 to 1 prior to the modelling. For global suitability factors, linear transformation method was applied to conduct the standardization. For spatial policy factors, the area where urban development is limited was assigned 0 and area that is designated for urban development was assigned 1. All the spatial data were registered to the same Universal Transverse Mercator (UTM) coordinate system and sampled to the same cell size of 100×100 m, which was sufficient to capture the detailed information about urban dynamics while keeping the volume of computation manageable.

1.2. Development of CA model

The definition of transition rule plays an important role in CA models. The key element of transition rule is the transition potential which determines the probability of a cell changing to a specific land use (Wu, Webster 1998). This involves a number of spatial variables that contribute to urban growth. In this study, the transition potential P_{ij} can be practically defined as a function of the global suitability value S_{ij}, neighborhood effects N_{ij}, constraints

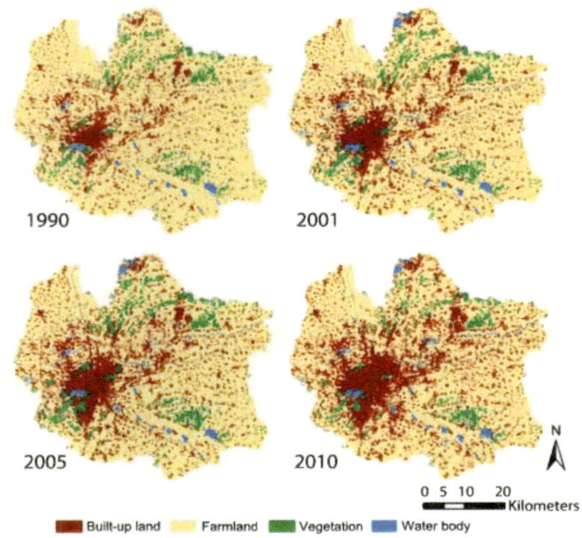

Fig. 2. Classified land cover maps of Xuzhou city from 1990 to 2010

$CONS_{ij}$ and stochastic perturbation V_{ij}. It can be expressed as follows:

$$P_{ij} = S_{ij} \times N_{ij} \times CONS_{ij} \times V_{ij} . \quad (1)$$

The global suitability value represents the intrinsic suitability of urban development. It was calculated as a function of global spatial variables:

$$S_{ij} = f(x_{l,ij}, w_l) , \quad (2)$$

where $x_{l,ij}$ (l = 0, 1, 2, …, n) represents the values of global factors for the cell (i, j), w_l represents the corresponding weight of the global factor.

Neighborhood effect was introduced by many studies to consider the effects of spatial interaction and neighborhood characteristics on urban growth. In this study, this neighborhood score was calculated according to following equation:

$$N_{ij} = \sum\nolimits_c W_{mn} \times I_{mn} , \quad (3)$$

where N_{ij} is the effect of neighborhood cells on the central cell (i, j) within the neighborhood space c; W_{mn} represents the weight indicating the impact of the interaction between the central cell and cell (m, n) within the neighborhood (Barredo *et al.* 2003). Following the first law of geography (Tobler 1970), a distance decay function was applied, so that cells closer to the central cell carry larger weight. I_{mn} represents the state of the cell (m, n) using binary value. I_{mn} = 1, when the cell is urban land, otherwise I_{mn} = 0. The neighborhood size, neighborhood type and weighting function have significant effects on the CA model results (Kocabas, Dragicevic 2006). As shown in Figure 3, three different neighborhood types (Moore, Moore Circular, and Von Neumann Circular) with different neighborhood size (radius of 1 to 6) were involved. In addition, three different weighting functions (Eq. 4) were applied to define the weights W_{mn} for cells within neighborhood:

$$W_{mn} = \exp(-\beta \times D_{mn}), \quad (4)$$

where D_{mn} is the distance between cell (m, n) to the central cell within a neighborhood. β is the exponent of the function. The higher of the value, the more abrupt is the function curve. In this study, β was assigned 0, 0.2 and 0.5, respectively.

The total constraint score was calculated as:

$$CONS_{ij} = \prod_{f=1}^{n} con_{ij,f}, \quad (5)$$

where $CONS_{ij}$ is the total evaluated constraint score representing natural constraints to urban expansion. If $CONS_{ij} = 0$, cell (i, j) is constrained by some constraint factors, and the cell cannot be converted to urban land use. Otherwise, $CONS_{ij} = 1$. $Con_{ij,f}$ represents the binary value of constraint factor f for the cell (i, j).

From a practical point of view, the related complexity of urban systems could be modeled as some degree of stochasticity (Barredo *et al.* 2003). Thus, a stochastic disturbance parameter was introduced into the model. It was calculated with Eq. (6):

$$V = 1 + (-\ln(rand))^{\alpha}, \quad (6)$$

where *rand* is a random value within the range from 0 to 1, and α is random variable which is used to control the degree of stochasticity. A higher value of α represents more random degree involved in this model.

Once the transition potential is calculated, decision rules need to be identified to spatially allocate the new urban area in order to simulate the historical and future urban growth process. At each iteration, the new urban pixels are allocated by selecting the non-urban pixels with the higher transition potential values. The non-urban pixels with lower values remain unchanged. The iteration continued until the total urban expansion area is reached.

1.3. Calibration and validation of CA model

In this study, the weights of global suitability variables, neighborhood size, neighborhood types, weighting function, and random variables need to be calibrated. An advantage of the logistic regression is its ability to estimate the weights of various spatial factors by developing statistical relationships between historical urban growth and spatial factors (Arsanjani *et al.* 2013; Ward *et al.* 2000). It can avoid subjectivity in determining the weights involved in transition rules of the CA model. However, it does not include all the relevant variables and cannot explain temporal dynamics of relationships (Hu, Lo 2007). The global factors which keep constant during each simulation period are involved into the logistic regression model. While the neighborhood effect and the random variable change with the running of the CA model. It is impossible to estimate these parameters using the logistic regression model. The trial and error method is a more rigorous calibration method. But its time cost for calibrating all parameters is not acceptable because trial and error method is implemented by running CA model many times with different parameter values. In this study, the hybrid method consisting of logistic regression and trial and error was used for the calibration in this study.

Validation is conducted by comparing the simulated results generated from calibrated CA models with observed maps in order to assess the simulation ability of CA models for different periods. With the consideration of the simulation purpose which is to make the simulated urban growth as close as the actual one in terms of location and pattern, a mixed measure based on the cell by cell and landscape pattern analysis was chosen in this study. The figure of merit (Eq. 7) (Pontius *et al.* 2007), and the relative difference of landscape metrics (Eq. 8) were computed to evaluate the fit of goodness between simulated and observed maps.

The figure of merit is the ratio of the intersection of the observed developed and simulated developed to the union of the observed developed and predicted developed (Pontius *et al.* 2008). The figure of merit can range from 0% to 100%. A higher value of figure of merit indicates a higher agreement in terms of cell by cell comparison. The figure of merit is calculated using the following equation

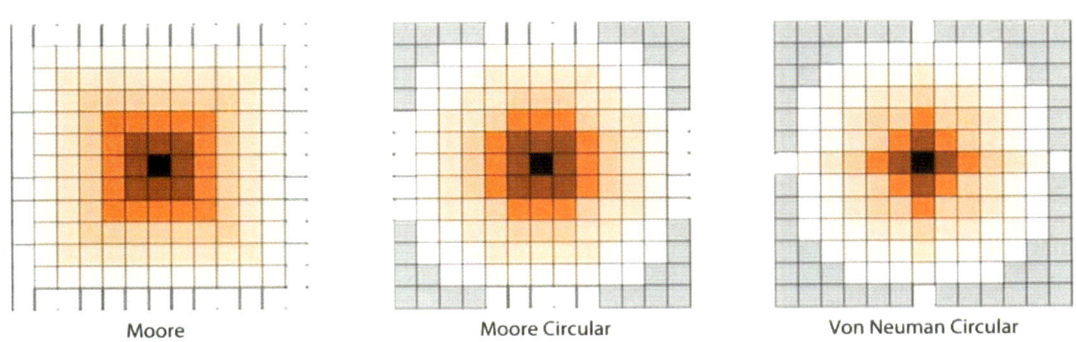

Fig. 3. Neighborhood types and sizes

(Pontius *et al.* 2008):

$$Fm = \frac{B}{A+B+C+D}, \quad (7)$$

where A is the area of error due to observed developed and simulated as persistence, B is the area of correct due to observed developed and simulated as developed, C represents the area of error due to observed developed and simulated as incorrect gaining category, and D is the area of error due to observed persistence and simulated as developed. Because the CA model only simulates the change of states from non-urban to urban, the value of C should be equal to 0.

The pattern similarity was incorporated, which was estimated through the comparison of landscape metrics between simulated patterns and observed ones. A total of four landscape metrics were selected to represent the landscape pattern from different aspects. These metrics are: number of patches (NP), largest patch index (LPI), Area Weighted mean shape index (SHAPE_AM) and Area Weighted mean Euclidean nearest neighbor distance (ENN_AM). The relative difference Rd can be calculated as follows:

$$Rd(\%) = \frac{1}{4} \times \sum_i \left| \frac{M_{s,i} - M_{o,i}}{M_{o,i}} \right| \times 100, \quad (8)$$

where $M_{s,i}$ and $M_{o,i}$ are the values of landscape metric i calculated from the simulated and observed urban land use maps, respectively. A smaller absolute value of Rd indicates that the simulated landscape pattern is closer to the observable pattern.

1.4. Simulation of future scenarios

In recent years, there has been an increasing interest for developing sustainable urban form. A compact development is necessary in order to improve sustainability. Although there is a strong agreement on this statement, a debate between compact city and dispersed city has never stopped. Both of positive and negative effects of each type of city have been reported. In order to provide an insight into the different urban development strategies, five urban growth scenarios were designed towards 2020 and were named according to the main themes that result from the scenarios (Fig. 4). The land demand during this period was estimated according to the urban plan of Xuzhou.

The business as usual scenario (BUS) assumes that the future urban growth follows historical trend without any adjustment when environmental and developmental conditions are similar to the ones observed from the historical data.

The planning-strengthened scenario (PSS) assumes that the future urban growth strictly follows the master plan of Xuzhou city. The plan influences new developed urban allocation, as it establishes the legal regulatory framework for future land use. This scenario provides a better understanding of the impact of planning on urban growth.

Considering the urban development challenges, we established the compact development scenario (CDS) that aims to prevent sprawl-like development and to create a more compact city. The development is mainly concentrated around the existing city center, providing a more compact urban form. Second, in order to increase land use efficiency, a major development policy is implemented to increase the development of high-density residential and to decrease the development of low-density residential, which can reduce the per capita demand for the occupied land.

Contrary, the dispersed development scenario (DDS) was developed to simulate the future urban pattern with an increase of urban sprawl but without any effective urban planning against this trend. Due to the rapid economic growth and widespread of private vehicles, people desire to move to the low density settlements in order to avoid the congestion and large pollution in the city center, as well as to pursue a better living environment. In addition, the cost of housing outside the city center could be lower. Hence, the aim of the scenario is to encourage developments of new urban patches and urban infrastructure outside the city center. This scenario reflects a lesser degree of environmental protection.

The debate concerning sustainable urban form could move towards more moderate position where agreements are easier to achieve. Closer link between the former city center and several developed regions in fringe area is established. In addition, due to the fact that the rural settlements in Xuzhou are small in size but numerous and scattered, coordinating urban and rural development is also involved in this scenario not only for intensive land use but also for economic growth in rural areas. Therefore, the scenario aims to promote the endogenous potential of city center, suburban center, and rural area and the cooperation between them in order to achieve a physically and functionally connected region.

Multi-Criteria Evaluation (MCE) is an important means of analysis in spatial decision support systems, as it allows weighted value to be assigned to spatial layers, and the sum of these values produces a final suitability map. However, determining factor weights is a complicated task

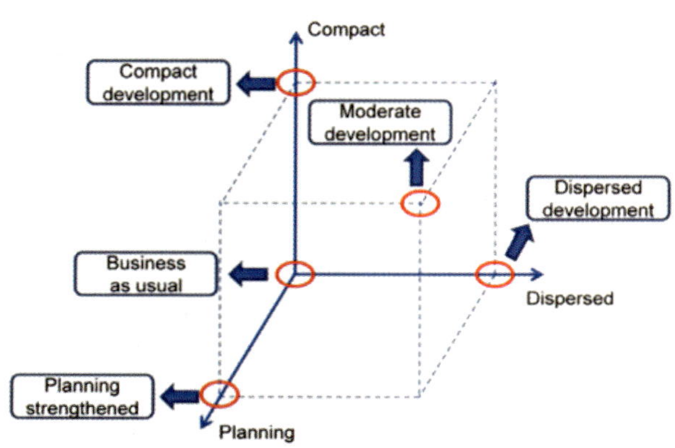

Fig. 4. Different development scenarios

in MCE. Analytic Hierarchy Process (AHP) originally developed by Saaty (1980) is one of the most commonly used approaches when analyzing complex decision problems. It can be used to derive behavior-oriented transition rules (Wu 1998). Basically, the pair-wise comparison of the relative importance is conducted to arrive at a scale of preference among a set of alternatives (Malczewski 1999).

By using AHP, the weights of global factors are identified and imported into the CA model for calculating global suitability value. AHP provides a comprehensive and rational framework for structural conceptualization of decision making, in which the relative importance of several variables can be compared (Vaz *et al.* 2012). Hence, the integration of MCE and AHP benefited this study in that it has capability to link scenario simulation with decision making processes and to translate the qualitative descriptions of scenarios into quantitative spatial analysis.

However, the weight of each global factor in AHP is usually identified by direct subjective assessment because preferences of decision makers determine the relative importance of each factor. In order to incorporate more realistic behavior into the simulation, the historical urban growth trend needs to be considered in decision making process. Therefore, the logistic regression coefficients for the period of 2005–2010 in CA model were used to identify the relative importance of each global factor in AHP for further modification. In addition, the discussion with five experts provided insights into the determination of the weights.

2. Results

2.1. Historical urban growth simulation

The specific procedure of calibration followed three steps. Take the period of 1990–2001 as an example:

- The historical urban growth (1 = changed and 0 = no change) was set as a dependent variable, and the global factors after standardization were set as independent variables. Based on the historical urban development trends, the weights for the global factors were accurately determined using a binary logistic regression model. The correlation analysis indicates that Dis2CBD and Dis2Cens are significantly correlated variables with the Pearson correlation coefficients of 0.680, 0.702 and 0.654 for three time points, respectively. In order to exclude redundant variable and select optimum set of variables, logistic regression was estimated for two possible variables sets, which considered all variables excluding Dis2CBD or Dis2Cens. The optimal set was determined using Relative Operating Characteristic (ROC) measure;
- The neighborhood configurations were calibrated through the trial and error method by running the model many times with different neighborhood configurations, while the random variable was set as 0, and held constant. The figure of merit value was calculated for each simulated result to measure the overall performance of the model (Fig. 5). The

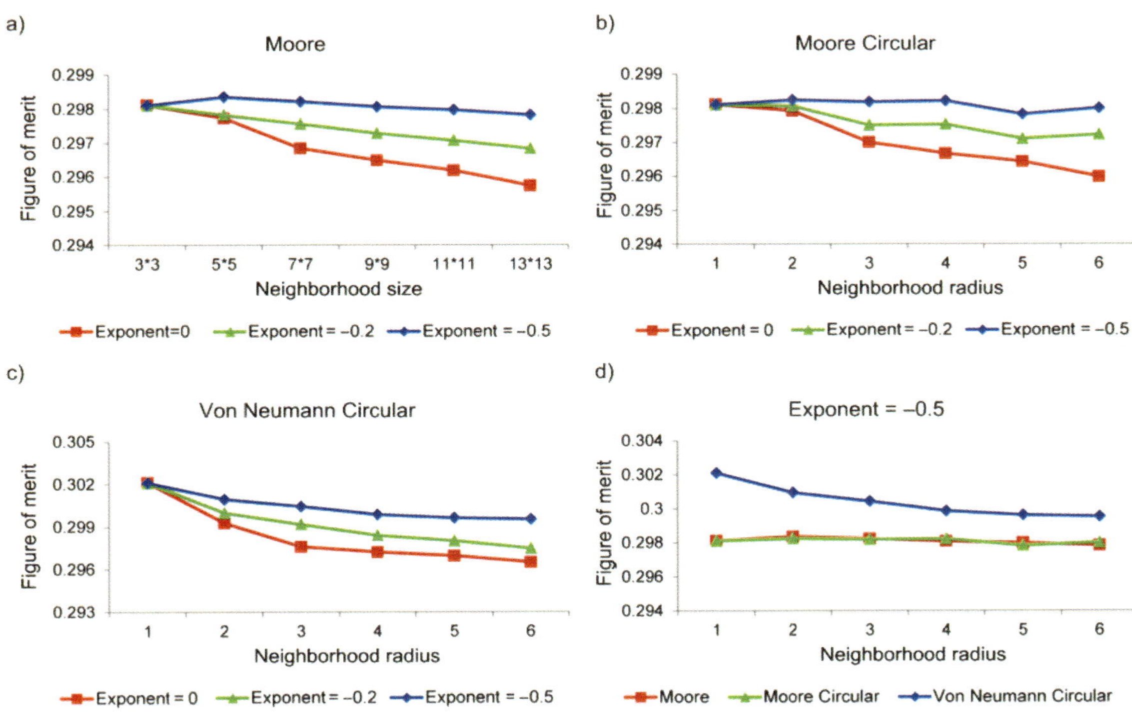

Fig. 5. Variation of figure of merit value response to neighborhood configuration variation. (a) The figure of merit value calculated for different exponent values and sizes using Moore type, (b) The figure of merit value calculated for different exponent values and sizes using Moore Circular type, (c) The figure of merit value calculated for different exponent values and sizes using Von Neumann Circular type and (d) The figure of merit value calculated for different neighborhood types and sizes when exponent is set as −0.5

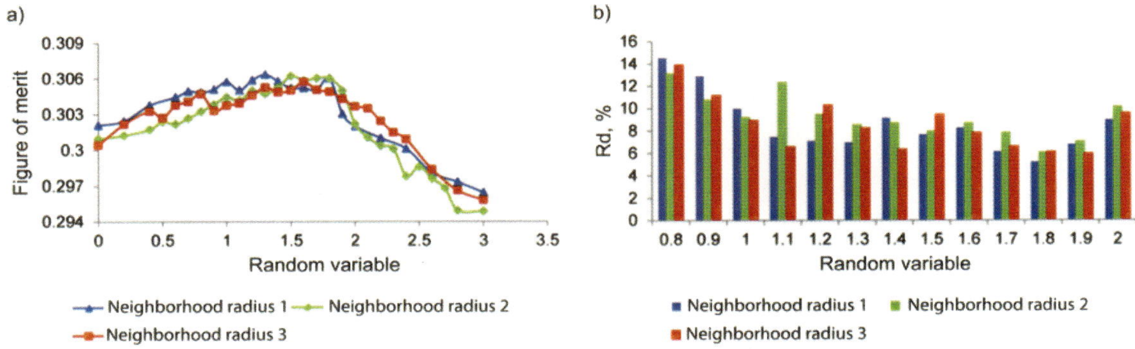

Fig. 6. Variation of figure of merit and Rd values response to neighborhood size and random variable variation: a) – the figure of merit value calculated for different neighborhood sizes and random variables and b) – the Rd value calculated for different neighborhood sizes and random variables

results show that the model with neighborhood type of Von Neumann Circular and exponent value of –0.5 generated the simulation result with the highest value. Therefore, they were used for further simulation;

- Various simulations were performed using random variables in the range 0–3 with 0.1 increment and different neighborhood sizes. Figure of merit value was calculated at each neighborhood size and random variable (Fig. 6(a)). Because of the involvement of random variable, each simulation generated different result with different value of figure of merit. However, the stochastic CA can maintain stability in landscape pattern (Yeh, Li, 2006). A range of random variables (0.8–2.0) were selected to ensure that CA model can generate relatively high figure of merit values. Focusing on this range, Rd value was calculated for each simulation result (Fig. 6(b)). The result indicates that random variable and neighborhood size should be set with values of 1.8 and 1 respectively so as to fit the observed urban land use map in terms of location and pattern.

The same procedure was used to calibrate the CA models for other periods. The estimated parameters for the periods of 1990–2001, 2001–2005 and 2005–2010 are listed in Table 1. The calibrated CA model was then used to simulate urban growth in Xuzhou during the periods of 1990–2001, 2001–2005, and 2005–2010, respectively.

As shown in Figure 7, the observed urban development map was overlaid with simulated map to identify the four groups of cells (observed change simulated as persistence, observed persistence simulated as change, observed change simulated as change, observed persistence simulated as persistence). Owing to the transition rules of CA models, such rules can in effect evenly locate new urban cells mainly in the city core and around the edge of initial urban patches. As a result, some of simulated urban cells in the city core were located where no

Table 1. Calibration results of the CA model during 1990–2001, 2001–2005 and 2005–2010

	Parameters	1990–2001	2001–2005	2005–2010
Factors	Dis2CBD	−1.422	–	–
	Dis2Cen	–	−1.697	−1.964
	Dis2MajR	−1.295	−1.114	−1.470
	Dis2MinR	−0.975	−1.002	−0.843
	Slope	−1.636	−1.392	−0.375
	Popden	0.846	0.254	−0.310
	Subsidence	0.413	0.435	0.343
	Environment	0.363	0.328	0.239
Neighborhood configurations	Type	Von Neumann Circular	Von Neumann Circular	Von Neumann Circular
	Function	exp(−0.5*D)	exp(−0.5*D)	exp(−0.5*D)
	Size	1	2	2
Stochastic disturbance	Variable	1.8	2.0	2.1

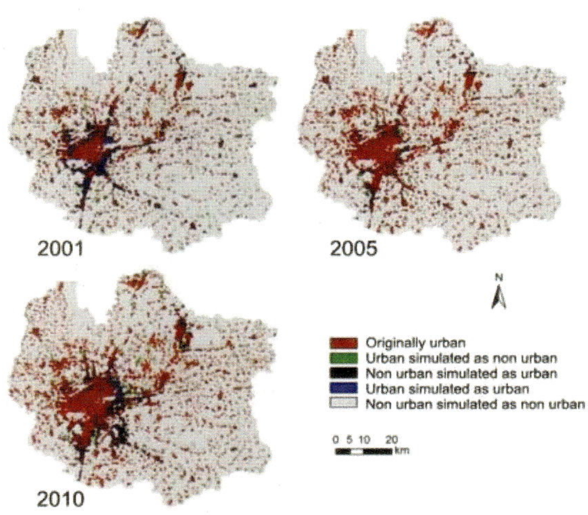

Fig. 7. Spatial distribution of corrects and errors of the simulation results

changes from non-urban to urban land uses took place. While some real developed cells in fringe and rural areas were underestimated by models. The errors found in simulation results also reveal that some errors are caused by issues not related to the model, such as the complexity of urban growth. Urban growth processes usually have some unpredictable features because of the complexity of nature. Although the uncertainty of urban growth can be represented by incorporating random variable, as demonstrated by Yeh and Li (2006), each simulation will generate different result when the inputs are the same because of the involvement of random variable. However, major uncertainties caused by random variable only existed in the fringe and rural areas, which can partly explain the errors in simulating new isolated urban cells outside the city core. Furthermore, some errors are observed due to the difficulties in considering all driving factors. The use of more or less number of variables will affect the outcome of CA simulation (Poelmans, Van Rompaey 2010). For example, the urban simulated as non-urban in the city core was mainly located in the southern and eastern part, where the development policies acted an accelerating factor to promote more new development. However, it was not involved into the transition rules, which made the accurate simulation of urban growth more difficult.

In addition to the description of visual comparison, the quantitative validation methods are required to quantify the degree of error of the simulation results. Figure 8 presents a summary of the error analysis according to Figure 7. The value represents the number of cells at the resolution of 100 m. The union sections of observed change simulated as persistence and observed change simulated as change represent the area of change according to the observed maps, and the union sections of observed change simulated as change and observed persistence simulated as change are the area of change according to the simulation maps. Table 2 presents the simulation accuracy at cell level. The figure of merit was calculated based on the quantitative error analysis. It enables to assess the cell to cell coincidence between simulated and actual maps in a more realistic way (Santé *et al.* 2010). The values of figure of merit are 30.6%, 33.2%, and 27.7%, respectively. The model for 2010 is the one that produced less matches. New development areas were promoted by planning policies, which cannot be involved in the transition rules. In addition, the increase in stochasticity of development indicated by random variables also influenced the performance of the CA model. The Kappa value was calculated to measure the overall performance of the model using the entire area with fixed land use. The kappa value for the simulated results of 2001, 2005 and 2010 are 0.82, 0.86 and 0.81, respectively.

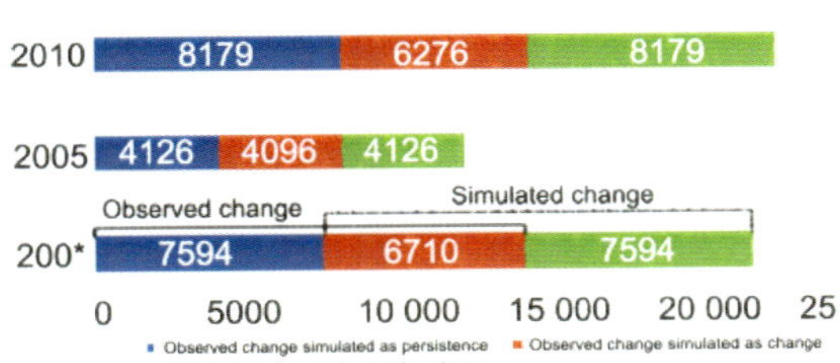

Fig. 8. Quantities of correct and errors values in the model validation

Table 2. Quantitative assessment of accuracy based on cell by cell comparison

Year	Figure of merit	Kappa
2001	30.6%	0.82
2005	33.2%	0.86
2010	27.7%	0.81

Besides the matching the exact location of urban land use change, the generating urban patterns similar to actual urban spatial patterns is also an important objective of CA models. Spatial metrics were used to objectively characterize the spatial pattern observed in the visual analysis in order to make quantitative comparison and to determine whether simulated patterns are similar to the actual patterns. Table 3 shows the comparison between observed and simulated landscape metrics values for 2001, 2005, and 2010. According to the landscape metrics, the models produced the landscape patterns substantially close to the observed ones. When looking at the relative error value calculated for the different landscape metrics, however, the models had relatively larger error in the simulated NP. The CA models generated lower number of patches, which were larger and more clustered than those in observed patterns. The isolated cells can be developed only by involving the random variables in this model, such that some of small new patches cannot be generated. Although

Table 3. Comparison between observed and simulated landscape metrics values for 2001, 2005, and 2010

		NP	LPI	SHAPE_ AM	ENN_ AM	Rd(%)
2001	Observed	2412	3.78	5.21	275.01	5.29
	Simulated	2195	3.93	5.53	280.63	
	Relative error (%)	9.00	3.97	6.14	2.04	
2005	Observed	2489	4.63	5.73	261.87	4.39
	Simulated	2307	4.89	5.72	273.57	
	Relative error (%)	7.31	5.62	0.17	4.47	
2010	Observed	2509	7.07	8.46	246.36	3.28
	Simulated	2354	7.36	8.56	250.48	
	Relative error (%)	6.18	4.10	1.18	1.67	

fewer patches generated by CA models, the lower relative error values of SHAPE_AM and ENN_AM indicate that the compaction and isolation were similar to the observed ones. The Rd values for the three simulated results are 5.29%, 4.39%, and 3.28%, respectively.

Overall, the validation results shown above reveal that the CA models for three time points have the ability to produce the multi-temporal simulation results which can be considered to be in line with the observed maps in terms of location and pattern similarity.

2.2. Future development scenarios

The parameters of CA model were modified according to the story-line of each scenario and the weights of BUS scenario. As illustrated in Figure 9, the elements of the story-line of each urban growth scenario were defined through the AHP process, in which the value represents the relative importance of global factors used for each scenario. Under the CDS scenario, Dis2CBD was considered to represent the distance to socioeconomic centers, while Dis2Cen was used for other scenarios instead of the Dis2CBD. A summary of the neighborhood configurations and random variables, and constraints for each scenario is given in Table 4. Using the modeling configuration, the input map of 2010, we performed simulations under the five scenarios aiming to project alternative spatio-temporal patterns of urban growth in 2020. Figure 10 shows the simulated landscape patterns of urban land use in 2020. Although all scenarios have the same urban land area as that of the 2010–2020 urban planning, the urban growth patterns differed.

To further evaluate and compare these scenarios, NP, LPI, SHAPE_AM and ENN_AM were used to further clarify the differences between the five scenarios by quantifying the landscape pattern under each scenario. When looking at these comparison results in Figure 11, we can gain the insight into the differences of the landscape patterns among different scenarios at global levels. The landscape patterns of scenarios differed from that of the observed urban pattern in 2010 due to the urbanization. If we look at the binary comparisons (Figs 12–16) between

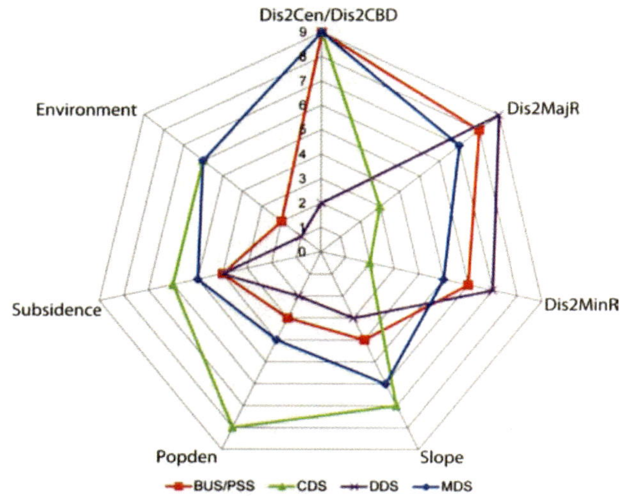

Fig. 9. Relative importance of the factors for each scenario

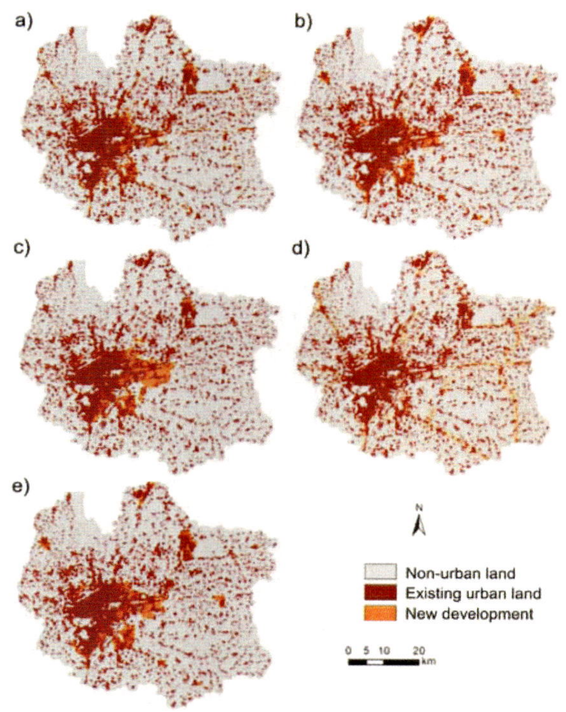

Fig. 10. The alternative urban maps of Xuzhou city for 2020 under different scenarios: (a) BUS, (b) PSS, (c) CDS, (d) DDS, and (e) MDS

Table 4. The configurations of CA model for each scenario

		BUS	PSS	CDS	DDS	MDS
Neighborhood configurations	Type	Von Neumann Circular	Von Neumann Circular	Von Neumann Circular	Von Neumann Circular	Von Neumann Circular
	Size	2	2	1	3	1
	Exponent value	–0.5	–0.5	–	–1.114	–
Stochastic perturbation	Random variable	2.1	2.1	1.5	3.0	1.5
Constraint	Nature	water	water	water	water	water
	Policy	–	Master planning	–	–	Master planning

scenarios in 2020 and observed landscape pattern in 2010 at local level, the urban growth pattern under each scenario can be discovered and located. In addition, the role of the transition rule can be better understood.

Under the BUS scenario, the increases in NP and LPI values are observed, which illustrate that urban growth in Xuzhou is focused on the development of new urban patches, as well as the expansion of the existing urban patches. The urban pattern becomes compact as reflected by the slight decrease in SHAPE_AM. As evidenced by the decrease in ENN_AM value, the individual urban patches get close to each other, becoming more connected with the city core. This is also reflected by the binary comparison in Figure 12, Xuzhou city expends outside its historical core. Furthermore, urban growth is constantly moving toward the urban fringe in the eastern part of Xuzhou city. Some of new urban land is situated adjacent to or near major roads with scattered and irregular patches.

In contrast to the historical urban growth trend, the PSS, CDS, and MDS scenarios have fewer urban patches compared to historical urban landscape pattern in 2010. Under these scenarios, the individual patches tend to be aggregated with increasing connection with previous individual urban patches already close to the city center and district centers, which is indicated by the decreases in NP and the increases in LPI values. The SHAPE_AM values decline suggests that urban pattern become more compact by locating continued growth in diffuse sprawl urban areas. Among these scenarios, the PSS scenario has highest NP, lowest LPI, highest SHAPE_AM and ENN_AM values,

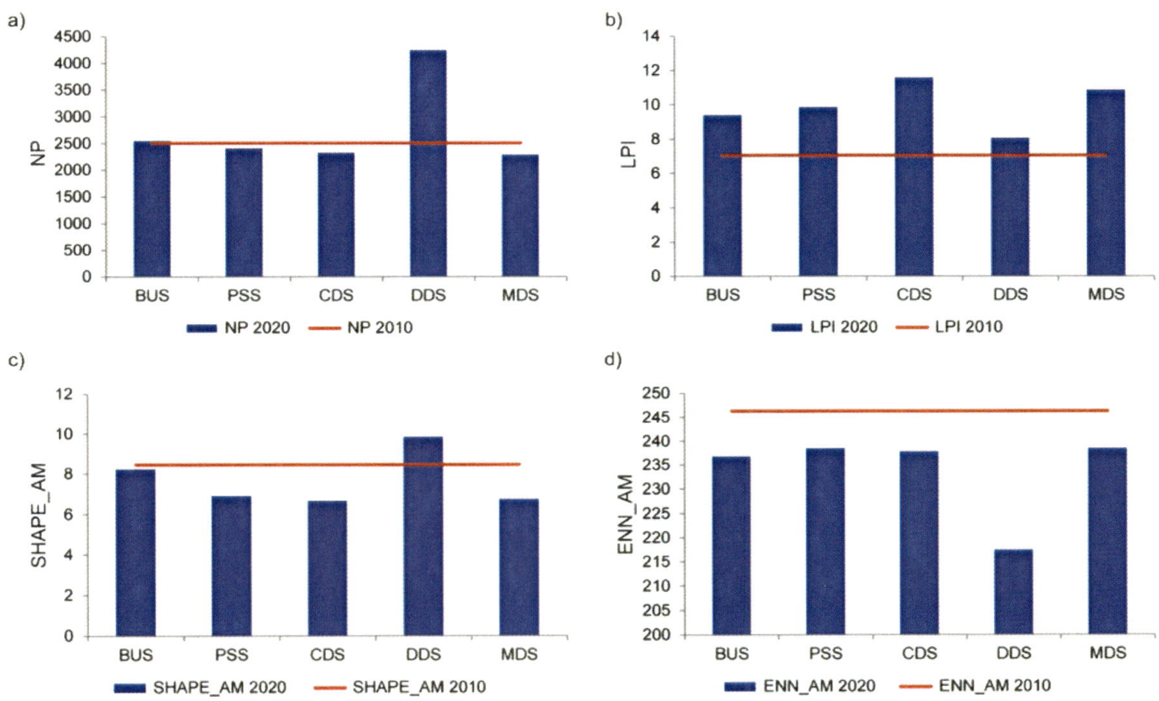

Fig. 11. Landscape metrics values of urban land use under different scenarios and observed urban landscape pattern in 2010: (a) NP value, (b) LPI value, (c) SHAPEAM value, and (d) ENN_AM value

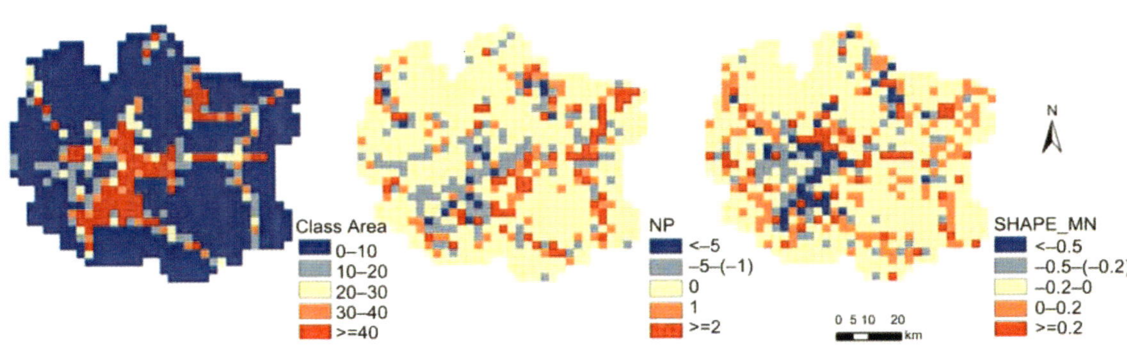

Fig. 12. Binary comparison between BUS scenario and observed urban landscape pattern 2010 in Class Area, NP, and SHAPE_MN values

which are more similar to the BUS scenario compared to other scenarios. It is also confirmed by the small number of blocks with significant increase in NP and SHAPE_MN values in Figure 13. By 2020, some of the sprawl areas develop into compact urban land by infill of vacant land between the existing urban patches. It is clear that in this scenario, some areas would become more attractive, since they are enforced as hotspots in fringe and rural areas that are potential for future urban development and the evolution of compact centers. Around the hotspots, the fragmentation and diffuse urban development slows down as evidenced by the significant decreases in NP and SHAPE_MN values. Despite the strict implementation of master plan of Xuzhou city leads to the slowing down of the urban diffuse sprawl around the hotspots, it does not change the landscape pattern significantly, which could be explained by the fact that only master plan is involved without considering other factors that have significant impacts on urban growth pattern.

With regard to the CDS, the compact urban pattern is observed which can be attributed to the considerable edge growth of historical urban patches. In specific, the highest LPI value in CDS scenario indicates that urban growth under this scenario has a preference to occur around the city center that is more attractive for development. Hence, the urban patches around the city center grow together to form larger patches, which is described as "dense-onion" model by Herold *et al.* (2003). The lowest value of SHAPE_AM suggests the urban areas are growing more compact. It can be seen from Figure 14 that almost all the vacant land suitable for development in the city core is used by 2020. The blocks in the city core have NP and SHAPE_MN values lower than 0. This indicates that the urban patches grow together to former larger and more compact urban patches. However, most of distant fringe and rural areas still remain unchanged or grow at a slow rate under CDS.

The significant increases in NP and SHAPE_AM values indicate the increasing fragmentation and irregularity of the landscape pattern with continued urbanization under DDS scenario. In concert with the increase in NP, the corresponding decrease in ENN_AM suggests that distance between urban patches dramatically declines. The intensive urban sprawl speeds up during the 2010–2020. The development centers appear to be less attractive for development compared to other scenario, which is reflected by the slight increase in LPI since 2010. Figure 15 shows that the urban areas spread outward from the city core and along the major road. Many blocks have high NP and SHAPE_MN values, which indicate that the new development creates many smaller and more fragmented patches in 2020. While the central urban area changes

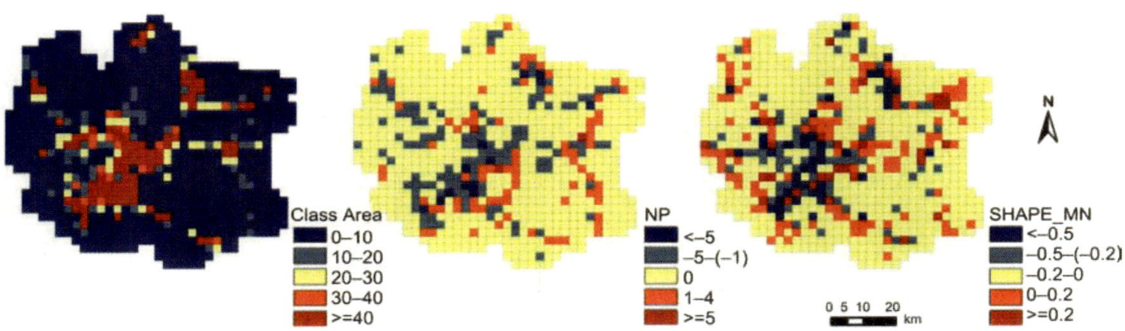

Fig. 13. Binary comparison between PSS scenario and observed landscape pattern 2010 in Class Area, NP, and SHAPE_MN values

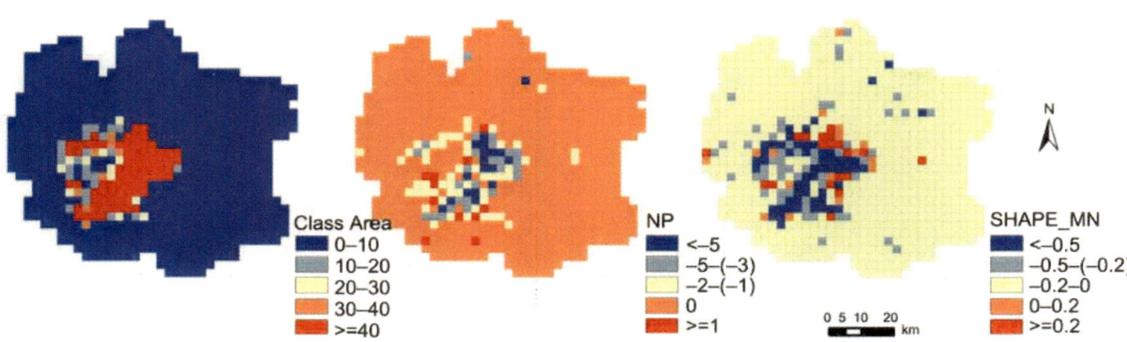

Fig. 14. Binary comparison between CDS scenario and observed landscape pattern 2010 in Class Area, NP, and SHAPE_MN values

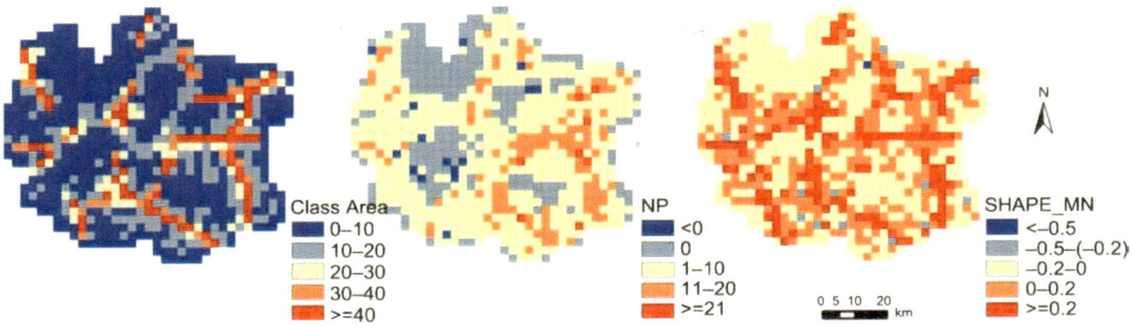

Fig. 15. Binary comparison between DDS scenario and observed landscape pattern 2010 in Class Area, NP, and SHAPE_MN values

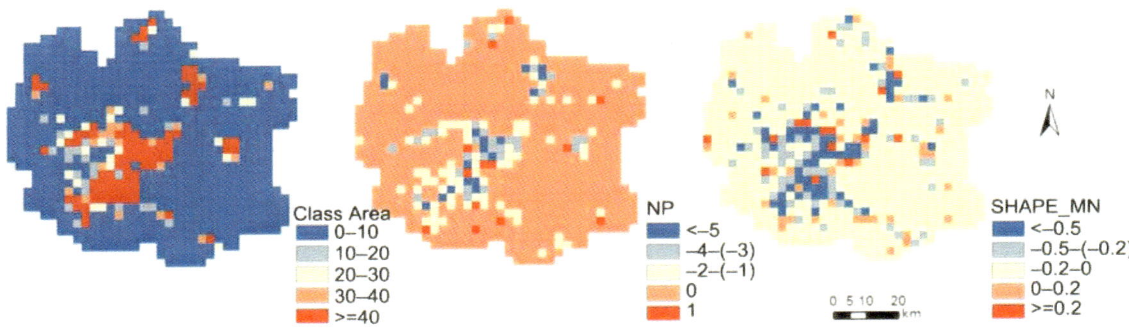

Fig. 16. Binary comparison between MDS scenario and observed landscape pattern 2010 in Class Area, NP, and SHAPE_MN values

slowly, there is a rapid increase in the new urban patches. The open spaces surrounded by developed urban land are created under this scenario.

Although the smallest number of urban patches is observed under MDS scenario, the LPI value is not the highest due to the development of several hotspots with relatively large size. Like the PSS scenario, the urban growth under MDS shows that the dominant trend of urban growth is the emergence of new development hotspots as shown in Figure 16. Most of land development is focused on the regions, where a large numbers of non-urban patches are encroached into urban land to form compact patches. Subsequently, the areas of diffuse sprawl are connected to the hotspots. Concomitant with the urbanization trend, however, spaces between the fragmented patches are further urbanized and enveloped on each other, which is similar to the CDS scenario.

3. Discussion

By involving natural and socioeconomic variables, the developed CA model has proved to be able to reproduce the historical urban growth process and assess the consequence of future urban growth. The hybrid calibration method combining logistic regression with trial and error was designed to calibrate the CA model, which can capture the complex interaction of various variables and promote the computational efficiency of the calibration. The existing validation method was improved by considering both the location and landscape pattern similarity to ensure that the CA model can produce more accurate result. Furthermore, five scenarios for 2020 were designed with focusing on specific urban development strategies. The study proposed the integration method of MCE and AHP that can be utilized to effectively translate the qualitative descriptions for scenarios into quantitative spatial analysis. Finally, the evaluation and comparison of the different scenarios presented in this paper provide an effective method for analyzing the impacts of different urban development strategies on landscape patterns at global and local scale and for supporting urban planning.

3.1. Methodology implication

The model can simulate the past urban growth and a wide variety of future scenarios based on the parameter values. Consequently, the method of parameter estimation is an important task. Furthermore, the calibration of CA models is difficult, particularly when there are many parameters to be considered in understanding spatial and temporal processes of urban growth (Cheng, Masser 2004). Therefore, selecting an appropriate method for the study is a challenge. Different from other calibration method used in previous studies, a hybrid calibration method consisting of the logistic regression and the trial and error has proved an effective and quick approach for calibrating the CA model in this study. The presented approach potentially captures the complex interaction of various environmental and socio-economic variables and promotes the computational efficiency of calibration. Moreover, it allows for sensitive analysis which demonstrates that the results of the CA model are sensitive to the parameter values, for example the neighborhood configurations and random variable. This is an important issue in CA models for understanding the urbanization process and its uncertainty.

Our study agrees well with the previously reported the usefulness of figure of merit and landscape metrics in the validation of CA model (García *et al.* 2012; Wang *et al.* 2013). While the study differed in that is the study focused on the effectiveness of combination of the two indicators in quantifying the agreement between simulated and observed urban land use maps. Firstly, figure of merit value was used to quantify the agreement using pixel by pixel comparison. It is a simple but promising way to measure location errors (Pontius *et al.* 2007). Secondly, the relative difference of landscape metric was utilized to objectively assess the goodness-of-fit of the outcomes with the actual urban patterns. For analyzing urbanization process, the landscape patterns are likely to be more important than the absolute locations of new urban pixels (Jenerette, Wu 2001). That is why landscape metrics were used to analyze the landscape patterns of model results. The rapid urbanization process may lead to the variation in the landscape patterns, which can be captured by a set of landscape metrics. Furthermore, each simulation will generate different results due to the involvement of random variables, but the stochastic CA can maintain stability in landscape pattern (Yeh, Li 2006). Consequently, the integration of figure of merit with landscape metrics can provide an effective way to identify the suitable random variables.

The establishment of connection between CA models and the urban decision making process needs to be considered as an important aspect of urban spatial models when such models are applied in the context of realistic cities. In this study, the efforts were made to illustrate a way in which CA models can be better linked with the decision making process. The challenge for a scenario simulation is to correctly define the relative importance of the global factors in qualitative terms, and then to translate the qualitative process description into quantitative scenarios of urban land use. This study proposed a combined methodology of translating the alternative futures into quantitative scenarios by integrating AHP, MCE and CA models. The pairwise function of different options quantified by AHP enables the decision makers to express their insights into the growth of Xuzhou city. The main advantage of this method is related to the structural conceptualization of decision making, in which several parameters may be compared, thus, bridging the gap between qualitative analysis and quantitative outputs.

The combination of scenario simulation and the landscape metrics has proved to be capable of making the processes and patterns of urban growth more prominent than using simulation on its own, and the landscape metrics also serves as a comparative platform to other cities. One important problem that has been often ignored by previous studies is the effects of scale on scenario evaluation. Scale effect refers to the variation in the results of statistical analysis caused by the variation of scale (Buyantuyev *et al.* 2010). In this study, two different scales were adopted to evaluate and compare the scenarios. The landscape metrics were calculated based on the block which makes it possible to discover and locate the patterns in different urban areas. Moreover, the local scale with a multi-temporal perspective enables us to better evaluate small-scale urbanization process, which cannot be detected at the global scale. The study presented here allows the integration of global and local scales and is able to highlight the consequences of urbanization at different scales.

3.2. Future development scenarios

The scenarios represent alternative policy, and the ways in which each policy could potentially unfold into the future. The scenarios may be of use for planners to better understand the consequences of drivers on urbanization (Aguilera *et al.* 2011; Fuglsang *et al.* 2013; Song *et al.* 2006). BUS scenario suggests that urban development will continue through both expansion of existing urban areas and outward diffuse sprawl in the future. If it continues as indicated by the BUS scenario, the conflict between rapid urban growth demand and the limitation of scarce land resources will intensify. The polycentric development was promoted as the development strategy since 2001. According to the historical growth trend, however, the polycentric development pattern is not significant by 2020. The development still focuses in the city core with rapid urban growth rate. A shift away from BUS might lead to significant alteration for Xuzhou city. There are still possibilities to enforce the polycentric development if the master plan is strictly implemented as shown in PSS scenario. The implementation of master plan leads to generation of new hotspots in fringe and rural areas for future development. Consequently, the closer linkage between the former city core and new development hotspots is established, which is necessary for solving the imbalance of development among the city core, fringe and rural areas. However, the development is also scattered across the study area when other factors are not considered. In the compact development scenario, the development continues through infill in the existing city core and edge-expansion growth. The compact urban pattern is generally considered to be more efficient in the use of natural resources. Therefore, it is regarded as a sustainable urban pattern (Li *et al.* 2008). However, the compact development within the city core makes it denser because the constraints within city core are not involved in the CA model. The densification of city core results in its limitless expansion and the loss of green open space, which influence the quality of urban life and urban environment. Concerning this shortcoming, DDS was simulated with the consideration of the demands of residents. As described in the storyline of dispersed scenario, the economy growth would result in increasing residents living in the fringe and rural areas and the development of new residential areas would be stimulated. DDS assumes the loss of agricultural and natural areas. The increasing residents would also encourage a large increase in road construction, and infrastructures. The economically oriented scenario presents more diffuse patterns (Reginster, Rounsevell 2006). The study also confirms that the DSS scenario presents a more diffuse sprawl pattern which is recognized to have a negative impact on environment and sustainable development. Considering this fact and urban development policy, in addition to the implementation of compact development, the scientific urban planning policies should also be required in order to avoid the limitless expansion of city core and to balance the conflicts among the inter-administrative regions. The demand of people for better residential environment also needs to be satisfied. This development strategy in MDS scenario optimizes the growth allocation in an environmentally and economically efficient way, which can support sustainable urban development in Xuzhou city.

Conclusions

The methodology framework proposed for this study has demonstrated to be useful in monitoring and analyzing urban growth in Xuzhou city and in providing a support for decision making processes towards a sustainable development. Some valuable results provide a better understanding of impact of urbanization process on landscape pattern.

Besides the factors involved in this study, urban growth is also strongly affected by political, cultural and other factors, which are difficult to incorporate into spatial model due to their aspatial characteristics and the lack of data. It is recommended that more potential variables should be included in the future studies to improve the performance of spatial models and to evaluate the effects of the factors on urban growth. The study only focused on the simulation of urban development without consideration of the detailed land use categories (commercial, industrial, and settlement land) due to the lack of detailed land use data. With taking into account the interactions among them, it would be interesting and valuable to simulate the change of several detailed land use categories within urban areas to provide a better understanding of the urban land use development.

References

Aguilera, F.; Valenzuela, L. M.; Botequilha-Leitão, A. 2011. Landscape metrics in the analysis of urban land use pattern: a case study in a Spanish metropolitan area, *Landscape and Urban Planning* 99(3–4): 226–238. https://doi.org/10.1016/j.landurbplan.2010.10.004

Arsanjani, J. J.; Helbich, M.; Kainz, W.; Boloorani, A. D. 2013. Integration of logistic regression, Markov chain and cellular automata models to simulate urban expansion, *International Journal of Applied Earth Observation and Geoinformation* 21: 265–275. https://doi.org/10.1016/j.jag.2011.12.014

Barredo, J. I.; Kasanko, M. K.; McCormick, N.; Lavalle, C. 2003. Modelling dynamic spatial processes: simulation of urban future scenarios through cellular automata, *Landscape and Urban Planning* 64(3): 145–160. https://doi.org/10.1016/S0169-2046(02)00218-9

Barron, O. V.; Barr, A.D.; Donn, M. J. 2013. Effect of urbanisation on the water balance of a catchment with shallow groundwater, *Journal of Hydrology* 485: 162–176. https://doi.org/10.1016/j.jhydrol.2012.04.027

Batty, M. 2005. *Cities and complexity: understanding cities with cellular automata, agent based models, and fractals*. Cambridge: The MIT Press.

Berling-wolff, S.; Wu, J. 2004. Modeling urban landscape dynamics: a review, *Ecological Research* 19(1): 119–129. https://doi.org/10.1111/j.1440-1703.2003.00611.x

Buyantuyev, A.; Wu, J.; Gries, C. 2010. Multiscale analysis of the urbanization pattern of the Phoenix metropolitan landscape of USA: time, space and thematic resolution, *Landscape and Urban Planning* 94(3–4): 206–217. https://doi.org/10.1016/j.landurbplan.2009.10.005

Cheng, J.; Masser, I. 2004. Understanding spatial and temporal processes of urban growth: cellular automata modelling, *Environment and Planning B: Planning and Design* 31(2): 167–194. https://doi.org/10.1068/b2975

Dewan, A. M.; Yamaguchi, Y. 2009. Land use and land cover change in Greater Dhaka, Bangladesh: using remote sensing to promote sustainable urbanization, *Applied Geography* 29(3): 390–401. https://doi.org/10.1016/j.apgeog.2008.12.005

Fuglsang, M.; Münier, B.; Hansen, H. S. 2013. Modelling land-use effects of future urbanization using cellular automata: an Eastern Danish case, *Environmental Modelling and Software* 50: 1–11. https://doi.org/10.1016/j.envsoft.2013.08.003

García, A. M.; Santé, I.; Boullón, M.; Crecente, R. 2012. A comparative analysis of cellular automata models for simulation of small urban areas in Galicia, NW Spain, *Computers, Environment and Urban Systems* 36(4): 291–301. https://doi.org/10.1016/j.compenvurbsys.2012.01.001

He, C.; Okada, N.; Zhang, Q.; Shi, P.; Zhang, J. 2006. Modeling urban expansion scenarios by coupling cellular automata model and system dynamic model in Beijing, China, *Applied Geography* 36(3–4): 323–345. https://doi.org/10.1016/j.apgeog.2006.09.006

Herold, M.; Goldstein, N. C.; Clarke, K. C. 2003. The spatiotemporal form of urban growth: measurement, analysis and modelling, *Remote Sensing of Environment* 86(3): 286–302. https://doi.org/10.1016/S0034-4257(03)00075-0

Hu, Z.; Lo, C. P. 2007. Modeling urban growth in Atlanta using logistic regression, *Computers, Environment and Urban Systems* 31(6): 667–688. https://doi.org/10.1016/j.compenvurbsys.2006.11.001

Jenerette, G. D.; Wu, J. 2001. Analysis and simulation of land-use change in the central Arizona-Phoenix region, USA, *Landscape Ecology* 16(7): 611–626. https://doi.org/10.1023/A:1013170528551

Kaufmann, R. K.; Seto, K. C.; Schneider, A.; Liu, Z.; Zhou, L.; Wang, W. 2007. Climate response to rapid urban growth: evidence of a human-induced precipitation deficit, *Journal of Climate* 20(10): 2299–2306. https://doi.org/10.1175/JCLI4109.1

Kocabas, V.; Dragicevic, S. 2006. Assessing a cellular automata model behavior using a sensitivity analysis approach, *Computers, Environment and Urban Systems* 30(6): 921–953. https://doi.org/10.1016/j.compenvurbsys.2006.01.001

Lambin, E. F.; Turner, B. I.; Geist, H. J. 2001. The causes of land-use and land cover change: moving beyond the myths, *Global Environmental Change* 11(4): 261–269. https://doi.org/10.1016/S0959-3780(01)00007-3

Li, C.; Thinh, N. X.; Zhao, J. 2014. Spatiotemporally varying relationships between urban growth patterns and driving factors in Xuzhou city, China, *Photogrammetrie Fernerkundung Geoinformation* 6: 535–548. https://doi.org/10.1127/pfg/2014/0246

Li, C.; Thinh, N. X. 2013. Investigation and comparison of land-cover change patterns in Xuzhou city, China, and Dortmund city region, Germany, using multitemporal Landsat images, *Journal of Applied Remote Sensing* 7(1): 073458. https://doi.org/10.1117/1.JRS.7.073458

Li, X.; Yang, Q.; Liu, X. 2008. Discovering and evaluating urban signatures for simulating compact development using cellular automata, *Landscape and Urban Planning* 86(2): 177–186. https://doi.org/10.1016/j.landurbplan.2008.02.005

Li, X.; Yeh, A. G. O. 2002. Neural-network-based cellular automata for simulating multiple land use changes using GIS, *International Journal of Geographical Information Science* 16(4): 323–343. https://doi.org/10.1080/13658810210137004

Liu, X.; Li, X.; Shi, X.; Wu, S.; Liu, T. 2008. Simulating complex urban development using kernel-based non-linear cellular automata, *Ecological Modelling* 211(1–2): 169–181. https://doi.org/10.1016/j.ecolmodel.2007.08.024

Malczewski, J. 1999. *GIS and Multicriteria Decision Analysis*. New York: John Wiley & Sons Inc.

Mitsova, D.; Shuster, W.; Wang, X. H. 2011. A cellular automata model of land cover change to integrate urban growth with open space conservation, *Landscape and Urban Planning* 99(2): 141–153. https://doi.org/10.1016/j.landurbplan.2010.10.001

Munshi, T.; Zuidgeest, M.; Brussel, M.; van Maarseveen, M. 2014. Logistic regression and cellular automata-based modelling of retail, commercial and residential development in the city of Ahmedabad, India, *Cities* 39: 68–86. https://doi.org/10.1016/j.cities.2014.02.007

Petrov, L. O.; Lavalle, C.; Kasanko, M. 2009. Urban land use scenarios for a tourist region in Europe: applying the MOLAND model to Algarve, Portugal, *Landscape and Urban Planning* 92(1): 10–23. https://doi.org/10.1016/j.landurbplan.2009.01.011

Poelmans, L.; Van Rompaey, A. 2010. Complexity and performance of urban expansion models, *Computers, Environment and Urban Systems* 34(1): 17–27. http://dx.doi.org/10.1016/j.compenvurbsys.2009.06.001

Pontius, R. G.; Boersma, W.; Castella, J. C.; Clarke, K.; de Nijs, T.; Dietzel, C.; Duan, Z.; Fotsing, E.; Goldstein, N.; Kok, K.; Koomen, E.; Lippitt, C. D.; McConnell, W.; Sood, A. M.; Pijanowski, B.; Pithadia, S.; Sweeney, S.; Trung, T. N.; Veldkamp, A. T.; Verburg, P. H. 2008. Comparing the input, output, and validation maps for several models of land change, *Annals of Regional Science* 42(1): 11–37. https://doi.org/10.1007/s00168-007-0138-2

Pontius, R. G.; Walker, R.; Yao-kumah, R.; Arima, E.; Aldrich, S.; Caldas, M.; Vergara, D. 2007. Accuracy assessment for a simulation model of Amazonian deforestation, *Annals of the Association of American Geographers* 97(4): 677–695. https://doi.org/10.1111/j.1467-8306.2007.00577.x

Reginster, I.; Rounsevell, M. 2006. Scenarios of future urban land use in Europe, *Environment and Planning B: Planning and Design* 33(4): 619–636. https://doi.org/10.1068/b31079

Saaty, L. T. 1980. *The Analytic Hierarchy Process*. New York: McGraw-Hill.

Santé, I.; Marcía, A. M.; Miranda, D.; Crecente, R. 2010. Cellular automata models for the simulation of real-world urban processes: a review and analysis, *Landscape and Urban Planning* 96(2): 108–122. https://doi.org/10.1016/j.landurbplan.2010.03.001

Shen, Q.; Chen, Q.; Tang, B.; Yeung, S.; Hu, Y.; Cheung, G. 2009. A system dynamics model for the sustainable land use planning and development, *Habitat International* 33(1): 15–25. https://doi.org/10.1016/j.habitatint.2008.02.004

Song, Y.; Ding, C.; Knaap, G. 2006. Envisioning Beijing 2020 through sketches of urban scenarios, *Habitat International* 30(4): 1018–1034. https://doi.org/10.1016/j.habitatint.2005.10.006

Sui, D.; Zeng, H. 2001. Modeling the dynamics of landscape structure in Asia's emerging Desakota regions: a case study in Shenzhen, *Landscape and Urban Planning* 53(1–4): 37–52. https://doi.org/10.1016/S0169-2046(00)00136-5

Tan, M.; Li, X.; Xie, H.; Lu, C. 2005. Urban land expansion and arable land loss in China-a case study of Beijing-Tianjin-Hebei region, *Land Use Policy* 22(3): 187–196. https://doi.org/10.1016/j.landusepol.2004.03.003

Thapa, R. B.; Murayama, Y. 2012. Scenario based urban growth allocation in Kathmandu Valley, Nepal, *Landscape and Urban Planning* 105(1–2): 140–148. https://doi.org/10.1016/j.landurbplan.2011.12.007

Tobler, W. R. 1970. A computer movie simulating urban growth in the Detroit region, *Economic Geography* 46: 234–240. https://doi.org/10.2307/143141

United Nations. 2012. *World urbanization prospects: the 2011 revision* [online], [cited 5 March 2014]. Population Division, Department of Economic and Social Affairs, UN. Available from Internet: http://esa.un.org/unup/

Vaz, E. d N.; Nijkamp, P.; Painho, M.; Caetano, M. 2012. A multi-scenario forecast of urban change: a study on urban growth in the Algarve, *Landscape and Urban Planning* 104(2): 201–211. https://doi.org/10.1016/j.landurbplan.2011.10.007

Wang, H.; He, S.; Liu, X.; Dai, L.; Pan, P.; Hong, S.; Zhang, W. 2013. Simulating urban expansion using a cloud-based cellular automata model: a case study of Jiangxia, Wuhan, China,

Landscape and Urban Planning 110: 99–112. https://doi.org/10.1016/j.landurbplan.2012.10.016

Ward, D.; Phinn, S. R.; Murray, A. T. 2000. Monitoring growth in rapidly urbanizing areas using remotely sensed data, *Professional Geographer* 52(3): 371–386. https://doi.org/10.1111/0033-0124.00232

Wu, F. 1998. SimLand: a prototype to simulate land conversion through the integrated GIS and CA with AHP-derived transition rules, *International Journal of Geographical Information Science* 12(1): 63–82. https://doi.org/10.1080/136588198242012

Wu, F. 2002. Calibration of stochastic cellular automata: the application to rural-urban land conversions, *International Journal of Geographical Information Science* 16(8): 795–818. https://doi.org/10.1080/13658810210157769

Yang, Q.; Li, X.; Shi, X. 2008. Cellular automata for simulating land use changes based on support vector machines, *Computers and Geosciences* 34(6): 592–602. https://doi.org/10.1016/j.cageo.2007.08.003

Yeh, A. G. O.; Li, X. 2006. Errors and uncertainties in urban cellular automata, *Computers, Environment and Urban Systems* 30(1): 10–28. https://doi.org/10.1016/j.compenvurbsys.2004.05.007

Zhang, Q.; Ban, Y.; Liu, J.; Hu, Y. 2011. Simulation and analysis of urban growth scenarios for the Greater Shanghai Area, China, *Computers, Environment and Urban Systems* 35(2): 126–139. https://doi.org/10.1016/j.compenvurbsys.2010.12.002

Zimmermann, P.; Tasser, E.; Leitinger, G.; Tappeiner, U. 2010. Effects of land-use and land cover pattern on landscape-scale biodiversity in the European Alps, *Agriculture, Ecosystems and Environment* 139(1–2): 13–22. https://doi.org/10.1016/j.agee.2010.06.010

Cheng LI. He is working at School of Resources and Geoscience, China University of Mining and Technology, Xuzhou, Jiangsu, China. From 2010 to 2014, he was a PhD student at TU Dortmund University. Now his research focuses on monitoring and analysis of urban growth by using remote sensing data, GIS and land use change modeling methods.

Jie ZHAO. She is working at Jiangsu Normal University, Xuzhou, China. She received PhD degree from TU Dortmund University in 2016. Her current research interests include monitoring urban growth by using remote sensing data, assessment and simulation of urban dynamics with a view to sustainable development and to eco-efficiency.

17

LANDSCAPE SPATIAL PATTERN INDICES AND SOUNDSCAPE PERCEPTION IN A MULTI-FUNCTIONAL URBAN AREA, GERMANY

Jiang Liu[a, b], Jian Kang[a], Holger Behm[b], Tao Luo[c]

[a]School of Architecture, University of Sheffield, Western Bank, Sheffield S10 2TN, UK
[b]Faculty of Agricultural and Environmental Sciences, University of Rostock, Justus-von-Liebig-weg-6, 18059 Rostock, Germany
[c]Key Lab of Urban Environment and Health, Institute of Urban Environment, Chinese Academy of Sciences, 1799 Jimei Road, 361021 Xiamen, China

Abstract. Soundscape research could provide more information about urban acoustic environment, which should be integrated into urban management. The aim of this study is to test how landscape spatial pattern could affect soundscape perception. Soundscape data on specifically defined spatial and temporal scales were observed and evaluated in a multi-functional urban area in Rostock, Germany. The results show that urban soundscapes were characterised by artificial sounds (human, mechanical and traffic sounds) overwhelming the natural ones (biological and geophysical sounds). Major sound categories were normally mutual exclusive and dynamic on temporal scale, and have different spatial distribution on spatial scale. However, biological and traffic sounds seem to be co-existing on both temporal and spatial scales. Significant relationships were found existing between perception of major sound categories and a number of landscape spatial pattern indices, among which vegetation density (NDVI), landscape shape index (LSI) and largest patch index (LPI) showed the most effective indicating ability. The research indicated that soundscape concepts could be applied into landscape and urban planning process through the quantitative landscape indices to achieve a better urban acoustic environment.

Keywords: landscape spatial pattern indices, multi-functional urban area, urban soundscape, human perception, landscape management.

Introduction

"Noise" has been considered as a global problem affecting the well-being of humans and organisms (André *et al.* 2011; Butkus, Januševičius 2011). The traditional approaches of noise control, such as noise mapping, noise zoning, noise monitoring and abatement have been pointed out by many researches as not effective enough (Paožalytė *et al.* 2012), as the physical parameters such as A-weighted sound pressure level (L_{Aeq}) ignore to a certain extent the physiological and psychological consequence on human (De Ruiter 2004; Raimbault, Dubois 2005).

The concept of soundscape, initiated by Schafer (1969), advocates treating "noise" in cities objectively and emphasises human experience. Thus it overcomes the limitations of quantitative approaches focusing only on the physical measurements of noise, and is advocated by many researches (Schulte-Fortkamp, Fiebig 2006; Kang 2007; Kang, Zhang 2010; Liu *et al.* 2013a, b). Soundscape research is also thought to be an integrated and multidisciplinary approach (Raimbault, Dubois 2005; Kull 2006; Brown *et al.* 2011), and concerted efforts have been made in this area (Genuit, Fiebig 2006; Schulte-Fortkamp, Fiebig 2006; Adams *et al.* 2006; Lavandier, Defreville 2006; Zhang, Kang 2007; Kang 2007; Brambilla, Maffei 2006, 2010; Joo *et al.* 2011; Pijanowski *et al.* 2011). Although using the same term "soundscape", the focuses of these researches mainly fall into two groups, i.e. one presented soundscape as the collection of all sounds that occur at a place (Schafer 1969), the other presented soundscapes as an ecological reflection of underlying ecological processes occurring in ecosystems (Pijanowski *et al.* 2011).

Corresponding author: Jian Kang
E-mail: j.kang@sheffield.ac.uk

Particularly in the area of urban acoustics, combining soundscape concept into the planning process is thought to be the most effective and practical way to realise a better acoustic environment (Raimbault, Dubois 2005; Guastavino 2006; Adams *et al.* 2006; De Coensel *et al.* 2010). In terms of planning, urban planners thought that "how to conceive and design desirable soundscapes" was more important than "a simple decrease of noise level or the elimination of noises" (Raimbault, Dubois 2005). However, how could planners and decision makers easily apply the soundscape concept into practical urban planning management is still to be considered, where an inter-disciplinary approach could be helpful.

In the research area of landscape ecology, landscape spatial patterns are mainly analysed in terms of ecological processes (Turner 1989), and a series of spatial metrics that quantitatively measure landscape structures have been developed (Turner *et al.* 2001). Based on the area and spatial distribution of different landscape patches, these indices could be used to characterise landscapes in two main aspects, i.e. landscape composition, namely the variety and abundance of patch types within a landscape, and landscape configuration, namely the spatial distribution of patches within a landscape (McGarigal, Marks 1994; De la Fuente de Val *et al.* 2006). Recently, landscape metrics have been widely applied in indicating landscape changes and different landscape functions such as habitation, regulation and information (Uuemaa *et al.* 2012). In particular, for the information function, numerous studies have revealed the close relationship between landscape visual attributes and landscape patterns (Palmer 2004; Dramstad *et al.* 2006). The visual aspect of landscape has also been, up to now, paying more attention in urban planning process (Adams *et al.* 2006).

On the other hand, researches on the aural aspect of landscape have been drawing more attention, but mainly focusing on aural-visual interactions in landscape perception, treating sound as an important factor affecting landscape preference by human beings (Pheasant *et al.* 2008). As an auditory correspondent of landscape(Schafer 1969), soundscape and its quality are closely related to characteristics of the underlying landscape. For example, Matsinos *et al.* (2008) revealed that spatial sound variability in a coastal rural area in Greece was mainly shaped by landscape attributes. In the same rural area, Mazaris *et al.* (2009) pointed out later the interactions between soundscapes and several landscape features. Liu *et al.* (2013b) found that urban soundscape perception also showed a relationship with underlying landscape characteristics. Compared with visual perception, which is related only to the landscape characteristics in viewshed, sounds can propagate and be perceived by humans through the local landscapes even the sound sources are visually hidden. Although the previous studies revealed at an early stage how landscape spatial patterns could affect soundscape perception, their relationship needs to be explored in a more detailed way.

The key aim of this research is therefore trying to do so by examining the soundscape-landscape relationship at a relatively large scale. Concepts from landscape ecology were applied into the soundscape approach, attempting to combine soundscape perception with urban ecological processes, aiming to reveal the relationships: 1) among the perceptual sound categories; 2) between each of the perceptual sound categories and landscape spatial pattern. Urban soundscapes in a multi-functional urban area were subjectively observed and evaluated on specifically defined spatial and temporal scales, in which soundscape was treated as an assembly of different sounds that were perceived by humans (Brown *et al.* 2011). Then based on the characteristics of the urban soundscapes, relationships between the major sound categories and the landscape spatial pattern indices were analysed. If landscape spatial indices could be indicators of soundscape perception, they could be an effective tool for planners and decision makers in creating a better acoustic environment.

1. Methods

1.1. Study area

The study was conducted in a coastal area of the Baltic Sea named Warnemünde, in Rostock, Germany, as shown in Figure 1. The area was chosen because of its rich landscape diversity and multiple land use types. As indicated in Figure 1, the local landscape includes 17

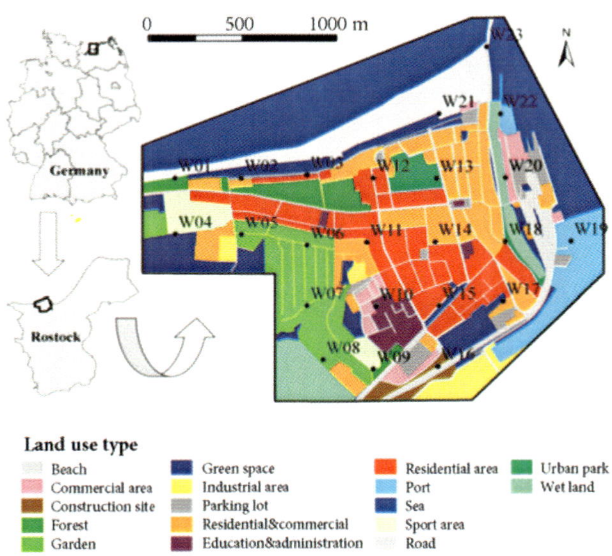

Fig. 1. Location of the study area and the 23 sampled sites (W01–W23), and land use structure of the study area

different types of land use according to the digital topographic-cartographic information system (ATKIS) 2010 from the Working Committee of the Surveying Authorities of the States of the Federal Republic of Germany (AdV), and actual on-site situation. Moreover, as a tourist resort, this area is subject to high leisure activity demands especially in summer time. As a result, on the one hand, different urban functions and human activities here generate diverse soundscapes; and on the other hand, different types of landscape appearing in this relatively small area can facilitate the analysis between soundscape perception and landscape spatial patterns on a human scale.

1.2. Soundscape observation and evaluation

Understanding the characteristics of urban soundscapes is the prerequisite of a soundscape-landscape relationship research. Urban sounds that form the urban soundscapes are rich and diverse. In soundscape researches, sounds were usually classified into different categories (Brown *et al.* 2011). In this research, according to the sources and nature of major sounds, urban sounds were decomposed into five major categories, i.e. human, mechanical, traffic, geophysical and biological sound (Table 1). Traffic sound was chosen to be independent from other mechanical sounds mainly because it is more concerned in urban areas.

Table 1. Classification of major sound categories in the study area

Sound category	Major sound	Code
Human sound (Hum)	Children voice	CS
	Adult voice	AS
	Footstep	FS
Mechanical sound (Mech)	Airplane flying	AF
	Bicycle riding	BC
	Bell ringing	BR
	Construction	CT
	Emergency	ES
	Grass mowing	GM
	Music	MS
	Ship moving	SM
	Other mechanical sounds	OA
Traffic sound (Traf)	Train moving	TM
	Traffic sound (foreground)	TSF
	Traffic sound (background)	TSB
	Motorcycle rumbling	MR
Geophysical sound (Geo)	Grass rustling	GR
	Raining	RS
	Sea wave	SW
	Tree rustling	TR
	Wind blowing	WF
	Water sound	WS
Biological sound (Bio)	Bird song	BS
	Chicken clucking	CC
	Dog barking	DB
	Frog	FR
	Insects	IS

Soundscape data was collected by a group of observers in terms of on-site perception and evaluation. Firstly, based on 350*350 m fishnet cells created in ArcGIS 9.1, 23 sites were randomly sampled in the study area (Fig. 1). During a pilot study, major sounds that frequently appear in this area were recognised and coded, as shown in Table 1, as a reference for the observers. A team of 12 people, including 10 students, one audio engineer and one musician conducted the investigation (7 male and 5 females; average age: 26 ± 2.8). All of them went through a training process one month before on-site investigation, to ensure a good inter-user reliability. The training included (a) getting familiar with all the major sounds on the list by presenting videos shot recorded on site; (b) remembering the code names to ensure a fast recording; and (c) performing pilot studies to test the investigation procedure and to minimise observation bias. After the training process, the observers could achieve an average inter-user reliability of 0.91 for loudness evaluation of the five major sound categories. Thus, in order to promote the observation efficiency, the observers were divided into 6 small groups of 2 to work together in responsible for 3 or 4 of the sampled sites (Matsinos *et al.* 2008; Mazaris *et al.* 2009).

Soundscape data were recorded in eight two-hour successive sampled periods in two days, covering main daily active periods in summer time (1st period: 06:00–08:00, 2nd period: 08:00–10:00, 3rd period: 10:00–12:00, 4th period: 12:00–14:00, 5th period: 14:00–16:00, 6th period: 16:00–18:00, 7th period: 18:00–20:00, 8th period: 20:00–22:00). Within each sampled period, a random 10 min sub-set was chosen to carry out the on-site observation-evaluation, which was further divided into twenty sequential time-steps, each of 30 s. The observers filled in a form with recognised sounds, evaluated their perceived loudness by using a five-point linear scale (1 = *very quiet*, 2 = *quiet*, 3 = *normal*, 4 = *loud*, 5 = *very loud*), respectively, and at the end of each time-step, evaluated soundscape preference by using a five-point linear scale (1 = *very pleasant*, 2 = *pleasant*, 3 = *normal*, 4 = *unpleasant*, 5 = *very unpleasant*). Overall 3680 samplings (8*20*23) of urban soundscapes and corresponding preference data were obtained. All the recognised major sounds were reclassified into five sound categories to simplify the analysis process, according to Table 1. The perceived loudness of each sound category and overall soundscape loudness were calculated by accumulating scores.

1.3. Landscape spatial pattern indices

Although landscape metrics are usually used to indicate ecological functions on a large scale, this research is more concerned with their ability to quantify landscape spatial patterns (Corry, Nassauer 2005). Considering that sound generation is more related to the activities in a certain area with particular functions, raster land use map (resolution 1 m) of the study area was used to calculate landscape metrics (Fig. 1). Among the several landscape metrics available, the following landscape metrics that may affect the composition and/or perception of soundscapes were selected. For the landscape composition, metrics used include largest patch index (LPI), patch density (PD) and evenness (SIEI). Landscape configuration was measured according to landscape shape index (LSI), contagion (CONT), and fractal dimension (FRAC). All the metrics mentioned above were calculated in Fragstats software (McGarigal, Marks 1994), based on a 175 m radius buffer area centred on each sampled site, according to some previous studies (Mazaris *et al.* 2009; Liu *et al.* 2013b).

Considering the land cover factors that may influence soundscape generation and/or propagation, another three indices reflecting landscape composition status were introduced, including construction density (CD), road density (RD) and vegetation density. For vegetation density, normalised difference vegetation index (NDVI) was chosen to be the indicator (Liu *et al.* 2013b), and the calculation was based on the Landsat TM image (30 m) on July 27, 2011 from U.S. Geological Survey (USGS). The three indices were also calculated based on the 175 m radius buffer area of each sampled site in ArcGIS 9.1. Table 2 lists all the indices and their descriptions according to McGarigal and Marks (1994).

1.4. Statistical analysis

Based on the daily accumulated perceived loudness of each sound category on each sampled site, interpolation method (regularised spline with tension in GRASS) was used to show the spatial patterns of them across the study area (GRASS Development Team 2008). Kruskal-Wallis independent samples non-parametric test was used to examine the possible differences on perceived loudness of each sound category among 23 sampled sites and among 8 sampled periods. Regression analysis between overall soundscape loudness and loudness of each of the five sound categories by time-step per-period was conducted to find out their relative contributions in different sampled periods. Spearman's roh correlation analysis between perceived loudness of different sound categories and corresponding soundscape preference, among perceived loudness of different sound categories, as well as between perceived loudness of different sound categories and each of the landscape indices were conducted by time-step per-period. Although the landscape indices chosen all have their specific intention, Pearson correlation coefficients among all the landscape pattern indices were calculated too, in order to detect colinearity in landscape indices (Turner, Ruscher 1988). All the statistical analyses were conducted in SPSS 16.0.

Table 2. Indices used to describe the landscape spatial patterns (McGarigal, Marks 1994)

Indices	Acronym	Description
Construction density	CD	Density of buildings of different uses
Road density	RD	Density of main traffic roads
Vegetation density	NDVI	Normalized Difference Vegetation Index. A simple graphical indicator that assess whether the target being observed contains live green vegetation or not
Patch density	PD	Patch density has the same basic utility as number of patches as an index, but facilitates comparisons among landscapes of varying size. It is used as a measure of landscape fragmentation
Landscape shape index	LSI	Landscape shape index can be interpreted as a measure of the total shape complexity of patches, indicating landscape fragmentation status
Largest patch index	LPI	Largest patch index quantifies the percentage of total landscape area comprised by the largest patch. It is a simple measure of dominance
Fractal dimension	FRAC	Area-weighted mean fractal dimension for all patches. It reflects shape complexity of patches. As polygons become more complex, the fractal dimension changes from 1 to 2
Contagion	CONT	It measures the aggregation extent of landscape patches. Higher values may characterise landscapes with a few large, contiguous patches. Lower values might be resulted by many small and dispersed patches formed landscapes
Evenness	SIEI	Simpson's evenness index indicates a structural component of diversity. An even distribution among landscape types results in maximum evenness, and low evenness indicates that either one or just a few elements are dominant

Table 3. Kruskal-Wallis independent samples non-parametric analysis for each sound category by sampled period and sampled site

Sound category	Sampled period		Sampled site	
	χ^2	P	χ^2	P
Hum	43.917	<0.001	85.859	<0.001
Mech	27.929	<0.001	77.525	<0.001
Traf	5.439	0.607	137.55	<0.001
Geo	25.5	0.001	78.555	<0.001
Bio	20.093	0.005	67.981	<0.001

Table 4. Spearman' rho correlation among different sound categories by time-step per-period (*p < 0.05, **p < 0.01)

Period	Sound	Hum	Mech	Traf	Geo
1	Mech	−0.003			
	Traf	−0.212**	−0.340**		
	Geo	−0.242**	0.102*	−0.128**	
	Bio	0.081	−0.245**	0.081	−0.125**
2	Mech	−0.098*			
	Traf	−0.051	−0.394**		
	Geo	0.033	−0.121**	−0.087	
	Bio	−0.05	−0.193**	0.269**	−0.067
3	Mech	0.099*			
	Traf	−0.283**	−0.419**		
	Geo	−0.104*	0.025	−0.069	
	Bio	−0.087	−0.316**	0.194**	0.05
4	Mech	0.043			
	Traf	−0.488**	−0.364**		
	Geo	0.002	−0.116*	−0.004	
	Bio	−0.335**	−0.269**	0.219**	0.086
5	Mech	−0.064			
	Traf	−0.340**	−0.354**		
	Geo	0.106*	0	−0.176**	
	Bio	−0.166**	−0.191**	0.163**	−0.132**
6	Mech	0.06			
	Traf	−0.266**	−0.347**		
	Geo	−0.011	−0.113*	−0.072	
	Bio	−0.201**	−0.261**	0.431**	0.003
7	Mech	0.102*			
	Traf	−0.397**	−0.430**		
	Geo	0.058	−0.031	−0.205**	
	Bio	−0.312**	−0.078	0.157**	−0.139**
8	Mech	0.403**			
	Traf	−0.441**	−0.333**		
	Geo	−0.041	−0.026	−0.269**	
	Bio	−0.293**	−0.332**	0.373**	−0.397**

2. Results

2.1. Urban soundscape characteristics

Correlation analysis showed that soundscape preference was significantly and positively correlated with geophysical sound (correlation coefficient: −0.155) and biological sound (correlation coefficient: −0.106), and negatively correlated with mechanical sound (correlation coefficient: 0.267) and traffic sound (correlation coefficient: 0.238), but there was no significant relationship with human sound. This, corresponding to previous research (Yang, Kang 2005a, b; Nilsson, Berglund 2006), suggests that soundscapes with more natural sounds (biological and geophysical sounds) are more preferable, while too much artificial sounds (mechanical and traffic sounds) could reduce soundscape quality.

Table 3 shows the results of non-parametric analysis of each sound category by sampled period and sampled site. It can be seen that perceived loudness of all five major sound categories had significant differences among both different sampled sites and different sampled periods, except that traffic sound had no significant difference across all sampled periods, indicating a dynamic characteristic of urban soundscape patterns.

Table 4 shows the correlation results among different sound categories by time-step per-period. It can be seen that human sound was negatively correlated with traffic sound and biological sound in seven and five periods, respectively. Mechanical sound was negatively correlated with traffic sound and biological sound in all and seven periods, respectively. Biological sound was positively correlated with traffic sound in seven periods except the first one. Geophysical sound was negatively correlated with biological sound and traffic sound both in four periods, but showed no consistent relationships with other sound categories. It seems that different sound categories were almost mutually exclusive, except that biological sound was not lessened by traffic sound.

Figure 2 shows the contributions of the five sound categories to the overall soundscape loudness during all sampled periods. In general, artificial sounds, namely human, mechanical and traffic sounds dominated the urban soundscapes, while contributions from natural sounds, i.e. geophysical and biological sounds were less to the overall soundscape loudness. It could also be seen from Figure 2 that, contributions of each sound categories to the overall soundscape loudness were changing in different sampled periods, indicating the dynamic temporal pattern of urban soundscapes again.

Figure 3 (map a–e) shows the interpolation results of human, mechanical, traffic, geophysical and biological sounds, respectively. It can be seen that, more human sounds appeared across the beach area, as many local people and tourist concentrated in this area. More mechanical sounds concentrated along the south-eastern and

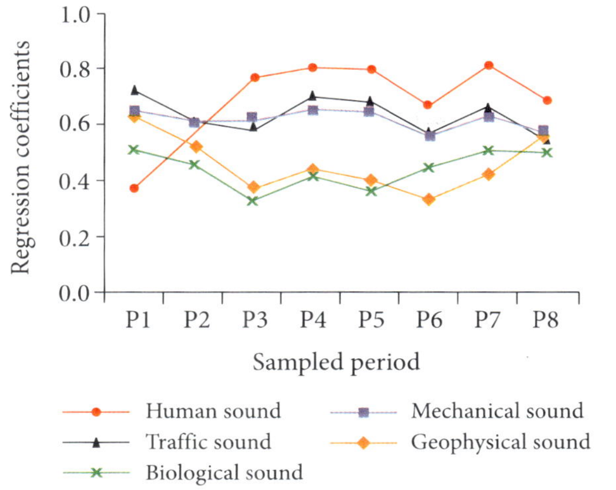

Fig. 2. Contributions of major sound categories to the perceived overall soundscape loudness during all sampled periods (P1–P8)

eastern boundary of the study area, where the rail way passes through and construction work was carried out at site W16. It is clear that traffic sounds concentrated almost along the direction of W09–W15, where the widest traffic road in the study area is located. Geophysical sounds were mainly perceived along the beach area, especially on the two ends, where the sea wave and wind were stronger. More biological sounds appeared in the central and south-eastern parts of the study area, mainly because dense constructions of the residential area form a quiet environment thus prevent fragile biological sounds from masking by other sounds, and dense vegetation of the urban park is ecologically good habitats for vocalising organisms such as birds and insects.

These maps indicate that artificial sounds, i.e. human, mechanical and traffic sounds permeated a larger part of

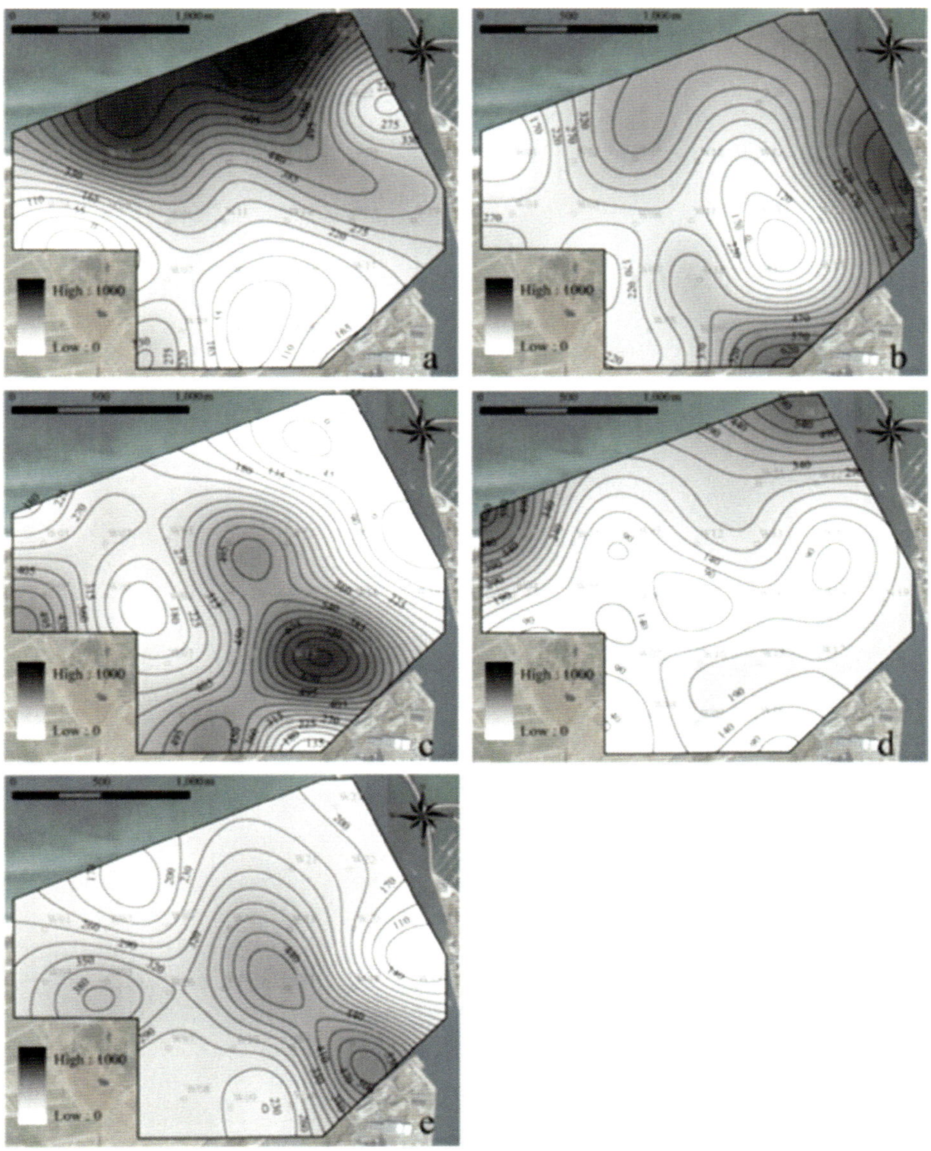

Fig. 3. Daily accumulated perceived loudness of human (a), mechanical (b), traffic (c), geophysical (d) and biological (e) sounds across the study area

the study area and were louder than natural sounds, i.e. geophysical and biological sounds, which show the dominating reality of artificial sounds in urban areas. Dominating areas of each sound category were usually not overlapping, or in other words, different sound category has different spatial arrangement, with an exception of biological and traffic sound, which indicates that biological organisms survived in urban areas may get used to the chronic traffic sound (Brumm 2004). From these maps, the underlying landscape characteristics show obvious relationships with soundscape perception.

2.2. Relationship between soundscape perception and landscape indices

Given the dynamic temporal characteristic of the urban soundscapes, and the stable characteristic of the landscape spatial patterns in the study area in a daily temporal period, the relationships between them were studied in each sampled period. The correlation results between landscape indices and each sound category were shown in Table 5. Though the relationship between the same landscape index and sound category may change in different period as supposed, some relatively stable significant relationships could still be found, i.e. mainly positively or negatively related with certain sound categories in most periods. The results showed that all the landscape indices correlated with at least one of the sound categories, except for patch density. Specifically, CD showed positive relationships with traffic and biological sounds, and negative relationship with geophysical sounds. RD only had stronger positive relationship with traffic sound, and possible negative relationship with geophysical sounds. FRAC showed similar effects with LSI, but stronger relationships were shown with human, traffic, and geophysical sounds. CONT and SIEI showed reverse effects on the same sound categories, namely human, traffic and biological sounds. In particular, NDVI, LSI and LPI showed the most significant relationships with all the five sound categories, making them the most potential "predictors" for soundscape characteristics.

3. Discussion

Landscape indices have several attributes, such as efficient, accessible, easily acquired, fully documented, and applicable to digital data, which make them attractive for planners and designers to apply to several alternative plans (Botequilha-Leitão, Ahern 2002; Corry, Nassauer 2005). Considering that landscape fragmentation is a common result of urbanisation process (Antrop 2004), all the landscape metrics used in this research could to some extent indicate landscape fragmentation pattern from different aspects, i.e. patch size (LPI), shape (LSI, FRAC), distribution (CONT), and heterogeneous status (PD, SIEI) (McGarigal, Marks 1994). The relationships between these indices and perception of different sound categories could be explained as following.

Construction density. Construction density is usually high in urban areas, which may form a kind of acoustic space termed "hi-fi area", a relatively quiet area formed by preventing a large part of sounds from outside (Schafer 1969), which makes it easier for human to recognise more sounds such as biological sounds. As most of the constructions are near traffic roads and form another kind of acoustic space termed "street canyons" to some extent (Table 6) (Kang 2007), this may explain the positive correlation between dense constructions and traffic sound (Table 5). Dense constructions also mean more barriers for geophysical sounds (e.g. tree rustling, grass rustling and wind blowing) before they are perceived, thus may impede the perceived loudness of this kind of sound.

Table 5. Spearman' rho correlation between each of the landscape indices and different sound categories by time-step per-period, where at each cell the number of periods when significant positive or negative correlations existed is given (*p < 0.05, **p < 0.01). See Table 1 and Table 2 for full names of the acronyms

Landscape indices	Positive/Negative				
	Hum	Mech	Traf	Geo	Bio
CD	1**/ 3**, 1*	—/3**	7**/—	—/5**	7**/—
RD	1**/ 3**, 1*	1**, 1*/1**	7**/—	1**/4**	3**/1**, 1*
NDVI	—/7**	—/8**	8**/—	—/8**	4**/—
PD	1**/1**	1*/1**	2*/1*	1**/ 3**, 1*	4**/2*
LPI	5**, 1*/1**	5**,1*/—	—/8**	7**/—	—/6**, 2*
LSI	1**/7**	—/ 4**, 1*	7**,1*/—	—/ 5**, 1*	4**,1*/1*
FRAC	1**/7**	1**/ 2**, 1*	6**/—	—/8**	3**/1**
CONT	7**/—	2**/1*	—/ 7**, 1*	2**/ 1**, 1*	—/ 5**, 1*
SIEI	—/ 2**, 4*	1**, 1*/2**	6**/—	2*/ 2**, 1*	3**, 2*/—

Road density. As the artery of urban areas, traffic roads are one of the main sound sources of urban acoustic environment. The result suggests that higher road density associates with more traffic sound. The possible negative effects on geophysical sound could be indirectly from masking by traffic sound, as they both contain low frequency components. Road effects on other sound categories have no obvious positive or negative trend.

Vegetation density. Vegetation density is indicated by NDVI value, and could act in two aspects in relationship to soundscape perception. On the one hand as sound

Table 6. Pearson correlation among landscape indices (*p < 0.05, **p < 0.01)

Indices	CD	RD	NDVI	PD	LPI	LSI	FRAC	CONTAG
RD	0.521*	1						
NDVI	−0.194	−0.176	1					
PD	0.482*	0.447*	−0.136	1				
LPI	−0.402	−0.274	−0.096	−0.37	1			
LSI	0.536**	0.617**	0.087	0.849**	−0.516*	1		
FRAC	0.365	0.546**	0.092	0.656**	−0.053	0.837**	1	
CONTAG	−0.263	−0.099	0.039	−0.318	0.776**	−0.323	0.059	1
SIEI	0.29	0.083	−0.059	0.271	−0.776**	0.242	−0.152	−0.972**

sources, areas with dense vegetation are usually ecologically good habitats for organisms such as birds and insects, so that dense vegetation may possess rich biological sounds (Gasc *et al.* 2013). As in urban areas traffic roads are usually planned with roadside trees, dense vegetation may also be related to traffic sound, as indicated in this study. On the other hand, dense vegetation could also act as barriers, thus affect sound propagation and perception. As revealed in this study, dense vegetation could minimise the perception of human sound, mechanical sound and geophysical sound.

Largest patch index. LPI indicates land use scale, and high value means existing of certain large scale and dominating land use patch in the landscape. As land use is related to certain activities happening in this area and further determines main sound sources, high LPI values may indicate more perception of certain sounds in the local landscape. In this study, the LPI values were mainly related to land use patches such as residential and commercial mixed area, beach, garden and urban park. As these areas are usually where human outdoor activities concentrate, e.g. chatting, shopping, relaxing and entertainment, it is reasonable that the positive relationships exist between LPI and the perception of human and mechanical sounds. The negative relationship between LPI and biological sound indicated that large areas with too many human activities may frighten off singing birds, for example. High LPI value could also mean less penetration by other land use types, such as traffic roads (Table 6), so that high LPI value is related to less traffic sound.

Landscape shape index and fractal dimension. Both LSI and FRAC could reflect shape complexity of land use patches. Although they are highly correlated, as shown in Table 6, LSI showed more indicating ability than FRAC. As indicated by both of them, complex land use shape may result in more traffic sound and less human and geophysical sounds. A possible reason of this pattern is that, in urban areas roads are usually the boundaries and/or connections of different functional areas, so that land use patches with complex shapes may be surrounded by more traffic roads, and a landscape penetrated by more roads could result in high landscape shape complexity. This point could also be verified by the positive relationships between road density and both LSI and FRAC (Table 6). As for human sound, since LSI (also FRAC) and LPI are negatively correlated and human sound was perceived more in certain large functional areas as indicated by LPI, it was reasonable in this case that less human sound was perceived in landscapes with high LSI and FRAC values and relatively small scale land use patches. For geophysical sound, either from far distance (e.g. sea wave) or sound caused by wind (e.g. tree or grass rustling), high LSI and FRAC means more "boundaries" during propagation before they are perceived, thus negative relationships between them were shown.

Contagion and Simpson's evenness index. CONT indicates the spatial distribution of land use patches, while SIEI indicates the distribution of diverse land use types in terms of structural component. They are highly negatively correlated, and both correlated with LPI (Table 6). It is suggested again that landscape with one large functional area could reduce the perception of traffic sound, and if there are a few large patches, they should be better contiguously and evenly distributed. This way of land use arrangement in planning could largely reduce the length of traffic roads as boundaries and/or connections among land use patches. However, biological sound require a total different pattern if more of them to be perceived, i.e. landscape with diverse and scattered land use types, since in this case biological organisms like birds could have more chance to find suitable habitats in a heterogeneous landscape (Andren 1994). Of course, landscapes with a large area of contiguous green areas would be definitely better for biological organisms, but at the scale of this research, this point could not be revealed.

It needs to be noted that, the relationships between the landscape spatial pattern indices and perceptual sound categories in this study were analysed at a relatively small scale. Thus, the effectiveness of certain landscape indices as soundscape indicators at larger scales still needs to be testified. Moreover, more landscape indices are available in the field of landscape ecology, and they could be further explored for soundscape evaluation. On the other hand, because parameters of soundscape quality have not yet come to a general accepted standard, the method of treating soundscape as an assembly of different meaningful sounds perceived by a certain user group as in this study could be adjusted in practical planning process by considering a more detailed user profile (Yu, Kang 2008, 2010; De Coensel *et al.* 2010).

Conclusions

The work shows how soundscape perception might be related to landscape spatial pattern. Based on the assumption that sound categories play a decisive role in soundscape perception, urban soundscapes were analysed in terms of five major sound categories. It has been found that:

1. Urban soundscapes were characterised by artificial sounds (human, mechanical and traffic sounds), overwhelming the natural ones (biological and geophysical sounds) which were more preferred by the observers.

2. Major sound categories were normally mutually exclusive and dynamic on a temporal scale, and have different spatial arrangement on a spatial scale. However, biological and traffic sounds seem to be co-existing on both temporal and spatial scales.

3. Significant correlations have been revealed between perception of major sound categories and a number of landscape spatial pattern indices, including vegetation density (NDVI), construction density (CD), road density (RD) and fragmentation status resulted from scale (LPI), shape (LSI, FRAC), composition (SIEI) and distribution (CONT) characteristics of different land use patches. Especially, NDVI, LSI, and LPI showed the most effective indicating ability.

4. As landscape spatial pattern indices are all quantitative, they could be adopted to compare different plans in terms of creating a better soundscape.

Acknowledgments

The authors would like to thank all the participants who were involved in the field work. The PhD research of the first author is supported by the German Academic Exchange Service (DAAD) and COST Action TD0804 (Short Term Scientific Mission). This project was also financed by Foundations in China (40971111; 51378139; 20112302110045; 2013BAJ12B02).

References

Adams, M.; Cox, T.; Moore, G.; Croxford, B.; Refaee, M.; Sharples, S. 2006. Sustainable soundscapes: noise policy and urban experience, *Urban Studies* 43(13): 2385–2398. http://dx.doi.org/10.1080/00420980600972504

André, M.; Solé, M.; Lenoir, M.; Durfort, M.; Quero, C.; Mas, A.; Lombarte, A.; van der Schaar, M.; López-Bejar, M.; Morell, M.; Zaugg, S.; Houégnigan, L. 2011. Low-frequency sounds induce acoustic trauma in cephalopods, *Frontiers in Ecology and the Environment* 9(9): 489–493. http://dx.doi.org/10.1890/100124

Andren, H. 1994. Effects of habits fragmentation on birds and mammals in landscapes with different proportions of suitable habitat: a review, *OIKOS* 71: 355–366.

Antrop, M. 2004. Landscape change and the urbanization process in Europe, *Landscape and Urban Planning* 67(1–4): 9–26. http://dx.doi.org/10.1016/S0169-2046(03)00026-4

Botequilha-Leitão, A.; Ahern, J. 2002. Applying landscape ecological concepts and metrics in sustainable landscape planning, *Landscape and Urban Planning* 59(2): 65–93. http://dx.doi.org/10.1016/S0169-2046(02)00005-1

Brambilla, G.; Maffei, L. 2006. Responses to noise in urban parks and in rural quiet areas, *Acta Acustica united with Acustica* 92(6): 881–886.

Brambilla, G.; Maffei, L. 2010. Perspective of the soundscape approach as a tool for urban space design, *Noise Control Engineering Journal* 58(5): 532–539. http://dx.doi.org/10.3397/1.3484180

Brown, A. L.; Kang, J.; Gjestland, T. 2011.Towards standardising methods in soundscape preference assessment, *Applied Acoustic* 72: 387–392. http://dx.doi.org/10.1016/j.apacoust.2011.01.001

Brumm, H. 2004. The impact of environmental noise on song amplitude in a territorial bird, *Journal of Animal Ecology* 73: 434–440. http://dx.doi.org/10.1111/j.0021-8790.2004.00814.x

Butkus, D.; Januševičius, T. 2011. Acoustic investigations of the exterior and interior wall of a log house, *Journal of Environmental Engineering and Landscape Management* 19(2): 140–147. http://dx.doi.org/10.3846/16486897.2011.580919

Corry, R. C.; Nassauer, J. I. 2005. Limitations of using landscape pattern indices to evaluate the ecological consequences of alternative plans and designs, *Landscape and Urban Planning* 72: 265–280. http://dx.doi.org/10.1016/j.landurbplan.2004.04.003

De Coensel, B.; Bockstael, A.; Dekoninck, L.; Botteldooren, D.; Schulte-Fortkamp, B.; Kang, J.; Nilson, M. E. 2010. The soundscape approach for early stage urban planning: a case study, in *Internoise 2010*, 13–16 June 2010, Lisbon, Portugal.

De la Fuente de Val, G.; Atauri, J. A.; de Lucio, J. V. 2006. Relationship between landscape visual attributes and spatial pattern indices: a test study in Mediterranean-climate landscapes, *Landscape and Urban Planning* 77: 393–407. http://dx.doi.org/10.1016/j.landurbplan.2005.05.003

De Ruiter, E. 2004. *Reclaiming land from urban traffic noise impact zones – the great canyon*: Dissertation. Technical University of Delft.

Dramstad, W. E.; Tveit, M. S.; Fjellstad, W. J.; Fry, G. L. A. 2006. Relationships between visual landscape preferences and map-based indicators of landscape structure, *Landscape and Urban Planning* 78: 465–474. http://dx.doi.org/10.1016/j.landurbplan.2005.12.006

Gasc, A.; Sueur, J.; Jiguet, F.; Devictor, V.; Grandcolas, P.; Burrow, C.; Depraetere, M.; Pavoine, S. 2013. Assessing biodiversity with sound: do acoustic diversity indices reflect phylogenetic and functional diversities of bird communities?, *Ecological Indicators* 25: 279–287. http://dx.doi.org/10.1016/j.ecolind.2012.10.009

Genuit, K.; Fiebig, A. 2006. Psychoacoustics and its benefit for the soundscape approach, *Acta Acustica united with Acustica* 92(6): 952–958.

GRASS Development Team. 2008. *Open Source Geospatial Foundation Project: Geographic Resources Analysis Support System (GRASS) Software* [online], [cited on 20 November 2011]. Available from Internet: http://grass.osgeo.org.

Guastavino, C. 2006. The ideal urban soundscape: investigating the sound quality of French cities, *Acta Acustica United Acustica* 92: 945–951.

Joo, W.; Gage, S. H.; Kasten, E. P. 2011. Analysis and interpretation of variability in soundscape along an urban-rural gradient, *Landscape and Urban Planning* 103: 259–276. http://dx.doi.org/10.1016/j.landurbplan.2011.08.001

Kang, J. 2007. *Urban sound environment*. London: Taylor & Francis incorporating Spon. 304 p.

Kang, J.; Zhang, M. 2010.Semantic differential analysis of the soundscape in urban open public spaces, *Building Environment* 45(1): 150–157. http://dx.doi.org/10.1016/j.buildenv.2009.05.014

Kull, R. C. 2006. Natural and urban soundscapes: the need for a multi-disciplinary approach, *Acta Acustica United with Acustica* 92(6): 898–902.

Lavandier, C.; Defreville, B. 2006. The contribution of sound characteristics in the assessment of urban soundscapes, *Acta Acustica United with Acustica* 92(6): 912–921.

Liu, J.; Kang, J.; Luo, T.; Behm, H. 2013a. Landscape effects on soundscape experience in city parks, *Science of the Total Environment* 454–455: 474–481. http://dx.doi.org/10.1016/j.scitotenv.2013.03.038

Liu, J.; Kang, J.; Luo, T.; Behm, H.; Coppack, T. 2013b. Spatiotemporal variability of soundscapes in a multiple functional urban area, *Landscape and Urban Planning* 115: 1–9. http://dx.doi.org/10.1016/j.landurbplan.2013.03.008

Matsinos, Y. G.; Mazaris, A. D.; Papadimitriou, K. D.; Mniestris, A.; Hatzigiannidis, G.; Maioglou, D.; Pantis, J. D. 2008. Spatio-temporal variability in human and natural sounds in a rural landscape, *Landscape Ecology* 23: 945–959. http://dx.doi.org/10.1007/s10980-008-9250-7

Mazaris, A. D.; Kallimanis, A. S.; Chatzigianidis, G.; Papadimitriou, K. D.; Pantis, J. D. 2009. Spatiotemporal analysis of an acoustic environment: interactions between landscape features and sounds, *Landscape Ecology* 24: 817–831. http://dx.doi.org/10.1007/s10980-009-9360-x

McGarigal, K.; Marks, B. J. 1994. *Fragstats– spatial analysis program for quantifying landscape structure*. USDA Forest Service, General Technical Report PNW-GTR-351, Corvallis, OR.

Nilsson, M. E.; Berglund, B. 2006.Soundscape quality in suburban green areas and city parks, *Acta Acustica United with Acustica* 92(6): 903–911.

Palmer, J. F. 2004. Using spatial metrics to predict scenic perception in a changing landscape: Dennis, Massachusetts, *Landscape and Urban Planning* 69: 201–218. http://dx.doi.org/10.1016/j.landurbplan.2003.08.010

Paožalytė, I.; Grubliauskas, R.; Vaitiekūnas, P. 2012. Modelling the noise generated by railway transport: statistical analysis of modelling results applying CADNAA and IMMI programs, *Journal of Environmental Engineering and Landscape Management* 20(3): 206–212. http://dx.doi.org/10.3846/16486897.2012.663090

Pheasant, R.; Horoshenkov, K.; Watts, G. 2008. The acoustic and visual factors influencing the construction of tranquil space in urban and rural environments tranquil spaces-quiet places?, *Journal of the Acoustical Society of America* 123(3): 1446–1457. http://dx.doi.org/10.1121/1.2831735

Pijanowski, B. C.; Farina, A.; Gage, S. H.; Dumyahn, S. L.; Krause, B. L. 2011. What is soundscape ecology? An introduction and overview of an emerging new science, *Landscape Ecology* 26: 1213–1232. http://dx.doi.org/10.1007/s10980-011-9600-8

Raimbault, M.; Dubois, D. 2005.Urban soundscapes: experiences and knowledge, *Cities* 22(5): 339–350. http://dx.doi.org/10.1016/j.cities.2005.05.003

Schafer, R. M. 1969. *The new soundscape: a handbook for the modern music teacher*. BMI, Don Mills, Canada.

Schulte-Fortkamp, B.; Fiebig, A. 2006. Soundscape analysis in a residential area: an evaluation of noise and people's mind, *Acta Acustica United with Acustica* 92(6): 875–880.

Turner, M. G. 1989. Landscape ecology: the effect of pattern on process, *Annual Review of Ecology and Systematics* 20: 171–197.

Turner, M. G.; Gardner, R. H.; O'Neill, R. V. 2001. *Landscape ecology in theory and practice: pattern and process*. NY, Berlin, Heidelberg: Springer-Verlag. 406 p.

Turner, M. G.; Ruscher, C. L. 1988. Changes in landscape patterns in Georgia, USA, *Landscape Ecology* 1(4): 241–251. http://dx.doi.org/10.1007/BF00157696

Uuemaa, E.; Mander, Ü.; Marja, R. 2012. Trends in the use of landscape spatial metrics as landscape indicators: a review, *Ecological Indicators* 28: 100–106. http://dx.doi.org/10.1016/j.ecolind.2012.07.018

Yang, W.; Kang, J. 2005a. Soundscape and sound preferences in urban squares: a case study in Sheffield, *Journal of Urban Design* 10(1): 61–80. http://dx.doi.org/10.1080/13574800500062395

Yang, W.; Kang, J. 2005b. Acoustic comfort evaluation in urban open public spaces, *Applied Acoustic* 66(2): 211–229. http://dx.doi.org/10.1016/j.apacoust.2004.07.011

Yu, L.; Kang, J. 2008. Effects of social, demographic and behavioral factors on sound level evaluation in urban open spaces, *Journal of the Acoustical Society of America* 123(2): 772–783. http://dx.doi.org/10.1121/1.2821955

Yu, L.; Kang, J. 2010. Factors influencing the sound preference in urban open spaces, *Applied Acoustic* 71(7): 622–633. http://dx.doi.org/10.1016/j.apacoust.2010.02.005

Zhang, M.; Kang, J. 2007. Towards the evaluation, description and creation of soundscape in urban open spaces, *Environment and Planning B: Planning and Design* 34(1): 68–86. http://dx.doi.org/10.1068/b31162

Jiang LIU finished his PhD at the faculty of Agricultural and Environmental Sciences, University of Rostock, Germany, and is recently a guest research at the School of Architecture, University of Sheffield. He has published 8 refereed journal papers. His research interest is on soundscapes in relation to landscape planning and design.

Jian KANG has been Professor of Acoustics at the University of Sheffield, School of Architecture since 2003. He obtained his PhD from the University of Cambridge. He has published 3 books, over 200 refereed journal papers and book chapters, and over 600 conference papers and technical reports.

Holger BEHM is an Apl. Professor at the Faculty of Agricultural and Environmental Sciences, University of Rostock, Germany. He has published more than 70 refereed journal papers. His research interest is in landscape planning and historical landscape management.

Tao LUO is an Associate Professor at the Institute of Urban Environment, Chinese Academy of Sciences. He obtained his PhD from the University of Rostock. His research interest is in regional sustainable development planning and landscape planning, landscape and environmental evaluation.

Permissions

All chapters in this book were first published in JEELM, by Vilnius Gediminas Technical University; hereby published with permission under the Creative Commons Attribution License or equivalent. Every chapter published in this book has been scrutinized by our experts. Their significance has been extensively debated. The topics covered herein carry significant findings which will fuel the growth of the discipline. They may even be implemented as practical applications or may be referred to as a beginning point for another development.

The contributors of this book come from diverse backgrounds, making this book a truly international effort. This book will bring forth new frontiers with its revolutionizing research information and detailed analysis of the nascent developments around the world.

We would like to thank all the contributing authors for lending their expertise to make the book truly unique. They have played a crucial role in the development of this book. Without their invaluable contributions this book wouldn't have been possible. They have made vital efforts to compile up to date information on the varied aspects of this subject to make this book a valuable addition to the collection of many professionals and students.

This book was conceptualized with the vision of imparting up-to-date information and advanced data in this field. To ensure the same, a matchless editorial board was set up. Every individual on the board went through rigorous rounds of assessment to prove their worth. After which they invested a large part of their time researching and compiling the most relevant data for our readers.

The editorial board has been involved in producing this book since its inception. They have spent rigorous hours researching and exploring the diverse topics which have resulted in the successful publishing of this book. They have passed on their knowledge of decades through this book. To expedite this challenging task, the publisher supported the team at every step. A small team of assistant editors was also appointed to further simplify the editing procedure and attain best results for the readers.

Apart from the editorial board, the designing team has also invested a significant amount of their time in understanding the subject and creating the most relevant covers. They scrutinized every image to scout for the most suitable representation of the subject and create an appropriate cover for the book.

The publishing team has been an ardent support to the editorial, designing and production team. Their endless efforts to recruit the best for this project, has resulted in the accomplishment of this book. They are a veteran in the field of academics and their pool of knowledge is as vast as their experience in printing. Their expertise and guidance has proved useful at every step. Their uncompromising quality standards have made this book an exceptional effort. Their encouragement from time to time has been an inspiration for everyone.

The publisher and the editorial board hope that this book will prove to be a valuable piece of knowledge for researchers, students, practitioners and scholars across the globe.

List of Contributors

Agnė Kazlauskienė and Vaida Brukštutė
Research Institute of Environmental Protection, Vilnius Gediminas Technical University, Saulėtekio al. 11, 10223 Vilnius, Lithuania

Long shao
School of Architecture, Harbin Institute of Technology, Harbin 150001, China

Jian xu
School of Architecture, Harbin Institute of Technology, Harbin 150001, China
School of Architecture, University of Sheffield, Sheffield S10 2TN, UK

Jian Kang
School of Architecture, University of Sheffield, Sheffield S10 2TN, UK
School of Architecture, Harbin Institute of Technology, Harbin 150001, China

Mei Zhang
School of East Asia Studies, University of Sheffield, Sheffield S10 2TD, UK

Chunyang Zhu
Huazhong Agricultural University, No. 1, Shizishan Street, Hongshan District, 430070 Wuhan Hubei Province, China

Peng Ji
Heilongjiang Bayi Agricultural University, No. 5, Xinfeng Road, Longfeng District, 163319 Daqing Heilongjiang Province, China

Shuhua Li
School of Architecture, Tsinghua University, Haidian District, 100084 Beijing, China

Shekoufeh Nematollahi, Sima Fakheran and Alireza Soffianian
Department of Natural Resources, Isfahan University of Technology, Isfahan, 84156-83111, Iran

Gintaras Jarašiūnas and Irena Kinderienė
Vėžaičiai Branch of the Lithuanian Research Centre for Agriculture and Forestry, Gargždų st. 29, 96216 Vėžaičiai, Klaipėda distr. , Lithuania

Rocío De Torre, María Dolores Jiménez and Miguel A. Casado
Department of Ecology, Complutense University, Madrid 28040, Spain

Álvaro Ramírez
Department of Zoology and Physical Anthropology, Complutense University, Madrid 28040, Spain

Ignacio Mola
Research, Development and Innovation (R&D) Service, Obrascón Huarte Lain, S.A. (OHL),Madrid 28046, Spain

Luis Balaguer
Department of Plant Biology I, Complutense University, Madrid 28040, Spain

Chuanan Liu, Tong Qi and Xiao Ma
College of Resource Environment and Tourism, Capital Normal University, 105 West Third Ring Road North, Haidian District, Beijing, 100048 China

Alexandre Marco Da Silva
Department of Environmental Engineering, Campus Sorocaba, São Paulo State University, Av. Três de Março 511, Sorocaba, SP, Brazil

Rodrigo Custodio Urban
Department of Sanitation and Environment, FEC, Campinas State University, Av. Albert Einstein, 951, Campinas, SP, Brazil

Luiz Augusto Manfré
Geoprocessing Laboratory, Polytechnic School, University of São Paulo, Avenue Professor Luciano Gualberto, Lane 3, 380 São Paulo, Brazil

Michel Brossard
IRD, UMR 210 Eco&Sols, BP 64501, 34394 Montpellier cedex 5, France

Marcelo Zacharias Moreira
Centre for Nuclear Energy in Agriculture, University of São Paulo, Av. Centenário, 303 - Piracicaba, SP, Brazil

Hakan Alphan and Laden Güvensoy
Department of Landscape Architecture, Cukurova University, Balcali Campus, 01330, Adana, Turkey

Huilin Liang and Qingping Zhang
College of Landscape Architecture, Nanjing Forestry University, No. 159, Longpan Road, Nanjing, Jiangsu, China

Di Chen
College of Architecture and Urban Planning, Tongji University, No. 1239, Siping Road, Shanghai, China

Qian Wu and Hyun-Kil Jo
Kangwon National University, Chuncheon, Republic of Korea

Maria Csete and Attila Buzasi
Department of Environmental Economics, Budapest University of Technology and Economics, Magyar Tudósok körútja 2, 1117 Budapest, Hungary

Nikrouz Mostofi
Department of Surveying Engineering, Islamic Azad University, South Tehran Branch, 1777613511 Tehran, Iran

Mahdi Hasanlou
School of Surveying and Geospatial Engineering, Norh Karegar, College of Engineering, University of Tehran, 1439957131 Tehran, Iran

Pietro Paolo Falciglia and Federico G. A. Vagliasindi
Department of Civil and Environmental Engineering, University of Catania, Viale A. Doria, 6-95125 Catania, Italy

Abir Al-Tabbaa
Department of Engineering, University of Cambridge, Trumpington Street, Cambridge CB2 1PZ, UK

Shyh-Chyang Lin
Department of Civil Engineering and Engineering Management, National Quemoy University Jinning, Kinmen, Taiwan

Cheng Li
Department of Geo-Information Science, School of Resources and Geosciences, China University of Mining and Technology, Daxue road 1, 221116, Xuzhou, Jiangsu, China

Jie Zhao
Huaihai Development Research Institute, Jiangsu Normal University, Heping road 57, 221009, Xuzhou, Jiangsu, China

Jian Kang
School of Architecture, University of Sheffield, Western Bank, Sheffield S10 2TN, UK

Jiang Liu
School of Architecture, University of Sheffield, Western Bank, Sheffield S10 2TN, UK
Faculty of Agricultural and Environmental Sciences, University of Rostock, Justus-von-Liebig-weg-6, 18059 Rostock, Germany

Holger Behm
Faculty of Agricultural and Environmental Sciences, University of Rostock, Justus-von-Liebig-weg-6, 18059 Rostock, Germany

Tao Luo
Key Lab of Urban Environment and Health, Institute of Urban Environment, Chinese Academy of Sciences, 1799 Jimei Road, 361021 Xiamen, China

Index